ECOPHYSIOLOGY OF METALS
IN TERRESTRIAL INVERTEBRATES

POLLUTION MONITORING SERIES

Advisory Editor: Professor Kenneth Mellanby

ECOPHYSIOLOGY
OF METALS
IN TERRESTRIAL
INVERTEBRATES

STEPHEN P. HOPKIN

Department of Pure and Applied Zoology, University of Reading, UK

ELSEVIER APPLIED SCIENCE
LONDON and NEW YORK

ELSEVIER APPLIED SCIENCE PUBLISHERS LTD
Crown House, Linton Road, Barking, Essex IG11 8JU, England

Sole Distributor in the USA and Canada
ELSEVIER SCIENCE PUBLISHING CO., INC.
655 Avenue of the Americas, New York, NY 10010, USA

WITH 41 TABLES AND 65 ILLUSTRATIONS

© 1989 ELSEVIER APPLIED SCIENCE PUBLISHERS LTD

British Library Cataloguing in Publication Data

Hopkin, Stephen P.
Ecophysiology of metals in terrestrial invertebrates.
1. Invertebrates. Physiology
I. Title
592'.01

Library of Congress Cataloging in Publication Data

Hopkin, Stephen P.
Ecophysiology of metals in terrestrial invertebrates.

(Pollution monitoring series)
Bibliography: p.
Includes index.
1. Invertebrates—Physiology. 2. Invertebrates—Ecology.
3. Metals—Environmental aspects. I. Title. II. Series.
QL364.H67 1989 592'.019214 88–37356

ISBN 1-85166-312-6

Printed in Great Britain by Page Bros (Norwich) Ltd

Prologue

THE MURDERED HITCHHIKER

On 19 August 1971, the badly decayed corpse of a young woman was discovered at Inkoo, south Finland, an area subject to mercury pollution. Final instar blowfly larvae collected from the cadaver were allowed to pupate and the concentrations of mercury were determined in the adults which emerged. The flies contained less than $0.15 \,\mu g \, g^{-1}$ (fresh weight) of mercury which indicated that the insects had developed in uncontaminated biological material (Nuorteva 1977). It was concluded that the murdered girl had not lived in the mercury-polluted area. Later when the corpse was identified, it was discovered that she had lived as a student in the city of Turku, an area not subject to mercury pollution. This provided the first indication that analysis of this kind may find application in forensic practice (Smith 1986).

Preface

Many reviews have been written on metals in terrestrial ecosystems but almost none of these texts have given invertebrates more than a passing mention. This is not due to a lack of research, for many hundreds of papers have been published on the subject. This book is an attempt to synthesise this wealth of information into a coherent review of the importance of metals in the ecology and physiology (hence 'ecophysiology') of terrestrial invertebrates.

I have tried throughout the book to consider the regulation of metals alongside the many other physiological and behavioural processes which invertebrates have had to evolve during their colonisation of the land. The underlying reasons for the different ways in which major taxonomic groups, species or even individual animals deal with metals, cannot be understood without such an approach.

Until recent years, most researchers (with a few honourable exceptions) have been content to report concentrations of metals in terrestrial invertebrates without considering the physiological implications of their findings (I speak with the authority of someone who has sacrificed many thousands of invertebrates on the altar of atomic absorption spectrometry!). This book inevitably reflects such bias. However, more physiological experiments are being conducted in which metals are considered as substances which invertebrates have to regulate to survive in terrestrial ecosystems rather than just as pollutants, an encouraging trend.

In terms of coverage, I have included multicellular invertebrates which live out of water for their whole life cycle, or where part of this is aquatic, the terrestrial stage. In a few cases, I have referred to the aquatic larval stages of insects where this helps to interpret metal dynamics in the adults. Microorganisms are dealt with only briefly, except where they form a symbiotic relationship in the gut of a terrestrial invertebrate. Free-living microorganisms are of great importance in terrestrial ecosystems but their inclusion would have doubled the size of this book were I to do the subject justice. Reviews have, in any case, been published recently on the ecology of microorganisms in terrestrial eco-systems by Richards (1987), and on metals in particular by Wood (1984b), Duxbury (1985) and Hughes & Poole (1988).

There is still very little known about the routes by which organisms assimilate metals from the environment. Even in an animal as well-studied as *Homo sapiens*, it has not yet been resolved what proportion of the lead in blood is derived from food or air (Elwood 1986). In terrestrial invertebrates, there are still whole Phyla about which next to nothing is known. I hope that the publication of this book will stimulate some researchers to attempt to fill this proverbial 'wide open field'.

Steve Hopkin
Reading

Acknowledgements

Many colleagues have helped in the production of this book. I am extremely grateful to Dr M.H. Martin, Professor K. Simkiss, Dr P.J. Coughtrey and Dr M. Taylor who commented extensively on the text, and to the research workers and publishers who gave their permission to reproduce figures and tables from their papers (in many cases loaning negatives or original prints of their diagrams and micrographs). Hugh Anger, Alan Jenkins and Liz Wheeler developed and printed many of the micrographs and the University of Reading Library Photographic Service helped in the production of the figures. June Hattrick undertook the tedious task of typing the tables. I am indebted to all these people without whose help the manuscript would have taken much longer to produce.

I have benefited from discussions with my long-suffering under-graduate and postgraduate students (especially Chris Hames and David Jones) who have acted as devil's advocates on many occasions to tone down some of my more outlandish theories! Other 'metalworkers' throughout the world have provided reprints of their publications for which I am grateful also.

My own research would not have been possible without the generous financial support of the Natural Environment Research Council, the Nuffield Foundation and the University of Reading Research Endow-ment Fund.

Finally, I would like to thank the staff at Elsevier Applied Science for their patience and help during the writing and production of this book.

S.P.H.

Contents

CHAPTER 1

Introduction

1.1 WHAT IS A METAL?

A metal is defined by chemists as being an element which has a characteristic lustrous appearance, is a good conductor of electricity and generally enters chemical reactions as positive ions or cations. Elements which are obviously metals are cadmium, copper, lead and zinc. However in some cases, the distinction between metals and non-metals is not sharp. Antimony, arsenic and tellurium, for example, have physical properties of metals and chemical properties of non-metals and are often referred to as metalloids. In biology, the distinction between metals and non-metals is even less clear and depends very much on the personal prejudice of the writer. Elements which are considered to be metals for the purposes of this book are shown in Fig. 1.1. Because special emphasis is placed on metals as environmental pollutants, coverage of metals which are major nutrients such as calcium, has been limited to where they are of direct importance in metal detoxification (e.g. Chapter 9).

Astute readers will notice the absence of the widely-used term 'heavy metal' from this book. Martin & Coughtrey (1982) argued for its retention because it was a term which invoked the concepts of permanence, toxicity and long residence time in the environment. However, the distinction between metals which are heavy and those which are not is blurred, to say the least. Many authors have defined a metal as heavy if it has a relative density of greater than five but there are numerous publications in which the term is used to describe elements which are neither heavy (e.g. aluminium), nor are metals in the strict sense of the word (e.g. selenium). One recent review on lead in the environment (Lansdown 1985) defines heavy metals as those with an atomic weight greater than sodium, and light metals as sodium, and metals lighter than

1

H																	He
Li	Be											B	C	N	O	F	Ne
Na	Mg											Al	Si	P	S	Cl	Ar
K	Ca	Sc	Ti	V	Cr	Mn	Fe	Co	Ni	Cu	Zn	Ga	Ge	As	Se	Br	Kr
Rb	Sr	Y	Zr	Nb	Mo	Tc	Ru	Rh	Pd	Ag	Cd	In	Sn	Sb	Te	I	Xe
Cs	Ba	La	Hf	Ta	W	Re	Os	Ir	Pt	Au	Hg	Tl	Pb	Bi	Po	At	Rn
Fr	Ra	Ac	Rf	Ha													

Ce	Pr	Nd	Pm	Sm	Eu	Gd	Tb	Dy	Ho	Er	Tm	Yb	Lu
Th	Pa	U	Np	Pu	Am	Cm	Bk	Cf	Es	Fm	Md	No	Lr

FIG. 1.1 Periodic table of the elements. Those considered to be metals are surrounded by bold lines. Metalloids (with properties of metals and non-metals) are shaded.

sodium. Under this definition, aluminium is a heavy metal, a conclusion that most researchers would find hard to reconcile.

An attempt to resolve these difficulties was made by Nieboer & Richardson (1980). They suggested that biologists should adopt the classification system originated by chemists, which uses Lewis acid properties (i.e. 'hardness' or 'softness' as acids and bases), to separate metals

TABLE 1.1

Separation of some essential and non-essential metal ions of importance as pollutants into class A (oxygen-seeking), class B (sulphur- or nitrogen-seeking) and borderline elements based on the classification scheme of Nieboer & Richardson (1980). This distinction is important in determining rates of transport across cell membranes (Section 9.2), and sites of intracellular storage in metal-binding proteins (Section 9.3) and metal-containing granules (Section 9.4).

Class A	Borderline	Class B
Calcium	Zinc	Cadmium
Magnesium	Lead	Copper
Manganese	Iron	Mercury
Potassium	Chromium	Silver
Strontium	Cobalt	
Sodium	Nickel	
	Arsenic	
	Vanadium	

into class A, class B and intermediate or borderline metals (Pearson 1963). Under this system, class A metals are predominantly oxygen-seeking and class B metals bind preferentially to nitrogen or sulphur-bearing ligands. Borderline metals exhibit characteristics of both class A and class B metals (Table 1.1). A classification system based on the chemical properties of metals is clearly preferable to one based on density so the scheme of Nieboer & Richardson (1980) is the one adopted in this book. Some researchers have used the hardness/softness concept when interpreting results of experiments on metal uptake in earthworms (Beyer 1981) and fruit flies (Williams *et al.* 1982).

The term 'trace metal' is also difficult to define. In studies on vertebrates, metals are described as being present in trace amounts if their concentrations in a tissue are less than that of iron (Underwood 1977). This definition is unsatisfactory for terrestrial invertebrates because the normal levels of iron in many species are much less than those of metals such as copper and zinc considered to be trace metals under the vertebrate definition (e.g. spiders, Hopkin & Martin 1985b). These differences may be even greater in the tissues of invertebrates from metal-polluted sites. The term trace metal is therefore not used in this book.

1.2 METALS AND EVOLUTION

In the primeval seas where life evolved some 3.5×10^9 years ago, the solubility of elements under the anaerobic conditions which existed at the time determined which metals could be incorporated into biochemical reactions (Wood 1984a). A 'natural selection of the elements' took place where the choice of one element rather than a similar one for a particular biochemical role was dictated by the energy cost of getting the element (i.e. its availability), the ability of the organism to retain it, and its functional advantages relative to those of other metals (Williams 1981; Esser & Moser 1982). Under these anaerobic conditions, free Fe^{2+} and ferrous sulphide were available to early life allowing iron-containing proteins to form, such as bacterial ferredoxin.

During the period 2.5×10^9 to 1.8×10^9 years ago, the activity of photosynthetic algae changed the atmosphere from being chemically-reducing with no oxygen, to one containing about 20% oxygen. Organisms which had developed under anaerobic conditions had to develop aerobic biochemistry and evolve ways of breaking down toxic by-products of oxidative metabolism. These included synthesis of the selenium-containing enzyme glutathione peroxidase which catalyses the breakdown of hydrogen peroxide in cells (Combs & Combs 1986). Copper, cobalt and zinc, which form highly insoluble sulphides under anaerobic conditions, became much more available under aerobic conditions and could therefore be incorporated into proteins. Iron in contrast, became almost completely insoluble and proteins for storage (ferritin) and transport (transferrin) of this metal had to be developed. The distribution of iron within the body of vertebrates, for example, is so closely controlled that the withholding of the element from wounds is a major inhibitor of infection by invading microorganisms (Weinberg 1984). It is interesting to note in the light of the endosymbiotic theory of the origin of the mitochondrion that eucaryotic cytoplasm yields a super-oxide dismutase containing copper and zinc while mitochondria and bacteria yield the presumably more ancient proteins containing iron or manganese.

To be successful, animals must maintain the concentrations of free metal ions in their cells within specific limits. An elaborate system of carrier and storage proteins has evolved to ensure that the correct distribution between internal organs, cells and intracellular compartments is maintained. Detoxification systems have developed which remove metals from circulation if they are surplus to requirements and store them in an unavailable form until such time as they can be excreted. During evolution of metal detoxification processes in marine invertebrates, the concentrations of metals in the ocean would have been fairly stable allowing metal-exchange processes at permeable surfaces exposed to seawater to operate at a constant rate. Concentrations in food would have been much less predictable. The levels of metal ions released into solution in the lumen of the gut by the action of digestive enzymes could have varied by more than an order of magnitude, particularly in animals with a heterogeneous diet. Thus, new species of marine invertebrates had to be evolved, presumably by mutation, which were able to restrict the amounts of these ions which were 'allowed' to pass across the epithelium of the digestive system into the blood to prevent them from interfering with biochemical reactions in the body. Ideally this would have been achieved by feedback mechanisms regulating the uptake of each metal in the gut at the lumen/cell interface. Unassimilated metals would simply pass through the lumen of the gut and be voided in the faeces. However, while a degree of regulation of uptake of some metals from the food must occur, it is clear that the passage of all metals into the epithelial cells of the digestive system under all dietary circumstances cannot be controlled. Detoxification of metals which did pass from the gut lumen into the cells in excess of physiological requirements, was achieved by sequestration within the cells in an unavailable form for subsequent excretion, or permanent storage within the tissues.

All the intracellular mechanisms for metal detoxification found so far in terrestrial invertebrates occur also in marine organisms and probably evolved long before the colonisation of the land took place. The strategies which terrestrial invertebrates have adopted for metal regulation have been modified by their need to conserve water and the fact that the epithelium of the gut has assumed almost total responsibility for metal exchange with the environment. The thesis which suggests that the dynamics of metal regulation in terrestrial invertebrates is intimately connected with their ecophysiology, forms the main theme of this book.

Numerous reviews have been published on the role of metals in the biology of animals (e.g. Miller & Neathery 1977; Underwood 1977; Bowen 1979; Mills 1979; Underwood 1979; Bronner & Coburn 1981; Anke *et al.* 1984; Frieden 1984; Adriano 1986; Friberg *et al.* 1986; Xavier 1986) but this book is the first to concentrate specifically on terrestrial invertebrates.

1.3 CONCENTRATION UNITS

If metals are to be considered as biologically active substances rather than material which is simply accumulated, their presence in organisms should be expressed in terms of numbers of atoms and molecules rather than weight. This is already the case in publications concerned with major nutrients, where levels of metals such as calcium are invariably expressed as molar concentrations rather than weight of metal per unit weight of tissue. However, almost all papers concerned with metals other than major nutrients in terrestrial invertebrates, have adopted the latter units and have expressed concentrations as $\mu g\, g^{-1}$ (or their equivalent). Some researchers have used both units in the same publication (e.g. Bertini *et al.* 1986). In this book, rather than convert all published figures to molar or weight concentrations, the units used by authors in their papers have been retained (except in a few cases where a change makes comparisons clearer). A table for conversion between molar and weight concentrations is given in Appendix 1.

Because of the variability in water content of invertebrates, it is preferable to express concentrations as amount of metal per unit dry tissue rather than fresh weight. In this book, unless otherwise stated, *all concentration values are given on a dry weight basis.*

1.4 DEVELOPMENT OF ANALYTICAL TECHNIQUES

Major advances in biological knowledge often follow the introduction of new analytical techniques. This has certainly been the case with research on metals in terrestrial invertebrates. The history of the subject can be divided broadly into three periods, pre 1960, 1960 to 1975 which saw increasing use of the electron microscope, and post 1975 when several new techniques became commercially available including flame and flameless atomic absorption spectrometry, X-ray microanalysis, and gel-permeation chromatography for separation of proteins (see Sections 4.4, 9.3.1 and 9.4.2 for details of these and other analytical methods).

Before about 1960, analytical techniques were relatively unsophisticated. The qualitative 'presence' of particular metals in invertebrate tissues was detected by simple chemical tests (Giunti 1879; Hogg 1895; Aronssohn 1911) and effects of metal-containing pesticides on molluscs were reported (Anderson & Taylor 1926). From 1940 to 1960, colourimetric and histochemical tests were employed to measure concentrations of metals in various insects to broadly localise the sites of storage within cells (Lennox 1940; Poulson & Bowen 1952; Waterhouse 1945a, 1945b). Radioisotopes were also used to study uptake rates of metals (Poulson & Bowen 1951).

From 1960 to about 1975, there was a major interest in the possible role of metals as insecticides. Work was carried out to determine whether particular metals were essential in the nutrition of insects, particularly aphids (Dadd & Krieger 1967; Srivestava & Auclair 1971a). The use of the electron microscope increased dramatically until by the early 1970s, most biological laboratories had begun to examine the ultrastructure of cells. Wright & Newell (1964) were among the first authors to observe metal-containing granules within the cells of a terrestrial invertebrate by electron microscopy but by the mid-1970s, these structures had been found in the digestive epithelia of a wide range of species (Simkiss 1976b).

Since 1975, there has been an exponential increase in the numbers of papers published each year on metals in invertebrates. This prodigious output has shown signs of stabilising only in the last few years. The sudden increase in research activity in the mid 1970s was due to an increased awareness of the importance of metals as environmental pollutants, and the availability of atomic absorption spectrometers which allowed rapid, routine analysis of the concentrations of a wide range of metals in biological samples to be conducted at relatively low cost. Towards the end of the 1970s, X-ray microanalysis became widely available. This extremely powerful technique enabled the ultrastructure and elemental composition of material to be examined simultaneously in the electron microscope and has made an enormous contribution to our understanding of intracellular metal detoxification systems.

The most exciting recent development is the inductively coupled plasma source mass spectrometer which has become commercially available only in the last couple of years. With this technique it is possible to determine simultaneously the concentrations of all elements and their isotopes to sub ng ml^{-1} levels in a sample in a matter of seconds (Eaton & Hutton 1988). Although an order of magnitude more expensive than atomic absorption spectrometry, the advantage of being able to obtain a complete spectrum of all elements in a sample must make this the most important technical development since the advent of X-ray microanalysis. It will be of particular use in environmental monitoring of metal contamination, especially radioisotopes.

CHAPTER 2

Essentiality and Toxicity of Metals

2.1 CRITERIA OF ESSENTIALITY

A vast number of experiments have been carried out on the mineral element requirements of vertebrates. Most of these studies have been conducted with rats and involve the use of highly purified diets. The requirement for a particular element is demonstrated by comparing the growth, survival and reproductive success of a control group, with a group fed on a diet which is identical, except for the absence of the element under test. A number of criteria have been established which must be satisfied before an element can unequivocally be considered to be essential (Muntau 1984).

(1) Removal of the element from the diet should result in disturbed growth and reproduction.
(2) These deficiency effects should be accompanied by pathological damage and/or changes in enzyme or hormone status of immunoactivity which can be reversed by supplementation with the element.
(3) Alleviation of deficiency effects should be dose-dependent.
(4) The element should form an essential component of an enzyme, hormone or other biologically active substance.

Many of the elements considered to be essential for vertebrates have satisfied only the first criterion. These elements are needed in such minute amounts that current analytical techniques have been unable to resolve their biochemical function. Several elements which are currently considered not to be essential, may eventually be shown to be required when the detection limits of analytical equipment are lowered and the purity of synthetic diets is improved. However, it may be impossible to

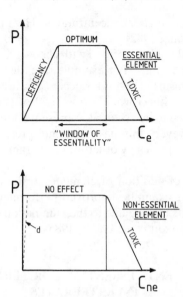

FIG. 2.1 Relationships between performance (P) (growth, fecundity, survival) and concentrations of an essential (C_e) or non-essential element (C_{ne}) in the diet of animals. Possible deficiency effects at ultra-trace levels (d) of an apparently non-essential element may be discovered as the sensitivities of analytical techniques are improved.

disprove an essential requirement for some elements as the theoretical lowest possible limit for essentiality is one atom for one specific gene in the chromosomes of a cell (Schwarz 1974). To put this into perspective, the concentration of zinc in the small red blood cells of cattle is about $4 \, \mu g \, g^{-1}$ (fresh weight). Individual blood cells contain more than 10^9 atoms of zinc. Even if the concentration of zinc in the cells was only $1 \, ng \, g^{-1}$, each cell would still contain about 250,000 zinc atoms (Miller & Neathery 1977). The distinction between essential and non-essential elements (Fig. 2.1) is therefore equivocal.

2.2 HORMESIS

Stimulation of growth following very small additions of an element to the diet should not necessarily be interpreted as a response to a deficiency. The rate of growth of a wide range of organisms may increase slightly when they are exposed to low levels of substances which are strongly inhibitory at higher concentrations. This phenomenon is known as hormesis (Stebbing 1982).

Hormesis should not be confused with changes in population structure which alters species ratios. For example, the 'decomposing activity' of certain microorganisms in leaf litter occurs at a faster rate (as measured by carbon dioxide efflux) when a low level of cadmium, a metal not generally considered to be essential, is present (Bond *et al.* 1976; Chaney *et al.* 1978). However, this has been interpreted as replacement of cadmium-sensitive species by other more tolerant ones which respire at a higher rate.

Indirect effects of addition of elements to the diet may also occur. In Collembola, for example, improved performance following dietary copper supplements may be due to the elimination of a copper-sensitive gut parasite (Bengtsson *et al.* 1983b, 1985a).

2.3 ELEMENTS ESSENTIAL FOR TERRESTRIAL INVERTEBRATES

In addition to carbon, hydrogen, oxygen and nitrogen, all animals need the seven major mineral elements calcium, phosphorus, potassium, magnesium, sodium, chlorine and sulphur for ionic balance, and as integral parts of amino acids, nucleic acids and structural compounds. Thirteen other so-called 'trace elements' are definitely required by higher vertebrates namely iron, iodine, copper, manganese, zinc, cobalt, molybdenum, selenium, chromium, nickel, vanadium, silicon and arsenic (Muntau 1984; Irgolic 1986). Other elements which may be essential at 'ultra-trace' levels include lithium, aluminium, tin, fluorine and lead (Anke *et al.* 1984) and cadmium (Schwarz & Spallholz 1977). However, definite biochemical roles have only been identified for iron, zinc, copper, selenium, molybdenum, cobalt and iodine (Table 2.1). Their presence is highly specific. Zinc, for example, is an essential component of carbonic anhydrase, the enzyme which catalyses the formation of calcium carbonate. If zinc is replaced in the protein by another element, the activity of the enzyme is greatly reduced (Table 2.2).

Very few laboratory experiments have been carried out to determine the trace metal requirements of terrestrial invertebrates. Most studies have been conducted with aphids and lepidopteran larvae because these are the only groups for which fully synthetic diets have been formulated. As well as the major mineral elements, iron, zinc, copper, manganese

TABLE 2.1

'Trace' elements essential for normal growth and reproduction in higher vertebrates for which definite biochemical roles have been identified. Iron, zinc and copper are involved in many more processes than are indicated in the table.

Element	Protein, vitamin or hormone	Example of process
Iron	Haemoglobin	Oxygen transport
Zinc	Carbonic anhydrase	Calcium carbonate formation
Copper	Cytochrome oxidase	Electron transport
Molybdenum	Xanthine oxidase	Purine degradation
Cobalt	Vitamin B_{12}	Nucleoprotein synthesis
Selenium	Glutathione peroxidase	Reduction of hydrogen peroxide
Iodine	Thyroxin	Growth hormone

TABLE 2.2

Level of activity of carbonic anhydrase relative to zinc, of different metals substituted in the protein (from Coleman 1967)

Metal	% Normal activity (hydration of CO_2)
Zinc	100
Cobalt	56
Nickel	5
Cadmium	4
Manganese	4
Copper	1
Mercury	0.05

and cobalt have also been shown to be essential for insects although many more elements are likely to be added to this list as analytical and dietary purification techniques are improved. These experiments are discussed in detail in Section 7.5. Some elements which are thought not to be essential for vertebrates may eventually be shown to be required by invertebrates.

Few essentiality experiments have been conducted on aquatic invertebrates although a requirement for selenium in *Daphnia* has been demonstrated recently (Keating & Dagbusan 1984).

2.4 TOXICITY OF METALS

Metals exert toxic effects on animals if they enter into biochemical reactions in which they are not normally involved. The threshold concentration at which such deleterious effects occur is usually higher for essential elements than for non-essential elements although the 'window of essentiality' for some elements such as selenium, is quite narrow (Shamberger 1983; Fig. 2.1). Unlike many organic chemicals, metals can not be broken down into less toxic components. When released into the environment by the industrial activities of humans, metals have long residence times in soils and may continue to exert harmful effects on the soil fauna long after the source of the pollution has ceased to operate.

The toxicity of metals is influenced by the chemical form. These can be classified into three categories depending on their complexity (Jorgensen & Jensen 1984).

(1) Simple aquated metal ions e.g. $Fe(H_2O)_6^{3+}$.
(2) Metal ions complexed by inorganic anions e.g. $CuCl^+$, $CuOH$.
(3) Metal ions complexed by organic ligands e.g. amino acids such as $Cu(NH_2CH_2COO)_2$.

There are numerous examples of how chemical form modifies the toxicity of metals. The toxicity of lead to moth larvae depends on the organic ligand to which the metal is bound (Hueck & La Brijn 1973). There is a ten-fold difference in the toxicity of organotin compounds to beetle larvae depending on the chemical formula of the insecticide (Becker 1978). The inhibitory power of vanadium on acid phosphatase activity of microorganisms in spruce needle litter decreases in the order $Na_3VO_4 > NaVO_3 > V_2O_5$ (Tyler 1976).

Methylation of metals may occur after the inorganic form has been released into the environment. This occurred in Minimata Bay in Japan where inorganic mercury discharged by an industrial concern was methylated by microorganisms in the sediments into methyl mercury which was much more bioavailable (Kudo *et al.* 1980). In terrestrial ecosystems, methylation may also occur but with arsenic (Mohan *et al.* 1982) and selenium (Allaway 1975) this has the effect of reducing toxicity. The 'speciation' of metals has been reviewed by Bernhard *et al.* (1986).

Most metals of concern as environmental pollutants are group B or borderline metals which form inorganic complexes in saline solutions that are very lipid soluble (Gutknecht 1981; Simkiss 1983a). This experimental finding may explain why metals such as cadmium and mercury pass so readily across the microvillus border of the gut of terrestrial invertebrates despite having no known biological function (see Section 9.2 for further discussion of this point).

CHAPTER 3

Sources of Metals in Terrestrial Ecosystems

3.1 GLOBAL DISPERSAL OF METALS

Metals are released continuously into the air and soil by volcanoes, natural weathering of rocks, and human activity. Natural fluxes of iron are similar to those resulting from industry (Fig. 3.1A) whereas anthropogenic sources of most other metals such as cadmium, far exceed those from natural processes (Fig. 3.1B). Environmental fluxes of metals have increased greatly since the industrial revolution (Wood 1974; Bowen 1979; Grandjean 1981; Galloway et al. 1982; Miller 1984a; Salomons & Forstner 1984; Sposito & Page 1984; De Jonghe & Adams 1986; Pacyna 1986a, 1986b; Salomons 1986). Metals released into the atmosphere as fine particles may remain aloft for several months and be carried over considerable distances (Rühling & Tyler 1968, 1969; Bowen 1975; Rolfe & Haney 1975; Lodenius & Tulisalo 1984; Wiersma 1986), as far even as the polar regions where levels in snow of different ages are correlated with the rise in industrial activity which has taken place over the past few hundred years (Wolff & Peel 1985).

If metal pollutants were distributed evenly throughout the biosphere after release, problems of toxicity would not arise. It is only when metals are localised near industrial plants, mine sites and major cities, that concentrations reach levels sufficient to upset normal ecological and physiological processes in living organisms (Peirson & Cawse 1979; Legge & Krupa 1986; Nriagu & Davidson 1986). The toxicity of metals is exacerbated by their permanence and long residence times in the environment. The half life of lead in undisturbed soil, for example, exceeds several hundred years in regions where the pH of the rainfall is neutral or only slightly acidic (Martin & Coughtrey 1987). Invertebrates which live in the soil/leaf litter ecosystem play an important role in the

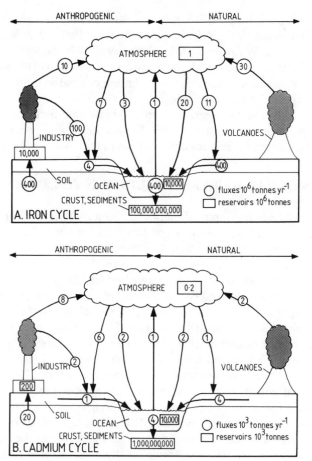

FIG. 3.1 Estimated annual global fluxes and reservoirs of iron and cadmium. Natural fluxes of iron (A) are similar to those released by mining. For cadmium (B), the flux is increased about 5 times by human activity. (After several authors.)

transport of metal contaminants from plant material to organisms higher in the food chain.

An enormous number of reviews on environmentally important metals have been published including articles and books on zinc (Nriagu 1980a; Jones & Burgess 1984), cadmium (Friberg *et al.* 1974; Webb 1979; Nriagu 1980b, 1981; Friberg & Kjellström 1981; Cole & Volpe 1983; Hutton 1983; Korte 1983; Elinder 1985; Friberg *et al.* 1985; Foulkes 1986; Mislin & Ravera 1986), lead (Nriagu 1978; Harrison & Laxen 1981; De Michele

1984), copper (Nriagu 1979), mercury (Mitra 1986), chromium (Allaway 1975; Anderson 1981), selenium (Shamberger 1981, 1983; Westermarck 1984), arsenic (Whitehead 1961; Watson *et al.* 1976; Fowler 1983), cobalt (Young 1979), vanadium (Robson *et al.* 1986) and nickel (Nriagu 1980c). The preponderence of studies on cadmium compared to other metals is a reflection of the concern which has surfaced in recent years as to the toxicity of this metal. Terrestrial invertebrates are hardly mentioned in these publications, with the notable exception of a paper by Wieser (1979) on copper in isopods and molluscs in the review by Nriagu (1979).

3.2 NATURAL SOURCES OF METALS IN THE ENVIRONMENT

The largest natural input of metals to the terrestrial environment on a global scale is from aerial fallout of dust particles emitted from volcanoes. Although the amounts of metals involved are huge, concentrations in the soil are rarely elevated to toxic levels due to the massive dilution which takes place in the atmosphere before the particles fall back to earth. Problems of metal contamination of soil are much more likely to occur where metal-rich rocks are exposed at the surface by natural weathering. Soils derived from shale, for example, may contain concentrations of cadmium of up to $7.5\ \mu g\ g^{-1}$, considerably greater than the normal background level of less than $1\ \mu g\ g^{-1}$ (Lund *et al.* 1981). Soils developed on serpentine rocks are characterised by their high content of cobalt, nickel and chromium which may be accumulated to very high levels by plants (Swaine & Mitchell 1960; Lyon *et al.* 1970; Proctor 1971; Proctor & Woodell 1975; Brooks *et al.* 1979; Anderson 1981; Vergnano-Gambi *et al.* 1982). Latex derived from the tree *Sebertia* in New Caledonia, an island built mostly of serpentine rock, contains 27% nickel on dry weight basis (Jaffre *et al.* 1976). Invertebrates such as termites (Wild 1975a), beetles (Wild 1975b) and earthworms (Schreier & Timmenga 1986) living in serpentine soils, may also contain high concentrations of cobalt, nickel and chromium.

3.3 ANTHROPOGENIC SOURCES OF METALS IN THE ENVIRONMENT

3.3.1 Introduction
There are many human activities which result in contamination of the

environment with metals. Such contamination may be deliberate in the case of metal-containing pesticides, or arise indirectly as a consequence of other agricultural or industrial processes. The amounts of metals accumulated by terrestrial invertebrates will be determined by the concentrations and chemical forms of the elements in their diet which depend to some extent on the source of contamination.

In heavily industrialised regions, metal contaminants of soil and vegetation may be derived from a number of sources and have an extremely complex composition which is difficult to define chemically. In such areas, point sources of contamination may be difficult to locate among the uniform blanket deposition of metals (Davies 1980; Hallet *et al.* 1982; Hosker & Lindberg 1982; Davies & Houghton 1984). Pacyna (1986a) recognised six main sources of metal contamination of the atmosphere, namely (1) stationary fuel combustion, (2) internal combustion engines, (3) non-ferrous metal manufacture, (4) iron, steel and ferroalloy plants and foundries, (5) refuse incineration and (6) cement production. Pollutants derived from such sources will eventually fall back to earth and contaminate soil and vegetation. Direct contamination of soil results from mining activity and the addition of metal-rich material to the land by agricultural practices, dumping of waste, and leaching of metals by rain from man-made structures such as electrical transmission lines and pylons (Hemkes & Hartmans 1973; Kraal & Ernst 1976; Jones 1983; Jones & Burgess 1984).

The major sources of metal pollutants in terrestrial ecosystems are described briefly below. More detailed information on the industrial uses of metals can be found in the reviews by Whitehead (1961), Berrow & Webber (1972), Barbour (1977), Stubbs (1977), Marples (1979), DoE (1980), Hiscock (1983) and Lansdown (1985).

3.3.2 Mining

Exposure of ore bodies by mining activity far exceeds exposure due to natural weathering. Serious pollution of the soil is usually restricted to the immediate area on and around spoil tips where metals are at too low a concentration to be extracted economically, but are at a high enough level to exert toxic effects on plants and animals. Most mine wastes are strongly acidic and it is often difficult to decide whether the absence of a particular soil organism is due to high levels of metals or the low pH. Metals may be transported away from mining areas *via* wind-blown particles or dissolved in acidic groundwaters. Marples (1979) and Khan

& Frankland (1983) have discussed contamination of soils with metal-rich mine waste.

3.3.3 Smelting

Once ores have been mined, the metals they contain must be purified by smelting, a process described in detail for the primary zinc, lead and cadmium works at Avonmouth, South West England, by Coy (1984). Metals may also be recovered from scrap by secondary smelting which accounts for about half of world production of lead bullion. The distance to which metals are carried from a smelting works (and indeed any industrial concern) and the degree of deposition, depend on the amounts of metals released, the height of the chimneys releasing the pollution, the particle sizes and the direction and strength of prevailing winds (Allen *et al.* 1974; see also Chapter 5).

In modern smelting works, waste gases are passed through fine filters and treated with electrostatic precipitators to remove most metal-containing particles from gases before they are released into the atmosphere. Unfortunately, even the most modern equipment working at removal efficiencies approaching 100% is unable to remove metals completely from these gases (Shendrikar & Ensor, 1986). At the Avonmouth plant which produces about 100,000 tonnes of zinc, 40,000 tonnes of lead and 300 tonnes of cadmium per year, some 6 kg of zinc, 4 kg of lead, 0.4 kg of cadmium and 0.1 kg of arsenic are released into the atmosphere each hour as fine particles with a diameter of a few μm or less (Harrison & Williams 1983; Coy 1984). These particles settle on the surrounding vegetation and soil and may be carried for more than 25 km from the plant by prevailing south-westerly winds (Hopkin *et al.* 1986). The effects of pollution from the Avonmouth works are described in Section 5.3.2.

The importance of effective removal of metals from waste gases was demonstrated vividly near a steelworks where symptoms of copper deficiency in cattle, induced by high molybdenum levels in vegetation, were removed entirely when more efficient bag filters were installed at the plant (Alary *et al.* 1983). Contamination of soil may still occur within the boundaries of a works by fugitive blowage of coarse particles containing primary ore minerals and from waste tips containing residues of the smelting process (Harrison & Williams 1983).

The environmental impact of pollutant metals from a smelting works can be alleviated if the emissions are spread over a wider area. In Sudbury, Canada, where nickel smelting activity had resulted in severe environmental damage (Costescu & Hutchinson 1972), the height of the chimney carrying waste gases was increased, toxic compounds were dispersed over a much greater area and the environment around the works recovered considerably (Chan & Lusis 1986).

3.3.4 Combustion of Fossil Fuels
The burning of coal, oil and natural gas for the domestic production of heat, or the generation of electricity in power stations, does not result in serious pollution of the environment with metals. Emissions of toxic gases such as oxides of sulphur and nitrogen are of more immediate concern. The amounts of metals released are large on a global scale, but they are dissipated away from their source in the same way as metals derived from volcanoes are diluted in the atmosphere. The burning of some oils, for example, is the main anthropogenic source of vanadium (Tyler 1976). In the United Kingdom, it has been estimated that about 50 tonnes of lead are released to the air each year from all fossil fuel consumption (excluding that derived from automobile exhausts; DoE 1974), an amount comparable to that emitted from the Avonmouth smelting works alone (Coy 1984). The main source of lead in the terrestrial environment is from the consumption of petrol to which the metal has been added as an 'anti-knock' agent (Hamilton *et al.* 1987). Lead is not added to diesel oil, the fuel used by most heavy goods and passenger vehicles.

Lead is added to petrol in the tetraethyl form to increase the octane rating and prevent 'pinking', the premature ignition of fuel in the engine cylinders (Farmer 1987). About 75% of the lead is emitted from the exhaust as fine particles. The remaining 25% is retained in the engine oil and internal surface of the exhaust. The chemical composition of lead compounds derived from traffic has been studied by Harrison and co-workers (Harrison & Johnston, 1985; Harrison *et al.* 1981, 1985, 1986). Huge quantities of lead are still added to petrol on a worldwide basis. In Venezuela in the early 1980s, for example, it was still legal to add 1.3 g of lead to each litre of petrol (Garcia-Miragaya *et al.* 1981).

Lead is undoubtedly toxic and there are still dozens of cases of poisoning reported each year in the U.K. resulting from ingestion of lead-based paints or contaminated water. However, controversy rages as to what proportion of the human body burden of this metal is derived from the release of lead additives in petrol (Lansdown 1985; Fergusson 1986). 'To be on the safe side', several countries have already removed this additive or have resolved to do so by the 1990s. In the U.K., the maximum permitted level of lead in petrol has been reduced gradually since 1970. The 1971 limit of $0.84 \, \text{g} \, \text{l}^{-1}$ was reduced to $0.64 \, \text{g} \, \text{l}^{-1}$ at the end of 1972 and to $0.55 \, \text{g} \, \text{l}^{-1}$ by 1975 (DoE 1974, 1979). This was reduced further to $0.15 \, \text{g} \, \text{l}^{-1}$ in the early 1980s with the ultimate aim of having the majority of cars running on lead-free petrol by 1990. Despite these reductions, the amount of lead released into the atmosphere in the U.K. in the 1970s remained fairly constant at about 10,000 tonnes per year due to the increase in road traffic (DoE 1979). Recently, however, the concentration of lead in air in the U.K. has shown a marked decline (Page *et al.* 1988) and this is expected to continue with greater use of lead-free petrol (especially as since the Budget of March 1988, it attracts less excise duty than petrol which contains lead!). North American and most European countries have followed a similar path towards lead-free petrol. Despite these moves, elevated concentrations of lead in soils are likely to persist for many thousands of years because of the long residence time of the metal in the environment (Siccama & Smith 1978).

Where traffic is fast-moving and the road is relatively exposed, the majority of particles of lead emitted from car exhausts are very small ($<0.5 \, \mu\text{m}$) and most are transported for considerable distances by the wind. Only 22% of all the lead released by cars on the M4 motorway in England since it was opened has fallen within 100 m of the road (Little & Wiffen 1977, 1978). In contrast, particles released from slow-moving cars in built up areas are much larger ($>5 \, \mu\text{m}$) and a greater proportion of the lead is deposited close to the road (Hughes *et al.* 1980). Lead deposited on the soil tends to remain near the surface due to its insolubility at neutral or slightly acidic pH's (Lagerwerff & Specht 1970). Consequently, concentrations of lead in some roadside soils may be more than two orders of magnitude greater than normal background levels which are themselves considerably greater than pre-industrial levels of lead in soil due to long distance transport (Davies & Holmes 1972; Quarles *et al.* 1974; Rolfe & Haney 1975; Smith 1976; Muskett & Jones 1980; Byrd *et al.* 1983).

3.3.5 Agriculture

Before the Second World War, pesticides containing arsenic, tin and mercury, were used extensively for the control of agricultural pests. Their long residence times in the environment made them apparently ideal for preventing fungal infections of crops (Hirst *et al.* 1961) and for treating attacks by molluscs (Lange & McLeod 1941; Lewis & La Follette 1942; Godan 1983), ticks (Matthewson & Baker 1975) and other invertebrate pests. However, the very factors which made metals

TABLE 3.1

Concentrations of metals in anaerobically digested sewage sludge used to 'reclaim' mine spoil in Chicago (mean ± standard deviation for 73 weekly composite samples taken from 1975 to 1977) (from Pietz *et al.* 1984, by permission of the American Society of Agronomy, Inc.; Crop Science Society of America, Inc.; Soil Science Society of America, Inc.)

Metal	Concentration ($\mu g \ g^{-1}$ dry weight)
Zinc	4 050 ± 683
Cadmium	309 ± 56
Chromium	3 820 ± 775
Copper	1 830 ± 269
Nickel	426 ± 71
Lead	917 ± 144

effective at long term control, namely their toxicity and long residence times in the environment, also resulted in serious pollution problems. In Iraq in the early 1970s, wheat seeds dressed with a mercury fungicide were accidentally made into bread. Several hundred people died before the source of the mercury poisoning was traced (Bakir *et al.* 1973; Mathew & Al-Doori 1976). Large areas of Vietnam were contaminated with metals due to the application by the American army of the defoliant 'Agent Blue' which contained 15% arsenic (Orians & Pfeiffer 1970). These and other similar incidents have led to the use of metal-based pesticides being discouraged.

Large quantities of metals are washed into drains from roads, industry and domestic households. Most of these metals are discharged into the sea in untreated sewage. However, concern over pollution of rivers and estuaries has led to an increase in the proportion of this waste which is treated in sewage works. The residue of the breakdown of organic material in sewage is a sludge rich in nitrogen which can be used as a fertiliser. Unfortunately, the sludge may contain considerable amounts of metals (Daum 1965; Berrow & Webber 1972; Hartenstein *et al.* 1980a; Lake *et al.* 1984; Table 3.1). Although generally not harmful in a single dose, repeated applications of contaminated sludge leads to accumulation of metals such as cadmium, copper, zinc and lead in the surface layers of soil (Soon 1981; Brown *et al.* 1983; Davis *et al.* 1988). These metals may then be assimilated to toxic levels by soil invertebrates (Furr *et al.* 1976; Berglund *et al.* 1984; Kruse & Barrett 1985). In the U.K., maximum permitted application rates of sewage sludge to farmland have been formulated to take account of this problem (DoE 1977).

A specific problem of copper contamination of agricultural soils may arise following application of pig slurry. The growth rate of pigs is substantially increased by the addition of copper to their diet at a concentration of 200 μg g^{-1}. This is an order of magnitude greater than the level required in their diet under normal circumstances. The physiological mechanisms which cause this effect are not fully understood. Most of the added copper is voided by the pigs in their faeces which, when applied as a fertiliser to soils, results in severe copper pollution which is particularly harmful to earthworms (Van Rhee 1975; Curry & Cotton 1980).

CHAPTER 4

Analysis of Metals in Biological Material

4.1 INTRODUCTION

Before going on to review the literature on the dynamics of metals in terrestrial ecosystems (Chapter 5), and invertebrates at the species, individual animal and organ level (Chapters 6 and 7), it is important to examine critically three aspects of the methodology of such studies. First, how are the results influenced by sampling procedures in the field and methods of preparation of material for analysis in the laboratory (Sections 4.2, 4.3), second, what are the relative advantages and disadvantages of the techniques available for the analysis of metals in biological material (Sections 4.4, 4.5, 4.6) and third, which statistical methods are most appropriate for the interpretation of the data (Section 4.7)? Methods of analysis for metals at the cellular level are covered in Chapter 9.

4.2 COLLECTION OF MATERIAL FROM THE FIELD

4.2.1 Climate and Season

Climatic factors have a considerable influence on the concentrations of metals in plants and animals in ecosystems polluted with metals. In sites subject to aerial deposition of metals, for example, much of the surface contamination of leaves is removed by heavy rainfall. Analysis of vegetation collected on the day after such precipitation will lead to a significant underestimation of the levels of metals to which herbivores are exposed at other times, particularly during periods of drought when metal-containing particles accumulate on the leaf surfaces (Martin & Coughtrey 1982).

Most groups of terrestrial invertebrates exhibit seasonal differences in concentrations of metals due to changes in population structure, diet and physiology (e.g. slugs, Ireland 1981, 1984a). Hunter *et al.* (1987a, 1987b, 1987c) collected plants and animals from a contaminated site on a monthly basis and showed that the concentrations of copper and cadmium in the biota varied substantially throughout the year (see Section 5.3.4 for a description of their work). It is clear that sampling programmes need to be designed to take account of seasonal variation.

4.2.2 Sampling Methods

No single collecting method is able to provide a representative sample of all species of invertebrate in a terrestrial ecosystem in the ratios in which they exist in the field. A comprehensive survey should include four sampling techniques (which are discussed further in Chapter 8).

1. *Collection by hand.* Some microsites such as rotting wood are difficult to sample and manual collection may be the only way to estimate abundance of species in these habitats. Handsorting of soil (after digging) is suitable for collection of earthworms if sampling by formalin extraction is not possible (see below). Invertebrates on above-ground vegetation can be dislodged into sweep nets or on to beating trays. Crude comparisons of densities of species in different sites can be achieved by searching for a specific time interval. In contaminated sites, these methods may of course catch invertebrates which have flown in from uncontaminated areas. This negates the use of light traps.

2. *Pitfall traps.* It is important to recognise that pitfall traps catch only surface-active invertebrates (mainly arthropods). Thus, not only do such traps only sample a restricted section of the population, but substantial differences also occur in the species composition of the animals collected at different times of the year because of seasonal variations in the activity (which may not be related to abundance) of different groups. The main advantage of pitfall traps is that they can be left in place for several weeks. However, the traps must contain preservative solutions to prevent the invertebrates from escaping or eating each other. Thus, there is the possibility that metals may leach out of the animals before they can be collected although where formalin and ethanol have been used, this does not seem to have been a problem (Clausen 1984a; Hunter *et al.* 1987b; Read 1988).

3. *Extraction of earthworms from soil in the field.* Lumbricid worms

comprise more than 70% of the biomass of soil and leaf litter invertebrates in uncontaminated temperate deciduous ecosystems. Worms can be induced to come to the surface by flooding their burrows with an irritant (usually a 1% solution of formalin) and are collected by hand (Raw 1959). Providing they are rinsed with water as soon as they emerge, worms collected in this way will survive unharmed. The density of worms in uncontaminated and contaminated sites can be compared by applying the same amount of formalin solution to the same area of soil and collecting emerging worms for a specific time interval. Providing several areas are sampled in each site to allow statistical comparisons to be made, the method is quite effective in demonstrating that soils contaminated heavily with metals support much lower numbers of earthworms than uncontaminated soils (Table 5.3).

4. *Leaf litter and soil extraction.* Unlike the previous three techniques, extraction of invertebrates from leaf litter and soil is conducted invariably in the laboratory. Most methods involve controlled drying by a source of heat (usually a light bulb) so that animals move out of the material, drop into a funnel and are directed into preservative solutions. This technique works reasonably well with loosely-packed soils but is less effective with dense soils such as clay from which the invertebrates have to be extracted by flotation. Such methods are good for comparing densities of the active leaf litter soil fauna, such as Collembola and mites, which are difficult to sample by any other method. Heat extraction may, however, seriously underestimate populations of soft-bodied invertebrates such as small Diptera larvae which are dehydrated before they can escape.

It may be necessary for invertebrates collected from the field to be kept alive in the laboratory for some days to allow the gut contents to be voided, for these may contribute substantially to the apparent amounts of metals which are 'in' the animals. Tissues in which the concentrations of metals are to be determined should be prepared for analysis immediately, or where this is inconvenient, they should be deep-frozen or dried, and stored in sealed containers. Metals may leach from material in preservative solutions so storage for lengthy periods in liquids is not recommended.

4.2.3 Dissection

The distribution of metals within individual animals is determined by dissection and analysis of the separate organs. Large soft-bodied invertebrates such as snails, slugs and earthworms may be anaesthetised with carbon dioxide if not already dead. Following dissection, each organ is placed in a separate pre-weighed container in which the tissues can be dried to constant weight at 70 °C in an oven. It is preferable to work on a dry weight basis as the organs lose water extremely rapidly by evaporation once they are removed from the body. Arthropods may be killed by rapid decapitation.

Small samples of tissue weighing a few mg are handled more easily when placed on to pre-weighed pieces of paper so long as blank analyses have shown that it does not contain any detectable metals. Filters manufactured from plastics (as opposed to plant fibre-based materials) are suitable and have been used successfully for weighing and transferring individual organs of isopods and centipedes (Hopkin & Martin 1982a, 1983).

4.3 PREPARATION OF MATERIAL FOR ANALYSIS

4.3.1 Introduction
The method adopted for the preparation of biological material for analysis depends on the analytical technique, cost, the metals which are to be determined, the sensitivity required and whether the analysis is to be destructive or non-destructive of the sample. Whatever the method, it is essential to prepare certified biological materials using the same procedures to ensure comparability between analyses conducted in different laboratories (Muntau 1984). The addresses of suppliers of such standard reference materials are given in Appendix 2.

Some methods for determining concentrations of metals in biological material do not require elaborate preparative procedures. The detection of gamma radiation from radioactive markers, for example, can be carried out on live animals (Section 4.6). However, most analyses of invertebrates at the species and organ level have been conducted by atomic absorption spectrometry (AAS) which requires that the metals are in solution. To get the material to this stage, organic matter has to be removed by ashing or wet digestion in acid. The problems inherent in such procedures are covered in this Section.

4.3.2 Wet Digestion
Sources of potential contamination of samples during wet digestion are

numerous (Stevenson 1985). In an ideal laboratory, the air would be filtered to remove all particulates from the atmosphere and all surfaces would be kept scrupulously clean. However, most researchers are not fortunate enough to work under such conditions and occasional contamination of samples is inevitable. This can be kept to an acceptable and *measurable* level by observing some simple precautions.

All glassware and plastic containers should be cleaned in a strong detergent before use, submerged in acidified water overnight (10% nitric acid is suitable) and rinsed at least three times in double-distilled water. The acids used for digestion of samples should be of analytical grade. Most manufacturers supply acids in several levels of purity and quote maximum values of metal contamination on the bottle. For example, BDH Chemicals (Poole, Dorset, U.K.) supply nitric acid in Analar and Aristar grades. The maximum level of lead in Analar nitric acid is quoted as 50 ng ml^{-1} whereas for Aristar grade nitric acid the figure is only 2 ng ml^{-1}. Analar grade is suitable for most environmental samples but where concentrations of metals are very low, Aristar acids should be used. Cost is an important consideration in this choice; Aristar grade nitric acid is an order of magnitude more expensive than Analar.

A variety of methods for wet digestion are available, some involving sealed units with elaborate devices for condensing and collecting acid fumes (Stevenson 1985). These may be essential for safety reasons when working with dangerous chemicals, or to prevent loss of the more volatile metals such as mercury, but with most elements all that is required is a method of heating the tissue samples together with some acid until the organic matter is broken down and driven off as a gas. Some researchers prefer to boil their digests to dryness and then make up to a standard volume with distilled water. Others allow the acid to cool after the organics have been degraded and add water so that the final acid concentration is quite high (10% to 20%). The method adopted depends very much on individual preference but its legitimacy should always be checked by taking standard reference materials containing certified levels of metals through the same procedures.

Nitric acid is the most frequently-used agent of digestion of organic material. Other reagents which are often employed include hydrochloric and perchloric acids but use of the latter has been discouraged in recent years because of its propensity to explode unpredictably! The presence of perchloric acid may also cause severe interference effects during analysis by flameless AAS for some elements, particularly selenium (Ringdal *et al.* 1984), whereas such problems arising from the use of nitric acid for digestion are rare.

Where the ratio of sample weight to acid volume is relatively high (i.e. more than about 1: 50), or the tissues have a high fat content, a residue may remain in the digests even after extended boiling. After making up to standard volume with distilled water it is possible to remove this residue by filtration, but care should be taken to use filters made from plastic rather than plant cellulose which contains substantial amounts of zinc. It may be acceptable, after dilution, to leave the samples overnight for residues to settle and to decant the liquid into a separate container when it has cleared.

For samples containing low levels of metals the final dilution volume is important. For flame AAS, a minimum of about 10 ml of solution are required for the determination of eight metals in the samples. For carbon furnace AAS, the required volumes are much smaller. Indeed, Bengtsson & Gunnarsson (1984) were able to measure the concentrations of lead in individual specimens of the collembolan *Onychiurus armatus* weighing only 25 µg because the final volume of the digests was only 50 µl.

Sources of aerial contamination are numerous so once digests have been diluted to standard volume and are ready for analysis they should be covered (clear plastic film is ideal). The talcum powder applied as a lubricant to disposable gloves contains a high concentration of zinc which may contaminate samples and it is not unknown for dust and fragments of paint to fall into digests from the roof of a fume cupboard!

Provided that samples are acidified and sealed, they can be kept almost indefinitely at room temperature. However, it is essential that blank solutions which have been through identical digestion processes are stored alongside samples in case metals leach from the walls of containers during storage. Indeed, the incorporation of blank digests in all runs is important so that the level of contamination of metals derived from the acids and other sources can be subtracted from the analytical signal. Blank readings of more than a few percent of the signal due to the sample should be a cause for concern and the source of contamination should be identified and removed.

4.3.3 Ashing

Removal of organic matter from biological samples can be carried out at room temperature by microwaves (Blust *et al.* 1985) or in an oxygen plasma created by radiofrequency (Mason 1983). However, most workers who have not adopted wet digestion methods have opted to ash biological material in a muffle furnace at about 600 °C (e.g. Ponomarenko *et al.* 1974; Beeby & Eaves 1983). This method is very effective at removing organics but the high temperatures involved may lead to loss of some metals. Zinc, chromium, cobalt and iron, for example, are stable at 600 °C but cadmium and mercury are removed at temperatures above 300 °C (Van Raaphorst *et al.* 1974; Koirtyohann & Hopkins 1976; Iyngar *et al.* 1978; Elinder & Lind 1985). Wet digestion is thus the preferred method if losses of the more volatile elements are to be avoided.

4.4 ANALYTICAL METHODS FOR MEASURING CONCENTRATIONS OF METALS IN BIOLOGICAL MATERIAL

4.4.1 Introduction
The vast majority of analyses of concentrations of metals in terrestrial invertebrates have been conducted by atomic absorption spectrometry and are likely to continue to be so until the cost of other methods decreases to a comparable level. Consequently, the technique is covered in some detail in this Section. Other analytical equipment which has been used for analysis of metals in invertebrate tissues is covered briefly in Section 4.4.3. Readers requiring a more comprehensive discussion of the relative advantages and disadvantages of the different methods available should consult the reviews by Kopp (1975), Willis (1975), Brätter & Schramel (1983, 1984), Jensen & Jorgensen (1984), Elinder & Lind (1985), Matsumoto & Fuwa (1986) and Hoffmann & Lieser (1987).

Concern as to the accuracy of published values of concentrations of metals in biological tissues has led to a number of interlaboratory comparability studies (Cornelis & Wallaeys 1984; Koh 1984; Pszonicki 1984; Elinder & Lind 1985; Stevenson 1985). The results of these studies have proved alarming to say the least! To take just one typical example, the International Atomic Energy Agency based in Austria organised an interlaboratory comparison in the early 1980s of the concentrations of elements in a standardised sample of 'animal blood' (Pszonicki 1984). Some 30 laboratories were involved in the study. About 80% of all values for 'minor' elements such as calcium and iron (concentrations above 100 μg g^{-1}), 50% to 70% of values for 'trace' elements such as copper and zinc (concentrations from 1 to 100 μg g^{-1}) and less than 30% of values for 'ultratrace' elements such as cadmium and lead (concentrations below 1 μg g^{-1}) were close enough to the true levels to be acceptable. Some laboratories reported concentrations for some metals which were more than two orders of magnitude greater than the true values in the sample.

Thus, it is highly likely that many published values for the concentrations of metals in invertebrate tissues are incorrect. The importance of calibrating instruments using standard reference materials cannot be overemphasised (Ihnat 1988). Perhaps editors of scientific journals should insist on authors providing evidence that they have carried out such a calibration before an article reporting metal concentrations in biological tissues can be accepted for publication.

The extent to which uncritical acceptance of analytical results can lead to erroneous conclusions was highlighted by Bowden *et al.* (1984) in their criticism of a paper by Turnock *et al.* (1980). The latter authors gave mean relative concentrations of 56 elements purported to have been found in field populations of the red turnip beetle *Entomoscelis americana* by energy-dispersive X-ray spectrometry. The relative concentrations given by Turnock *et al.* (1980) showed potassium concentrations several times lower than dysprosium, erbium, rhenium or tellurium, and phosphorus and sulphur no greater than thorium and uranium. Such data are of doubtful worth.

It is important to understand the difference between *precision* and *accuracy* of results (Fig. 4.1). Precision and accuracy of metal concentrations determined by a particular analytical technique depend on the relative sensitivity for each element. The sensitivity of flame atomic absorption spectrometry (AAS) for cadmium, for example, is about ten times greater than for lead (Fig. 4.2). Consequently, analyses of concentrations of lead near to the detection limit are much less precise *and* accurate than determinations of cadmium at the same concentration (Table 4.1). These comparisons are also valid for carbon furnace AAS but accuracy and precision are much better at low concentrations of lead in the samples due to the greater sensitivity of this technique (Fig. 4.2).

4.4.2 Atomic Absorption Spectrometry

In atomic absorption spectrometry (AAS), the absorbance of photons by atoms in the path of a light beam is measured. The light is provided by a lamp with a hollow metal cathode. The radiation emitted by the lamp depends on the metal contained in the cathode and is in the form of discrete line spectra which are unique to each element. For absorption of light to take place, metals in the analyte must be converted to the atomic state. In this form, the atoms are receptive to photons with

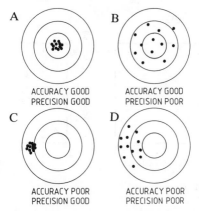

FIG. 4.1 The concepts of accuracy and precision of determinations of metal concentrations in a sample envisaged by 12 shots at a target. Situation A is ideal but B is acceptable since a mean of all 12 results gives a value close to the true level represented by the 'bull's-eye'.

energies corresponding to differences between energy levels of specific electron orbitals (the concept of energy levels of electron orbitals is discussed more fully in Section 9.4.2.3). Atomisation of the sample is achieved by heat decomposition in either a chemical flame or elec-trothermal furnace. The absorbance of light by atoms in the path of the beam is measured by a photoelectric cell and the levels of metals in the samples are quantified by comparison with absorbances of standard solutions containing known concentrations of the element.

The choice as to whether to use flame or furnace AAS depends on a number of factors which are discussed below. Several reviews have been published which go into the chemical principles and methodology of AAS in much greater detail than is possible here (Willis 1975; Thompson & Reynolds 1978; Tsalev & Zaprianov 1983; Varma 1984; Welz 1985; Matsumoto & Fuwa 1986).

4.4.2.1 Flame atomisation

In flame AAS, the analyte solution is sucked *via* a narrow-bore tube into a chamber where it is aspirated by impaction on to a glass nebuliser. The droplets pass into a flame, usually comprising a mixture of air and acetylene which burns at a temperature of about 2000 °C, or nitrous oxide and acetylene which burns at about 2700 °C. Light from the hollow cathode lamp passes through the flame and absorbance by metals which

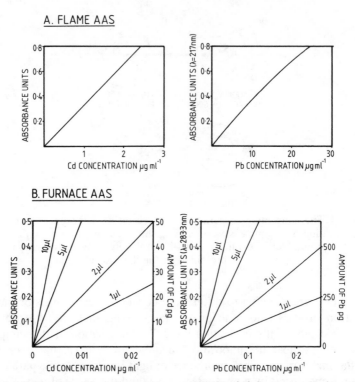

FIG. 4.2 Calibration curves for cadmium and lead in (A) flame atomic absorption spectrometry (AAS) and (B) electrothermal AAS with four injection volumes (μl). The amount of each metal injected into the furnace (pg) for each injection volume is indicated also. Wavelength of hollow cathode lamp for cadmium = 228.8 nm in both cases. For lead analyses on the furnace, a wavelength of 283.3 nm is usually used in preference to 217.0 nm. This reduces the sensitivity by about 50% but improves the signal/noise ratio.

have dissociated into the atomic state is measured by a photoelectric cell on the opposite side. Background absorbance due to non-atomic events in the flame is measured with a continuum source of radiation, usually a deuterium lamp. Modern machines are able to subtract the background signal from the total absorbance readings during the analysis. The sensitivity of flame AAS for hydride-forming elements such as mercury and arsenic, is much greater if the metals are introduced into the flame in the form of a vapour. Automatic vapour-generation accessories are available for most modern spectrometers.

TABLE 4.1
Comparison of the sensitivity of flame atomic absorption spectrometry (AAS)
for cadmium and lead. For a comparison with the sensitivity of electrothermal
furnace AAS, see Fig. 4.2.

Concentration of Cd in tissue sample (dry weight) = $10\ \mu g\ g^{-1}$
Concentration of Pb in tissue sample (dry weight) = $10\ \mu g\ g^{-1}$

1 g of tissue is digested in 20 ml of boiling concentrated nitric acid and made up
to 100 ml with distilled water to give concentrations of Cd and Pb in the diluted
digests of $0.1\ \mu g\ ml^{-1}$.

The following results are typical of those which might be obtained by flame
AAS of this digest analysed on three separate days after calibration with standard
solutions (lamp wavelength Cd, 228.8 nm; Pb, 217.0 nm; air–acetylene flame):

Day	Absorbance value	Calculated concentration in digest ($\mu g\ ml^{-1}$)	Calculated concentration in tissue ($\mu g\ g^{-1}$)	% of true value
Cd 1	0.024	0.096	9.6	96%
2	0.025	0.100	10.0	100%
3	0.026	0.104	10.4	104%
Pb 1	0.001	0.050	5.0	50%
2	0.002	0.100	10.0	100%
3	0.003	0.150	15.0	150%

Variation of 0.001 absorbance units from the true value is typical for flame AAS
and can result from slight contamination of blank solutions and day-to-day
variability in machine parameters. For cadmium in this example, this results in
a possible deviation of only 4% either side of the true mean, whereas for lead
the variation is 50%. Digests giving absorbance readings of <0.005 should be
analysed by flameless AAS, which is much more accurate (Fig. 4.2).

The main advantages of flame AAS are the relatively low cost, ease
of use, and ability to analyse large numbers of samples in a short time.
Indeed, the availability of such an inexpensive yet versatile technique
has probably done more to stimulate research on environmental aspects
of metal pollution than any other factor. It is quite possible to determine
the concentrations of eight elements in 50 digests in a single working
day. The great advances which have been made in automation of flame
AAS equipment over the past ten years have meant that the rate at
which analyses can be conducted is no longer limited by the equipment,
but is controlled by the time taken to collect and digest samples.

The principal disadvantages of flame atomisation are the presence of interference effects and poor sensitivity for some metals compared with electrothermal atomisation. The extent of interference can be measured by employing the method of standard additions (shown for electrothermal atomisation in Fig. 4.4). This should always be carried out when a new type of biological material is to be analysed, even if the method has been certified with standard reference materials. Changing the concentrations of calcium, for example, has a profound effect on the apparent concentration of lead in acid digests (Coughtrey & Martin 1976b). For this reason, published values of concentrations of lead in calcareous materials (such as bone or shell) should be treated with caution. A considerable number of chemical modifiers have been discovered in recent years which when added to the sample, are able to suppress some of these interference effects.

4.4.2.2 Electrothermal atomisation in a graphite furnace

In flameless AAS, small aliquots of sample (typically 10 μl) are dispensed into a furnace, usually a cylinder of graphite, through which the light beam from the hollow cathode lamp passes. The furnace is heated electrothermally and its temperature can be controlled very accurately. A typical heating programme involves a drying stage which evaporates liquid slowly from the sample, ashing stages which burn off any organic matter, and atomisation when the sample is taken to temperatures in excess of 2000 °C and absorbance is measured (Fig. 4.3). A typical heating programme is shown in Table 4.2. A constant flow of inert gas (argon or oxygen-free nitrogen) is maintained during the drying and atomisation stages to prevent oxidation of the graphite.

There are a number of advantages of flameless AAS over the flame method. During flame AAS, when the analyte impacts against the glass bead of the nebuliser in the spray chamber, some of the solution does not pass into the flame but is lost as waste. Furthermore, following introduction of the aspirated sample, the flame must be allowed to stabilise for a few seconds before a reading can be taken. Atoms which pass through the flame are in the path of the light beam for only a fraction of a second before they are expelled. Thus, less than 10% of the sample which is sucked into the machine is actually used for the concentration determination. In an electrothermal carbon furnace, all the atoms in the sample are held in the path of the light beam for the entire duration of the analytical reading. Because of this, flameless AAS is more than 100 times more sensitive for some elements than flame

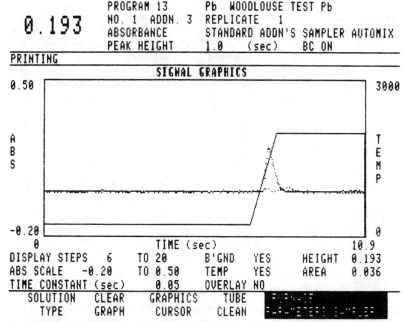

FIG. 4.3 Typical absorbance signal for lead (as part of a standard additions programme) produced during atomisation of a nitric acid digest of isopod tissue in an electrothermal graphite furnace. ('Screen dump' from Varian DS 15 data station fitted to Spectra AA30 and GTA 96.)

AAS (Fig. 4.2). Since such small amounts of analyte are dispensed, samples can be digested and diluted to final volumes of less than 1 ml leading to even greater sensitivity (Bengtsson & Gunnarsson 1984). Detection limits can be further improved by multiple injection of aliquots of the same sample. Ten injections of 20 µl of sample with a drying step between each would be equivalent to one injection of 200 µl. This is too great a volume to dispense into the furnace in a single step.

The facility to be able to inject a number of different solutions into the furnace prior to atomisation has other advantages. The sampler can be programmed to dispense standards of a range of concentrations by adjusting the amount of a single standard solution which is injected. Furthermore, standard additions are much easier to perform as these can be made up by the sampler 'on the furnace' (Table 4.3; Fig. 4.4).

Interference effects between some metals are reduced in flameless AAS because the elements atomise at different temperatures, unlike

TABLE 4.2

Typical heating programme for electrothermal atomisation of an acid digest of
animal tissue during flameless atomic absorption spectrometry (AAS). Volume
of sample injected into furnace = 10 μl. The ashing stage can be omitted if all
organic matter has been removed during digestion.

Stage	Step	Temperature (°C)	Time (s)	Gas flow (min^{-1})	Read
Dry	1	75	2	3.0	—
	2	95	15	3.0	—
	3	140	10	3.0	—
ASH	4	300	8	3.0	—
	5	450	7	3.0	—
	6	450	10	3.0	—
	7	450	1	0	—
ATOMISE	8	2 300	1	0	√
	9	2 300	1	0	√
	10	2 300	2	3.0	—

flame AAS where all atoms are heated simultaneously to the same
temperature. Thus, one metal may be completely atomised before
another metal with which it normally interferes comes off the wall of the
furnace. Careful control of the temperature and duration of the ashing
stage can also remove interference effects. Automatic sample dispens-
ation allows chemical modifiers to be mixed with the analyte in the
furnace. There are a very large number of these modifiers and more are
being discovered at a rapid rate. One of the most widely-used is
ammonium nitrate which is particularly useful when analysing for metals
in solutions with a high chloride and sodium concentration (e.g.
seawater). The ammonium nitrate combines with the salts to form more
volatile compounds which come off the furnace during the ashing rather
than the atomisation stage.

The main disadvantage of electrothermal atomisation is the much
greater time taken for analysis of samples. Determination of the concentrations of a single element in 50 samples by flame AAS may take as
little as 30 minutes whereas the same analyses conducted by flameless
AAS may take a whole working day depending on whether standard
additions or chemical modifiers are employed. However, the greater
sensitivity and precision of flameless AAS (which is particularly notice-
able for lead analyses — Fig. 4.2) are essential for the accurate determination of low concentrations of metals in invertebrate tissues. Analysis

TABLE 4.3

Volumes of standard solution (containing 0.050 μg ml^{-1} Pb), sample and blank injected into furnace to give readings for the standard addition calibration shown in Fig. 4.4 (lamp wavelength 217.0 nm)

Reading	Addition concentrations (μg ml^{-1})	Dispensed volumes (μl)			Total volume dispensed into furnace	Mean absorbance value obtained
		Standard	Sample	Blank		
1 BLANK	(0.000)	0	0	10	10	0.000
2 ADDITION 0	(0.000)	0	4	6	10	0.066
3 ADDITION 1	(0.025)	2	4	4	10	0.103
4 ADDITION 2	(0.050)	4	4	2	10	0.146
5 ADDITION 3	(0.075)	6	4	0	10	0.189

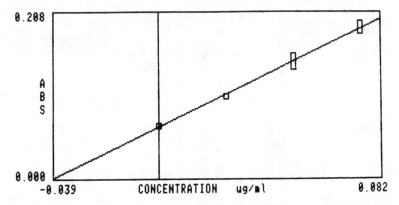

FIG. 4.4 Standard additions curve for lead concentration in Standard Reference Material 'Lobster Hepatopancreas' (TORT-1, National Research Council of Canada, Marine Analytical Chemistry Standards Programme). 1 g of the material was digested in 20 ml of concentrated nitric acid and made up to 250 ml with distilled water. Furnace and autosampler parameters for the analyses are shown in Tables 4.2 and 4.3. The 'curve' intersects the x-axis at 0.039 μg ml^{-1}, a value close to the certified concentration of lead in the sample of 0.041 μg ml^{-1}. When this sample is analysed without standard additions, a value of only 50% of the true level is obtained. The 'suppression' of the analytical signal (revealed by using the standard additions technique) is due probably to chemical interference when the lead is atomised from the furnace wall into the relatively cooler argon gas in the light path.

by flameless AAS has, for example, enabled the concentrations of metals to be determined in individual Collembola (Van Straalen & Van Meerendonk 1987) and internal organs of centipedes (Hopkin & Martin 1983) which would not have been possible using flame AAS.

4.4.3 Other Analytical Techniques

A small number of authors has used analytical techniques other than atomic absorption spectrometry to determine the concentrations of metals in the tissues of terrestrial invertebrates. The methods used most frequently are outlined briefly below. While all three techniques have the advantage of simultaneously displaying the concentrations of all elements in the samples, they have been used far less often than AAS because of poor sensitivity (e.g. energy dispersive X-ray fluorescence — Table 4.4) or much greater cost (plasma emission spectrometry and

TABLE 4.4
Sensitivity of a range of analytical techniques for cadmium (after Elinder &
Lind 1985; Matsumoto & Fuwa 1986)

Method	Detection limit
Atomic absorption spectrometry: flame	$1\,\mathrm{ng\,g^{-1}}$
flameless	$0.01\,\mathrm{ng\,g^{-1}}$
Inductively coupled plasma emission spectrometry	$1\,\mathrm{ng\,g^{-1}}$
Neutron activation analysis	$1\,\mathrm{ng\,g^{-1}}$
X-ray fluorescence: energy dispersive	$50\,\mathrm{\mu g\,g^{-1}}$
wavelength dispersive	$5\,\mathrm{\mu g\,g^{-1}}$

neutron activation analysis). All these techniques are described more
fully in the review by Matsumoto & Fuwa (1986).

4.4.3.1 Plasma emission spectrometry

This technique employs a high temperature plasma (c. 6000 °C) of rare
gases such as argon or helium into which the sample is aspirated. The
excited atoms emit photons of specific energies allowing the concentrations of most elements to be determined simultaneously. In inductively coupled plasma emission spectrometry (ICPES), the rare gas
atoms are excited by electrons accelerated by an alternating magnetic
field. ICPES has been used to determine the concentration of a range
of metals in larvae of the silkworm Bombyx mori (Suzuki et al. 1984).
An alternative method of generating the plasma is to use a direct
electrical current. Direct current plasma emission spectrometry has been
used for the analysis of lead and cadmium in earthworms (Pulliainen et
al. 1986).

4.4.3.2 Neutron activation analysis

In neutron activation analysis (NAA), a sample is irradiated with thermal
neutrons in a nuclear reactor. The gamma rays emitted by the radioactive
isotopes produced are measured by a suitable detector. The main advantage of this technique is that because the sample does not have to
be decomposed, chemical interferences are negligible and under some
circumstances it is possible to analyse live material. The main disadvantage is the requirement for a nuclear reactor! NAA has been used
for the analysis of arsenic in Orthoptera (Watson et al. 1976) and a range
of elements in earthworms (Helmke et al. 1979).

4.4.3.3 X-ray fluorescence

If the atoms in a sample are excited sufficiently they emit X-rays with characteristic energies corresponding to quantum levels. The excitation energy may be provided by bombardment of the sample with electrons or protons. There are two systems for analysis depending on whether the wavelength or energy of the X-rays is measured. In wavelength dispersive X-ray fluorescence (WDXF), the X-rays are separated by a crystal and a detector is moved to particular parts of the spectrum to measure the intensity of a narrow range of wavelengths. In energy dispersive X-ray fluorescence (EDXF), all the X-rays fall onto a detector which measures the whole range of energies. A spectrum is produced from which the concentrations of all elements in the sample can be calculated simultaneously. Although WDXF is limited in that it only provides information on a small part of the X-ray spectrum, it is 10 to 100 times more sensitive than EDXF. The sensitivity of both X-ray fluorescence methods is, however, very poor compared with other analytical methods (Table 4.4). WDXF has been used for the detection of metals in moths (Bowden *et al.* 1984; Sherlock *et al.* 1985) and a range of other insects (Udevitz *et al.* 1980), and EDXF for analysis of metals in fire ants (Levy *et al.* 1974a), aphids (Bowden *et al.* 1985a) and cockroaches (Van Rinsvelt *et al.* 1973). Applications of WDXF and EDXF in electron microscopy are covered in Section 9.4.2.3.

4.5 CHEMICAL SPECIATION OF METALS

The chemical speciation of metals in the environment is extremely complex and difficult to quantify (Cantillo & Segar 1975; Turner 1984). In terrestrial ecosystems, the proportions of metals bound to different sites in soil and leaf litter have been estimated by subjecting samples to a series of successively stronger chemical extractants (Martin *et al.* 1976; Mohan *et al.* 1982). The concentrations of metals are measured in each solution after shaking with the sample, and the sites from which the metals were removed are inferred from the chemical nature of the extractants used.

In terrestrial invertebrates, the forms in which metals are present in tissues can be assessed in homogenates separated into fractions by centrifugation. Proteins can be further separated on the basis of molecular weight and their metal contents measured by atomic absorption spectrometry or similar technique (see Section 9.3.1 for a detailed discussion of these and other methods).

The chemical form of metals in digestive fluids of terrestrial invertebrates has not been measured directly *in vivo* but has been inferred on the basis of how the metals should behave under the conditions of pH, temperature and presence of potential ligands derived from the food. Since this is vitally important in controlling the availability of metals at the gut/lumen interface where most metals are assimilated, much more attention should be devoted to research on the chemistry of digestive processes.

4.6 RADIOACTIVE ISOTOPES

The concentration of a metal in a liquid sample can be determined using a method known as isotope dilution (Matsumoto & Fuwa 1986). The isotope ratios are measured before and after a sample is spiked with the element to be analysed. The added element must possess a different isotope composition from that of the sample. The isotope ratio measurement is carried out either by a mass spectrometer when a stable isotope is added, or by radioactivity measurement when a radioactive isotope is used as a 'spike'. This method is very sensitive for most metals. The detection limit for cadmium, for example, is about $10 \, \text{ng g}^{-1}$ which compares favourably with other analytical techniques (Table 4.4).

The more traditional use of radioactive isotopes is as tracers for following pathways of metals. The metal in question is 'spiked' with a known amount of a radioactive isotope of the same element. Since the unradioactive and radioactive forms are treated in the same way by organisms, the pathways taken by the metal can be followed accurately. The great advantage of using isotopes in this way is the ability to measure fluxes of metals. With techniques such as atomic absorption spectrometry this is not possible because only net accumulation or movement can be determined. Furthermore, because radioactivity can be measured in solid samples in air, it is possible to analyse a living animal repeatedly over an extended period.

The range of metals which can be used as tracers is limited to those which possess half lives of a reasonable length. Experiments with ^{65}Zn (half life 245 days), ^{54}Mn (280 days) and ^{109}Cd (470 days) are fairly straightforward but most of the activity of ^{64}Cu (half life 12.8 hours) has decayed by the time it reaches the laboratory from the suppliers.

Examples of the use of isotopes to follow pathways of metals in terrestrial invertebrates include studies on snails (Williamson 1975), beetles (Dismukes & Mason 1975; Mason *et al*. 1983a, 1983b) and moths (Engebretson & Mason 1980). Other examples are described in the relevant sections of Chapters 7 and 9.

4.7 ANALYSIS AND PRESENTATION OF DATA

4.7.1 Introduction
The previous sections of this chapter have discussed the factors which affect whether the concentrations of metals determined in samples are an accurate reflection of the amounts present. Even if the values are correct, conclusions based on such data are influenced by the ways in which results are interpreted and presented. These considerations are examined in this section.

When interactions between metals are being considered, it is important to consider the elements in terms of their molar concentrations (even if figures are expressed on a weight basis). A sample containing 100 μg g^{-1} of zinc and copper, 200 μg g^{-1} of cadmium and 400 μg g^{-1} of lead contains about the same number of atoms of each metal (see Appendix 1). It is also important to state clearly whether concentrations values are expressed on a dry or wet tissue basis (for further discussion of these points see Section 1.3).

4.7.2 Amounts, Concentrations and 'Uptake' of Metals
The *amount* of a metal present in an organism is the weight of the element in the tissues. The *concentration* of a metal in an individual at a particular point in time (the most important factor for a predator) depends on the amount of the element the animal contains relative to its biomass. If the amount of a metal in an individual increases with age at the same rate as its biomass, then the concentration remains constant (Fig. 4.5). However, if the amount of the metal accumulates at a faster rate than biomass then the concentration increases. The concentrations of some metals in juveniles are often lower than in the adults in a population because the mature animals have ceased to grow but continue to assimilate metals.

These considerations are also important when assessing the relative importance of the different internal organs for storage and transport of metals in terrestrial invertebrates. For example, the kidney of a typical

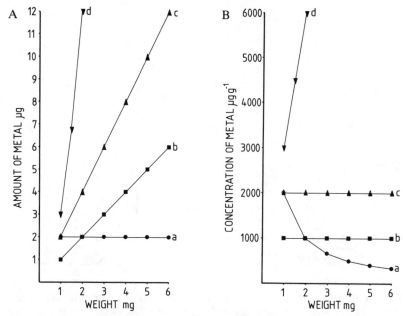

FIG. 4.5 Comparison between expressing levels of metals in terrestrial invertebrates as (A) amounts or (B) concentrations on the y-axis against weight of individuals on the x-axis. The same four groups of animals (a, b, c, d) are represented in both graphs.

fully-grown specimen of the snail *Helix aspersa* weighs 30 mg (dry weight) whereas the hepatopancreas is some ten times heavier at 300 mg. In snails from an uncontaminated site, these organs contain about 1 µg and 10 µg of copper respectively. Consequently, both organs maintain the same concentration of copper (*c.* 30 µg g^{-1}) but the hepatopancreas contains ten times the amount (Coughtrey & Martin 1976a).

The figure of 30 µg g^{-1} for the level of copper in the kidney and hepatopancreas of *Helix aspersa* represents the concentration in the organs at the moment the animals were killed. Fluxes of copper through the kidney are likely to be rapid whereas in the hepatopancreas, the atoms will have a much longer residence time. Techniques such as atomic absorption spectrometry, which measure only the amounts of metals in samples, can not evaluate these fluxes. Net accumulation of metals by individual animals or internal organs is not synonymous with 'uptake' in the strict sense of the word. Quantification of metal uptake *sensu stricto* is only possible using radioisotopes.

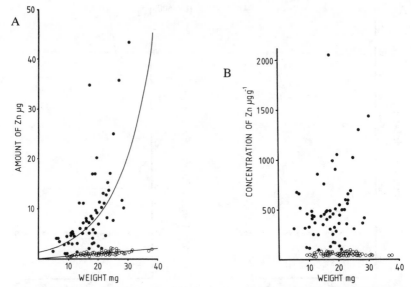

FIG. 4.6 Relationships between (A) amounts and (B) concentrations of zinc and dry weight of individual *Oniscus asellus* ($n = 60$). The isopods were collected from Shipham (disused zinc mine, solid circles) and Midger Wood (uncontaminated site, open circles). The three woodlice with more than 30 µg of zinc in the body were moribund. (A, after Hopkin & Martin 1984a, by permission of the Zoological Society of London; B, original).

Regression parameters for graph A:

Midger: Zinc content (µg) = 0.0477·body weight (mg) + 0.132 ($r = 0.724$)
Shipham: Zinc content (µg) = 1.366·$e^{(0.091 \cdot \text{body weight (mg)})}$ ($r = 0.658$)

(the exponential function gives a more significant correlation coefficient than a linear relationship, although it is likely that the animals represent two year classes—see text and Fig. 7.7).

The concentrations of metals in terrestrial invertebrates are usually expressed as a mean of several individual analyses, together with the standard error which gives a measure of the variability of the data. It is useful also to state the range which may be important when the extreme concentrations of metals in the diets of predators are being considered. Possible cumulative errors introduced during preparation and analysis of samples must be considered. It is not legitimate to express concentration values to seven or eight significant figures if repeated analyses of the same sample give values which differ even by as little as 1% (Table 4.1).

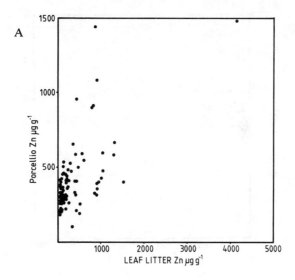

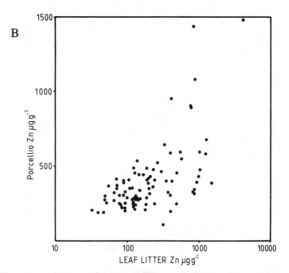

FIG. 4.7 The same data set (concentrations of zinc in 86 pooled samples of 12 *Porcellio scaber* from Avon and Somerset, England, against concentrations in leaf litter at the same sites) plotted with $\log_{10}$ or non-log axes. (Log × log plot (D), after Hopkin *et al.* 1986.)

(*contd*)

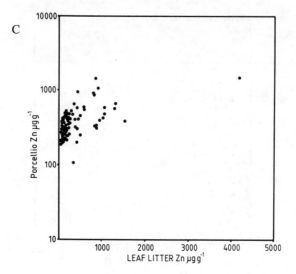

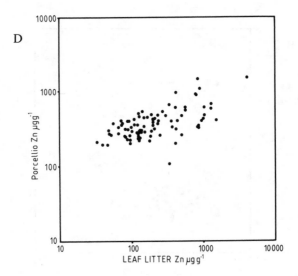

FIG. 4.7—*contd.*

The significance of differences in mean concentrations between different populations is determined with an appropriate statistical test. The correct use of statistics for the analysis of data on metal concentrations in invertebrates will not be discussed here (it would require several books), suffice to say that an appropriate text on the subject should be consulted before such analyses are undertaken.

4.7.3 Graphs

The concentrations and amounts of a metal in animals of different weight in a population can be displayed graphically (Fig. 4.6). If the isopods in this example put on weight at a constant rate then the x-axis is synonymous with age. However, the growth rates of isopods within a population are so variable that individuals of one year class may overtake those of a previous year (Fig. 7.7). It is important to recognise that such graphs only represent a 'snapshot' picture of the distribution of a metal within a population at a specific moment in time.

It is strongly recommended that scatter plots are always drawn when correlations between sets of data are examined. The reporting of regression equations and correlation coefficients only can lead to misinterpretation of relationships. The use of logarithmic rather than arithmetic axes on graphs may be more desirable if the data are spread over several orders of magnitude (Fig. 4.7).

CHAPTER 5

Metal Pollution and Terrestrial Ecosystems

5.1 TRANSPORT OF METALS THROUGH FOOD CHAINS

5.1.1 Introduction

The amounts of metals assimilated by invertebrates in terrestrial ecosystems are controlled by two factors. First, the concentrations of 'available' metals in their food and second, the physiological mechanisms which they possess for uptake and excretion of elements. For primary consumers (i.e. herbivores and detrivores), the degree of metal assimilation will depend on the concentrations and chemical forms of metals in the living or dead plant material on which they are feeding. For secondary consumers (i.e. carnivores), the degree of assimilation will depend on the availability of metals in the species of primary consumers which form their diet.

In this section, the main factors which control the amounts of metals transported between trophic levels are discussed. Uptake of metals by individual species is covered in greater detail in Chapters 6 and 7.

5.1.2 Accumulation of Metals by Plants

In terrestrial ecosystems, metals may be associated with plants as a surface deposit of particles, or contained within the plant tissue bound to organic ligands.

Surface deposition is a passive process over which the plant has little control. The amounts of metals retained on leaf surfaces depend on abiotic factors such as the ability of leaves to retain metal-containing particles of different diameters under particular climatic conditions. Laboratory experiments have shown, for example, that retention of plutonium and americium on leaves of bush bean (*Phaseolus vulgaris*) varies from 20% to 92% depending on particle size, chemical form of the metals and environmental conditions such as humidity, wind speed, and acidity and amount of the rainfall (Cataldo *et al.* 1981). Plants with 'hairy' leaves retain particles much more effectively than those with smooth leaves (Little 1977; Little & Wiffen 1977; Cataldo *et al.* 1981). Small particles have longer residence times on leaves than large particles because they are more strongly adsorbed onto the leaf surface (Little 1973).

Assimilation of metals into the cells of the plant occurs at the soil/root interface. The rates of uptake depend on the concentrations and chemical forms of elements in the interstitial water between soil particles and the relative efficiencies of active or passive mechanisms of transport for each metal into the root *via* the symplastic or apoplastic pathways. The availability of metals is affected by moisture content, pH, temperature and particle size of the soil, and differences in concentrations of metals with depth (Doelman & Haanstra 1979a).

Experiments on the uptake of cadmium by rape plants have shown that amounts assimilated are correlated with the exchangeable fraction of the metal and not total concentration (Singh & Narwal 1984). The most important factor which determines the proportion of the total metal content of soil which is available to plants is the pH. Hydrogen ions compete with metal ions for exchange sites on soil particles so that in general, the solubility of metals increases as the pH decreases (Harmsen 1977; Lawrey 1978; Herms & Brummer 1980; Esser & Bassam 1981; Tyler & McBride 1982; Martin & Coughtrey 1987). Consequently, residence times of metals are particularly long in calcareous soils with their alkaline pH. The pH can be quite critical. In loamy soils, the availability of cadmium increases markedly at around pH 6 but this does not occur with zinc until the pH falls below about 5 (Scokart *et al.* 1983).

The pH of leaf litter tends to decrease naturally with time due to the release of organic acids during decomposition (Bolter & Butz 1975). The concentrations of relatively immobile metals such as lead and copper, will increase relative to more mobile elements such as cadmium and zinc, as decay of organic matter proceeds (Staaf 1980).

Tyler (1978) conducted laboratory experiments to mimic the effects of 'acid rain' on the mobility of a range of metals in two mor spruce forest soils, one uncontaminated, and one similar soil heavily polluted with emissions from a brass foundry. Artificial rainwater acidified to pH 4.2, 3.2 or 2.8 was used as a leaching agent and applied at a rate equivalent to an annual precipitation of 150 l m^{-2}. Estimated 10% residence times in the contaminated soil (i.e. the time which it would have taken for concentrations of metals to decrease by 10%) were 10 to 20 years for zinc and cadmium, about 100 years for copper and more than 200 years for lead. Residence times for all metals were lowest in soils to which the most acidic precipitation had been applied.

Metals form highly stable organo-mineral complexes in soils rich in organic material (Kiekens 1984). Brown *et al.* (1983) applied sewage effluent containing 1 $\mu g\,g^{-1}$ of added cadmium, copper, nickel, lead and zinc to soil columns weekly for a year. All the added metals remained bound to organic components of the sewage in the top 25 cm and none were detected in leachates from the columns at a depth of 1.5 m.

Concentrations of metals in soils over ore bodies, or on spoil tips in mining areas, usually increase with depth where there is no source of aerial contamination (Watson 1970) whereas in aerially polluted regions, concentrations of metals are much higher at the surface (Buchauer 1973; John *et al.* 1976a, 1976b; Balicka *et al.* 1977; Brown *et al.* 1983; Scokart *et al.* 1983). The depth to which the roots of a plant penetrate will therefore be a critical factor in determining the concentrations and forms of metals to which the uptake sites on the roots are exposed (Fig. 5.1; Martin & Coughtrey 1981).

The chemical form of metals in the soil has a considerable influence on the degree of uptake by plants. The uptake of methylated mercury by bean plants, for example, is ten times greater than inorganic mercury at the same concentration (Huckabee & Blaylock 1973). Metals applied in an inorganic form may be methylated by soil microorganisms (Mohan *et al.* 1982).

Two metals present in the soil in the same chemical form and concentration may be assimilated at different rates if they follow different biochemical routes into the plant. However, metal ions may be antagonistic if they compete for the same uptake site on the roots, or synergistic if the presence of one metal increases uptake of another by stimulation of transport mechanisms which are common to both.

Considerable differences in metal accumulation may be exhibited by different species of plants, even if they are closely-related taxonomically (Guha & Mitchell 1966; Van Hook *et al.* 1977). For example, pot trials with different species of *Alyssum* (Cruciferae) collected from the same site, showed that *Alyssum serpyllifolium* barely survived 300 $\mu g\,g^{-1}$ nickel in the substrate and only contained 65 $\mu g\,g^{-1}$ nickel in its leaves. *Alyssum pintodasilvae*, on the other hand (a so-called 'metallophyte' — a plant confined to metal-contaminated sites), showed no adverse reaction when grown in soil containing 3000 $\mu g\,g^{-1}$ nickel and accumulated the metal

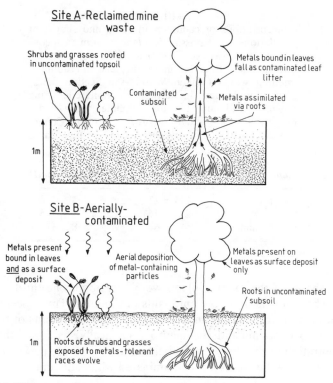

Fig. 5.1 Schematic diagrams comparing the distribution of metals in (A) a disused mine site 'rehabilitated' by application of uncontaminated topsoil, and (B) a site subject to aerial contamination.

in the leaves to a concentration of 14 000 μg g^{-1} (Brooks *et al.* 1979).

Tolerant races of several plant species ('pseudo-metallophytes') have evolved which are able to grow successfully in soils contaminated with metals. At Shipham, South West England, a tolerant race of the grass *Holcus lanatus* grows in mining waste that contains nearly 1000 μg g^{-1} of cadmium (Davies & Ginnever 1979; Marples 1979; Wigham *et al.* 1980; Khan & Frankland 1983). Tolerance to cadmium in *Holcus lanatus* is accompanied by restriction of transport of cadmium to the shoots. Plants from an uncontaminated site grown in nutrient solutions containing 1 mg l^{-1} of cadmium accumulated the metal to the same total concentration as tolerant plants from a site contaminated with cadmium (Coughtrey & Martin 1978a). However, the concentration of cadmium

TABLE 5.1

Concentrations of cadmium in roots and shoots of
populations of the grass *Holcus lanatus* from an uncontaminated site (Midger Wood) and a site contaminated
with emissions from a primary zinc, lead and cadmium
smelting works (Hallen Wood—see Section 5.3.2)
grown for 24 days in nutrient solutions containing
$1\ \mu g\ ml^{-1}$ cadmium (as the sulphate) ($\mu g\ g^{-1}$ dry
weight, mean $\pm$ standard error, $n = 5$) (from Coughtrey & Martin 1978a, by permission of Munksgaard
Ltd)

Plant part	Midger Wood	Hallen Wood
Shoot	145 ± 22	43 ± 6
Root	1 211 ± 161	1 799 ± 185
Whole plant	545 ± 64	538 ± 31

in the shoots of the tolerant plants were only about one third of those
in the shoots of the non-tolerant plants (Table 5.1). The concentration
of cadmium in the food of an animal which eats the shoots of tolerant
plants in a contaminated site may, paradoxically, be lower than the
concentration in the diet of an animal eating the same species of plant
in a site where the soil is not contaminated with cadmium.

5.1.3 Accumulation of Metals by Primary Consumers

The term 'concentration factor' is often used to demonstrate relative
differences in the transfer rates of metals in ecosystems. It is calculated
by dividing the concentration of a metal in an organism by the concentration of the same metal in its presumed diet. Such an approach has
demonstrated that cadmium is of great concern as a pollutant because
as well as being extremely toxic, it exhibits higher concentration factors
in food chains than most other metals including lead and zinc (Roberts
& Johnson 1978; Hunter & Johnson 1982).

Differences in the concentration factors for some metals between
species of invertebrates in the same site may be accounted for by
differences in their uptake, storage and excretory mechanisms (see
Chapter 6). This can only be categorically proved by comparing metal
uptake in different species reared on identical diets under laboratory
conditions. Where such experiments have been carried out, differences
have been demonstrated, even between closely related species. Some of

the variation between species may be accounted for by subtle differences in diets in the field.

If we are to understand the precise pathways that metals follow in terrestrial food chains, it is vitally important to know the specific organisms on which animals are feeding, and the precise parts which are ingested. With primary consumers, measuring the concentrations of metals in pooled samples of vegetation disguises a major source of variation in the metal content of the diet. Many herbivores exhibit distinct preferences for the leaves of some plant species over others (Jennings & Barkham 1975, 1979). Measuring the total concentration of a metal in the leaves of a plant is clearly inappropriate for assessing uptake rates in an insect which feeds on plant sap. Many soil arthropods which live in leaf litter do not feed directly on dead plant material but graze on fungal hyphae on the leaf surfaces (Reichle & Crossley 1965; Anderson & Healey 1972; Reichle 1977; Standen 1978; Parkinson *et al.* 1979; Ineson *et al.* 1982). Concentrations of metals in fungi may be an order of magnitude higher than in the leaf material on which they are growing (Ross 1975; Gadd 1981).

The availability to a primary consumer of a metal associated with plant material at a particular concentration will depend on its chemical form. The assimilation of mercury from food by dipteran larvae, for example, is much greater when the metal is in the methylated rather than the inorganic form (Lodenius 1981). In polluted ecosystems, the source of metal contamination is extremely important. Consider two contrasting sites (Fig. 5.1). In Site A which is a mining area, the soil is contaminated to a considerable depth. Trees in this site contain metals transported to the leaves from the soil. The metals are bound firmly to organic ligands in the leaf tissues and may be relatively insoluble in the gut of an animal consuming them. In Site B which is aerially polluted, the trees contain very little metal contamination internally because most of the roots are below the level to which pollutant metals have penetrated. Contamination of leaves in Site B is predominantly as a surface deposit of metal-containing particles which are removed easily as leaves pass through the gut of consumers. Near a copper refinery, for example, 25% of the copper associated with grass was incorporated in the plant tissues as protein-bound or ionic species whereas 75% was present as superficial deposits of metal-rich particles derived from fallout (Hunter *et al.* 1984a). Furthermore, such particulate matter may be distributed unevenly on the leaf surface, and depending on the diameter and purity, relatively few particles may give the impression that the vegetation is highly contaminated and that the tissues contain high concentrations of metals.

Metals accumulated by the larval stage of flying insects are dispersed from ecosystems by the adult stage. Adults of the fly *Psychoda alternata* (Diptera) are thought to be particularly important when they emerge in removing metal contaminants from sewage sludge applied as a fertiliser to the soil (Redborg *et al.* 1983).

5.1.4 Accumulation of Metals by Secondary Consumers

Metals in primary consumers are in three forms. First, gut contents (which may be relatively unaltered plant material), second, metals in solution or in a chemical form from which they are released easily into solution by the digestive enzymes of a predator and third, metals bound firmly in insoluble complexes such as intracellular granules (see Chapter 9). The ratio and amounts of each metal in these fractions will differ between species of primary consumer and between organs within their bodies depending on the food eaten and the uptake, storage and excretory mechanisms possessed. Thus, in the same way as it is important to know the precise component of the plant community which is ingested by primary consumers, it is important to know the precise species eaten by secondary consumers.

The relative importance of each species of primary consumer in transferring metals to secondary consumers depends on their population density and biomass, the concentrations of available metals in individuals of different ages, and the pattern of predation on the population. Take for example two animals which are known to predate on terrestrial isopods in temperate ecosystems, lithobiid centipedes (Sunderland & Sutton 1980) and spiders of the genus *Dysdera* (Hopkin & Martin 1985b). Centipedes consume large numbers of whole juvenile isopods but are reluctant to eat the adults unless they are injured or moulting (Hopkin & Martin 1984b). *Dysdera crocata* on the other hand, attacks adult isopods, injects digestive enzymes and extracts the soft tissues from the body where most of the metals are stored (Hopkin & Martin 1985b). In a heavily contaminated site where the concentrations of metals are higher in adult than in juvenile isopods (e.g. zinc in *Oniscus asellus* from Shipham, Fig. 4.6), the diet of the spider will contain much greater concentrations of metals than the centipede.

Furthermore, animals like *Dysdera* whose food consists of large individual prey items will, on occasion, be exposed to metal concentrations in its diet which are substantially greater than the mean levels in the prey population. The mean concentration of copper in a population of the isopod *Oniscus asellus* from a copper mining area, for example, was about 350 µg g^{-1} whereas the most contaminated individual contained more than 1000 µg g^{-1} of copper (Hopkin & Martin 1982a). This effect may be exacerbated even further if the most contaminated individuals become moribund due to the toxic effects of metals and are more easily captured by predators (Hopkin & Martin 1982b).

5.2 RESPONSES OF TERRESTRIAL ECOSYSTEMS TO METAL POLLUTION

5.2.1 Introduction

Contamination of terrestrial ecosystems with metals may upset normal ecological processes by altering the relative abundance and diversity of species of microorganisms, plants and animals. Indirect effects may follow such as a reduction in the rate of decomposition of leaf litter due to a decrease in the numbers of invertebrate detrivores, greater susceptibility of vegetation to attack by defoliating insects (Nuorteva *et al.* 1987), increased erosion of soil due to a lack of plant cover (Jordan 1975) and a further decrease in the abundance and diversity of invertebrates due to the absence of their normal food plants or prey items.

An overview of the direct and indirect effects of metal contamination on terrestrial ecosystems is given in this section. More detailed coverage is reserved for Section 5.3 where specific polluted sites are examined. Several reviews have been published on the dynamics of metals in terrestrial ecosystems and these should be consulted for further information (Roberts 1975; Martin & Coughtrey 1981; Miller 1984b; Sheehan 1984a, 1984b, 1984c; Coughtrey *et al.* 1987).

5.2.2 Population Structure

A reduction in abundance and diversity of terrestrial invertebrates is well documented in sites which have been heavily contaminated with metals emitted by smelting industries (Jackson & Watson 1977; Strojan 1978a, 1978b; Bisessar 1982; Bengtsson & Rundgren 1984; Hopkin *et al.* 1985a) and steelworks (Dmowski & Karolewski 1979), and soils to which metal-containing pesticides have been applied (Van Rhee 1967). Primary consumers of dead plant material are particularly vulnerable in aerially contaminated sites because metals tend to accumulate in the surface soil and leaf litter layers. Earthworms may be completely wiped out in orchards following ingestion of plant material and soil contaminated with copper-based fungicides (Van de Westeringh 1972; Niklas & Kennel 1978). This leads to even greater accumulation of metals at the surface of mull soils since under normal circumstances, earthworms are by far the most important agents in the mixing of soil horizons (Heath & Arnold 1966; Swift *et al.* 1979; Seastedt & Crossley 1980; Meyer *et al.* 1984). In temperate ecosystems, populations of larger species may deposit a layer of faeces of about 5 mm in thickness at the surface of the soil each year containing soil brought from depths of up to 2 m (Wallwork 1983).

Deleterious effects of metal pollution on microorganisms are much more difficult to quantify due to the huge natural variation in abundance and diversity of populations in uncontaminated sites (Martin *et al.* 1980). In the field, statistically significant reductions in overall numbers of bacteria, fungi and actinomycetes compared to control sites, are rarely detected except in grossly contaminated areas where metals such as zinc and copper comprise more than 1% of the dry weight of the soil (Jordan & Lechevalier 1975; Balicka *et al.* 1977; Babich *et al.* 1983).

5.2.3 Decomposition

One of the indirect effects of metal pollution in terrestrial ecosystems is the accumulation of a layer of partially decomposed leaf material on the soil surface (Hirst *et al.* 1961; Van Rhee 1967; Tyler 1972; Van de Westeringh 1972; Watson 1975; Inman & Parker 1978; Niklas & Kennel 1978; Strojan 1978a, 1978b; Coughtrey *et al.* 1979; Freedman & Hutchinson 1980; Killham & Wainwright 1981; McNeilly *et al.* 1984; Hopkin & Martin 1985a). At first sight, this effect would be expected to be due to direct inhibition of the activities of fungi and bacteria since in soil ecosystems, microorganisms constitute more than 90% of the biomass and are responsible for more than 90% of the chemical decomposition of organic matter (Swift *et al.* 1979). However, numerous field experiments have shown that where detrivorous invertebrates are excluded from litter by retention of the plant material in 'litter bags' made of a fine mesh, the rate of decomposition of dead leaves is reduced drastically.

The main effect of macroinvertebrates such as earthworms, millipedes and isopods in stimulating decomposition, is the conversion of the dead plant material in the litter layer to faeces which increases the surface area available for microbial attack (Edwards & Heath 1963; Heath *et al.* 1966; Nicholson *et al.* 1966; Kaplan & Hartenstein 1978; Anderson &

Bignell 1982; Cisternas & Mignolet 1982; Anderson & Ineson 1983, 1984). Absence of these invertebrates due to metal pollution of their food leads to accumulation of fragments of leaf material on the soil surface. It follows that the decomposition of animal remains in soil and leaf litter (Seastedt & Tate 1981) may also be inhibited by metal pollution if saprophages are unable to eat the contaminated tissues, but this possibility has not been studied to date.

Gut passage of material can be of much greater ecological significance in metal transport in food chains than accumulation of metals by the organisms. Food chain transfer of lead by the collembolan *Orchesella cincta* is very minor and only accounts for about 11 times the standing pool of lead in the population per year. The flux of lead through the animals by consumption and defaecation, however, is about 10,000 times the standing pool per year (Van Straalen *et al.* 1985).

Microbial activity may also be stimulated when material is coated with digestive enzymes and innoculated with components of the gut microflora of invertebrates. In addition, deposition of faecal pellets in areas deeper in the litter effects transport of surface fungal spores and bacteria to moist conditions where they can proliferate (Bocock 1963; Hanlon & Anderson 1980; Hassall *et al.* 1987).

Low levels of metal contamination may have subtle effects on microbial physiology which are easier to detect than changes in population structure (Bischoff 1982). Their activity is monitored in the laboratory by measuring rates of carbon dioxide production (Rühling & Tyler 1973), nitrogen immobilisation (Tyler 1975; Giashuddin & Cornfield 1978) and production of enzymes (Tyler 1974, 1976; Stott *et al.* 1985). At moderate levels of contamination, the density of bacteria in the soil increases at the expense of fungal hyphae which are more sensitive to metal pollution. Nutrients released by the dead fungi are utilised by the bacteria (Ausmus *et al.* 1978). In heavily contaminated sites, populations of all micro-organisms are reduced and immobilisation of mineral elements is inhibited leading to increased leaching of essential plant nutrients from the soil (Tyler *et al.* 1974; Jackson *et al.* 1978a, 1978b; Doelman & Haanstra 1979b; Ross *et al.* 1981; Brookes *et al.* 1986).

5.3 SPECIFIC EXAMPLES OF CONTAMINATED TERRESTRIAL ECOSYSTEMS

5.3.1 Introduction

Many sites throughout the world have been sampled to determine concentrations of metals in soils, plants and animals. However, very few of these studies have been long-term and most have concentrated on at most a few species of organism with little attempt to assess the overall ecological impact of metal pollution. For example, of the numerous reports on enrichment of roadsides with metals derived from automobiles (Giles *et al.* 1973; Maurer 1974; Price *et al.* 1974; Rolfe & Haney 1975; Goldsmith & Scanlon 1977; Zhulidov & Emets 1979; Beyer & Moore 1980; Wade *et al.* 1980; Udevitz *et al.* 1980; Robel *et al.* 1981), none have examined changes in concentrations in the invertebrates over a period of years, or have analysed more than a few components of the contaminated ecosystems.

In most sites studied, contamination with metals has been due to smelting activity. This reflects concern as to the environmental impact of aerial metal pollution from these point sources which may contaminate large areas of soil and vegetation with potentially toxic particulate material. Specific examples include the effects of contamination by a zinc smelter at Palmerton, U.S.A. (Buchauer 1973; Beyer *et al.* 1984, 1985b) and the huge nickel-copper smelting works at Sudbury, Canada (Costescu & Hutchinson 1972; Hutchinson & Whitby 1974; Freedman & Hutchinson 1980; Jeffries 1983; Chan & Lusis 1986). However, long-term studies which include extensive analysis of the terrestrial invertebrate fauna, have been conducted in only three areas, deciduous woodlands near to a primary zinc, lead and cadmium smelting works at Avonmouth near Bristol, England, coniferous forests surrounding a brass mill near Gusum, Sweden and grasslands close to a copper refinery near Liverpool, England.

5.3.2 Primary Zinc, Lead and Cadmium Smelting Works, Avonmouth, South West England

The primary zinc, cadmium and lead smelting works at Avonmouth near Bristol, South West England, started production in 1928 in a region which has been used for smelting since Roman times (Martin *et al.* 1979; Fig. 5.2). The works has the capacity to produce annually 100,000 tonnes of zinc, 40,000 tonnes of lead and 300 tonnes of cadmium. An elaborate system of pollution abatement devices are fitted to the plant which extract in excess of 98% of the metals present in waste gases before they are released into the environment. However, some 50 tonnes of zinc, 30 tonnes of lead and 3 tonnes of cadmium escape as fine particles from the chimneys each year (Coy 1984). These rates of release are accepted by the local factory inspector as being within 'permissible levels' which

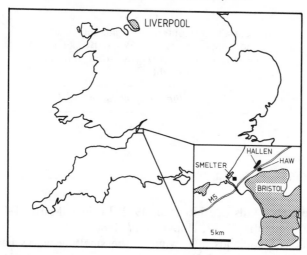

FIG. 5.2 Map to show locations of metal-contaminated sites in England mentioned in the text. Inset: positions of the Avonmouth smelting complex and Hallen and Haw Woods.

could not be reduced without making the smelting works uneconomic to run (Barbour 1977). Metals are also released from the smelting works into the Severn Estuary (for references see Hopkin *et al.* 1985b). The smelting works is the major source of metals in the region but the Avonmouth industrial area includes a number of other industrial concerns which cause pollution, notably a domestic rubbish incinerator and sulphuric and nitric acid-producing plants, emissions from which may increase substantially the acidity of local precipitation (Martin & Coughtrey 1987).

Soils close to the smelting works are severely contaminated with particles containing zinc, cadmium, lead and copper blown by the wind from unsmelted ore and tips of waste slag which are stored permanently on site. The diameter of most of these particles is greater than 10 μm and contamination from this source is restricted to within 1 km of the works (Harrison & Williams 1983). Most of the emissions from the chimneys contain particles of metal sulphates, sulphides and oxides with diameters of less than 2.5 μm. These may be transported for considerable distances (Fig. 8.1). Contamination of soil, vegetation and invertebrates by metals derived from the smelting works has been detected up to 25 km to the north-east of Avonmouth in the direction of the prevailing

TABLE 5.2
Total annual inputs of zinc, cadmium, lead
and copper to Hallen Wood by aerial
deposition during 1977–78 (from Martin
& Coughtrey 1981)

Metal	Input (mg metal $m^{-2} yr^{-1}$)
Zn	600
Cd	9.2
Pb	285
Cu	26

south-westerly winds (Burkitt *et al.* 1972; Little & Martin 1972, 1974; Martin & Coughtrey 1982; Hopkin *et al.* 1986).

Deciduous woodlands downwind of the smelting works have been studied intensively since the early 1970s by researchers at the University of Bristol (reviewed by Hutton 1984 and Read 1988). Research was concentrated initially on Hallen Wood, an oak-hazel woodland on clay about 3 km downwind of the smelting complex (Fig. 5.2), which is subject to considerable metal pollution (Table 5.2). Preliminary observations on the concentrations of metals in different components of the woodland demonstrated that soil, litter, plants and invertebrates were heavily contaminated with zinc, cadmium, lead and copper (Martin & Coughtrey 1975, 1976). Cadmium was considered potentially to be the most damaging pollutant because it was assimilated by detrivorous invertebrates from leaf litter to a greater relative extent than zinc or lead, leading to greater food chain transport (Martin *et al.* 1976). The kidneys of a thrush (*Turdus philomelos*) collected from the Avonmouth area at this time contained 387 µg g^{-1} of cadmium (Martin & Coughtrey 1975), a level high enough to cause pathological damage in human kidneys.

The greater affinity for cadmium by detrivorous invertebrates can be partly explained by the greater availability of this metal when compared with zinc and lead. Some 63% of the cadmium in leaf litter from Hallen Wood was released into solution by extractants designed to mimic digestive enzymes of detrivores whereas the corresponding figures for zinc and lead were 45% and 20% respectively (Martin *et al.* 1976). However, these differences were not large enough to account for the large differences in concentration factors for the three metals, so physiology must play a part also.

In the late 1970s, cattle were allowed into Hallen Wood and severely disturbed the leaf litter/soil profile. However, an experimental plot had been fenced prior to this and has since been maintained in an undisturbed state. Following the intrusion of the cattle, some aspects of the research were moved instead to the nearby Haw Wood which has remained undisturbed (Fig. 5.2). Haw Wood is similar to Hallen Wood (it is an oak-hazel wood on clay) and is also subject to substantial aerial deposition of metals from the smelting works (Martin *et al.* 1982).

The most obvious effect of metal contamination on woodlands near to the smelting works is the presence of a considerable layer of unde-composed leaf material on the surface of the soil (Coughtrey *et al.* 1979). In Hallen Wood, this effect is thought to be due to the almost complete absence from the soil and leaf litter layers of earthworms and millipedes which have been poisoned by the synergistic effects of low pH and high levels of metals in the diet (Hopkin *et al.* 1985a; Table 5.3). These macroinvertebrates normally stimulate decomposition by converting dead leaves into smaller fragments which are more easily decomposed by microorganisms. Nevertheless, the persistence of this thick layer of leaf material provides a stable habitat for some groups of invertebrates such as Collembola and mites, which reach higher population densities per unit area in Hallen Wood than in uncontaminated woodlands (Table 5.3). It has been suggested that the thick layer of leaf litter provides a favourable environment for the phoretic mite *Bakerdania elliptica* when it is not attached to its isopod host, as rates of infestation are higher on woodlice in the Avonmouth area than elsewhere in England (Colloff & Hopkin 1986).

Populations of bacteria and fungi in the Avonmouth area appear to be unaffected by concentrations of zinc, cadmium, lead and copper similar to those pertaining in soil and leaf litter in Hallen Wood, although further research is required to substantiate this claim (Gingell *et al.* 1976; Bewley 1980, 1981; Coughtrey *et al.* 1980; Martin *et al.* 1980). Grasses growing in contaminated soil in the region have evolved tolerance to metals (Coughtrey & Martin 1977b, 1978a, 1978b) but this has not been demonstrated conclusively in invertebrates.

A detailed budget of the distribution of metals in soil and vegetation in Haw Wood (Table 5.4) has confirmed that the 'final sink' for metal contaminants is the soil/litter microcosm (Martin *et al.* 1982). Analysis of concentrations of metals with depth of soil in Haw Wood has revealed the 'classic' profiles found in most aerially contaminated sites (Fig. 5.3). However, within the experimental plot in Hallen Wood, monitoring of the distribution of cadmium, zinc and lead in the soil over a ten-year

Ecophysiology of Metals in Terrestrial Invertebrates

TABLE 5.3

Densities of major groups of invertebrates (numbers of individuals m^{-2}) in leaf litter and soil from a contaminated woodland 3 km downwind of the Avonmouth smelting works (Hallen Wood) and a similar but uncontaminated woodland (Wetmoor Wood). Earthworms were collected by application of a 1% solution of formalin to the soil. All other invertebrates were heat-extracted in the laboratory from leaf litter and the top 3 cm of soil in Tullgren funnels (from Hopkin *et al.* 1985a, by permission of the University of Amsterdam, Commissie voor de Artis-Bibliotheek)

		Hallen Wood	*Wetmoor Wood*
Isopoda	*Oniscus asellus*	56	20
	Trichoniscus pusillus	0	151
Diplopoda	Polydesmidae	8	79
	Julidae	0	11
	Glomeridae	0	21
Chilopoda	Lithobiidae	112	116
	Geophilomorpha	328	263
Arachnida	Acari	129 000	19 400
	Aranae	248	81
	Pseudoscorpionidae	200	67
Insecta	Collembola	20 800	8 690
	Coleoptera (adult)	902	48
	Coleoptera (larvae)	120	4
	Diptera (larvae)	4 590	291
Annelida	*Lumbricus rubellus*	17	3
	Lumbricus terrestris	0	29
	Allolobophora longa	0	4
	Allolobophora caliginosa	0	30
	Octolasium cyaneum	0	9
Litter standing crop (kg m^{-2} dry weight)		14.28	1.35

period has revealed substantial changes which do not conform to the typical pattern (Martin & Coughtrey 1987). To take cadmium as an example (Fig. 5.4), in 1975 the metal was present in much higher concentrations in the leaf litter and surface layers of the soil than at depth, the situation one would expect. By 1983 however, the profile was very different and it appeared that a 'pulse' of cadmium was moving down through the soil.

Computer modelling of metal transfer rates suggested that a 'classic'

TABLE 5.4

Amounts of zinc, cadmium, lead and copper in different 'compartments' in Haw Wood (g m^{-2}), a site 3 km downwind of the Avonmouth smelting works (from Martin *et al.* 1982)

	Zn	Cd	Pb	Cu
Tree				
Wood	0.1481	0.0179	ND	0.0966
Bark	2.0416	0.02738	2.9287	0.1601
Branches	2.651	0.03119	7.1842	0.3827
Leaves	0.0639	0.00051	0.0319	0.0045
Shrub				
Wood	0.0031	0.00023	ND	0.0020
Bark	0.0115	0.00016	0.0777	0.0010
Branches	0.0566	0.00052	0.0966	0.0052
Leaves	0.0172	0.00017	0.0067	0.0013
Ground flora	0.1078	0.00279	0.0271	0.0052
Litter				
Twigs and wood	0.1411	0.00173	0.1344	0.0146
L1 layer	0.1139	0.00110	0.0519	0.0078
F1 layer	5.1335	0.07646	3.5151	0.2281
H1 layer	16.3578	0.36280	12.5924	0.6153
Total	21.7463	0.44209	16.2938	0.8658
Soil (0–30 cm)	93.4063	0.94510	16.7840	13.0121
Total metal burden	120.2565	1.4609	43.4407	14.5368

ND, not detected.

concentration–depth profile would be re-established in about ten years but with concentrations of cadmium substantially lower than those pertaining in 1975. Martin & Coughtrey (1987) proposed that the increased mobility of cadmium, zinc and lead was due to the reduction in pH of litter and soil which had taken place in Hallen Wood since 1975. The increase in acid precipitation in the area coincided with the erection in the mid-1970s of a 91 metre stack within the smelting complex which emits 400 kg h^{-1} of sulphur trioxide from the acid plant (Coy 1984). These findings demonstrate the importance of calculating fluxes of metals through the soil from repeated observations rather than deducing immobility of pollutants from measurements of concentrations with depth made on a single occasion.

Several papers have been published on the concentrations of zinc, cadmium, lead and copper in a wide range of invertebrates collected at

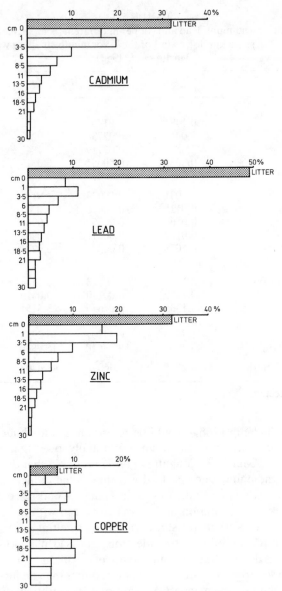

FIG. 5.3 Percentages of total cadmium, lead, zinc and copper content m^{-2} (to 30 cm depth) held in different layers of the soil profile in Haw Wood, a site 3 km downwind of the Avonmouth smelting complex. (Redrawn from Martin *et al.* 1982.)

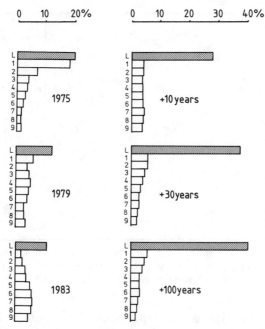

FIG. 5.4 Percentage of total cadmium content m^{-2} (to 9 cm depth) held in different layers of the leaf litter (L, shaded) and soil profile in Hallen Wood (a site 3 km downwind of the Avonmouth smelting complex) in 1975, 1979 and 1983. Computer predictions for 10, 30 and 100 years after 1983 assume a constant input of 10 mg Cd m^{-2} yr^{-1}. (Redrawn from Martin & Coughtrey 1987, by permission of the British Ecological Society.)

different distances from the smelting complex at Avonmouth. These studies have included analysis of isopods (Martin *et al.* 1976; Coughtrey *et al.* 1977; Hopkin & Martin 1982a, 1982b, 1984a, 1985a; Hopkin *et al.* 1986), the snail *Helix aspersa* (Coughtrey & Martin 1976a, 1977a), millipedes (Hopkin *et al.* 1985a; Read & Martin 1988), earthworms (Wright & Stringer 1980) and the centipede *Lithobius variegatus* (Hopkin & Martin 1983). The concentrations of zinc, cadmium, lead and copper in these invertebrates increases with proximity to the smelting works although this effect is much more pronounced in isopods (Hopkin *et al.* 1986) and snails (Coughtrey & Martin 1976a) than in centipedes (Hopkin & Martin 1983) and millipedes (Read 1988).

Laboratory experiments have been conducted on the uptake of zinc, cadmium, lead and copper by isopods and millipedes fed on leaf litter

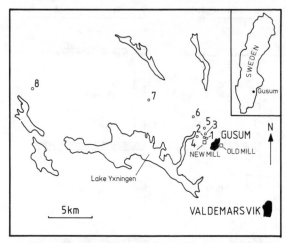

FIG. 5.5 Map to show positions of sampling sites near the Gusum brass mill, Sweden. (Redrawn from Bengtsson & Rundgren 1984, by permission of the Royal Swedish Academy of Sciences.)

from Hallen and Haw Woods (Hopkin & Martin 1984a; Hopkin *et al.* 1985a; Read & Martin 1988), and centipedes and spiders fed on metal-contaminated isopods (Hopkin & Martin 1984b; 1985b). However, it is extremely difficult to determine the precise routes of uptake of metals in contaminated field sites because of the lack of information on the exact diets of the animals in question. Some of the more subtle ways in which contamination with metals affects the life histories of terrestrial invertebrates in the Avonmouth area are only just beginning to be understood. Recent research in Haw Wood has shown that the maturation of carabid beetles is delayed in comparison to similar less-contaminated woodlands more distant from the smelting works due possibly to an affect on their normal prey items (Read *et al.* 1986, 1987; Read 1988). Further research in the Avonmouth area is required before the full effects of the zinc, cadmium, lead and copper pollution can be elucidated.

5.3.3 Brass Mill, Gusum, Sweden

An extensive study was conducted in the late 1970s by researchers at the University of Lund, into the effects of metals emitted from a brass mill on local plants, microorganisms, fungi and invertebrates. The mill is situated near the town of Gusum, S.E. Sweden (Fig. 5.5). It has been in operation since the late 1960s, replacing an older mill on a site on which smelting has been conducted since the fifteenth century (Bengtsson 1986). No figures for quantities of metals released into the atmosphere are available but about 98% of the emissions are zinc and copper with the remainder being mainly cadmium and lead. All the sites studied were in mature spruce forest. Levels of zinc and copper in leaf litter at different distances from the mill were somewhat lower than at comparable distances from the smelter at Avonmouth. However in their sampling programme, the Lund researchers included sites much closer to the source of pollution than those studied by the Bristol workers and analysed a wider range of soil/leaf litter organisms. The Swedish research has been summarised recently by Tyler (1984), Tyler *et al.* (1984) and Bengtsson (1986).

Cup-shaped lichens and the common wavy-hair grass *Deschampsia flexuosa* were dominant near to the mill where competition from reindeer lichens and carpet-forming mosses has been reduced due to the pollution. The most frequent fungal taxa of normal coniferous forests have been replaced by several 'new' species in the most contaminated sites and decreases in the total active fungal biomass and rates of soil respiration at sites less than 1 km from the mill have been detected (Tyler *et al.* 1984).

Lumbricids and enchytraeids were among the most sensitive animal groups, exhibiting a reduction in the numbers of species and individuals with increasing degree of pollution (Bengtsson & Rundgren 1982, 1984; Bengtsson *et al.* 1983b; Table 5.5). The reduction in earthworm feeding and burrowing, together with reduced microbial activity, has led to an accumulation of partially decomposed organic matter on the soil surface near to the mill, similar to the effect seen in woodlands at Avonmouth. The only species of earthworm found at sites 275 m and 650 m from the Gusum mill was *Allolobophora caliginosa* which survived, presumably, because it is a deep-soil dwelling annelid which does not feed directly on contaminated leaf litter at the surface (Bengtsson *et al.* 1983b). In contrast, the total number of species and density of Collembola were not significantly affected by zinc and copper pollution (Table 5.5) although there were some differences in the ratios of numbers of individuals of each species present. Carnivorous beetles were not found close to the mill whereas ants were plentiful. Invertebrates surviving near to the mill contained substantial amounts of copper providing considerable potential for food chain transfer of the pollutant (Table 5.6).

TABLE 5.5

Concentrations of metals in leaf litter of pine forest and number of species and density of lumbricid and enchytraeid worms and Collembola at different distances from a brass mill near Gusum, Sweden (from data in Bengtsson & Rundgren 1982, 1984; Bengtsson *et al.* 1983b; Tyler *et al.* 1984)

Site	Distance from mill (m)	Concentration of metals ($\mu g\ g^{-1}$ dry weight)			No. of species			Faunal density (numbers of individuals m^{-2})		
		Pb	Cu	Zn	*Lumbricids*	*Enchytraeids*	*Collembola*	*Lumbricids*	*Enchytraeids*	*Collembola*
1	175	960	16 500	18 000	0	2	21	0	300	21 000
2	275	650	10 800	11 000	1	3	20	1.1	600	21 000
3	650	220	2 600	3 000	1	4	21	0.3	11 000	47 000
4	1 000	80	500	1 500	3	6	28	9	800	45 000
5	1 250	120	600	1 000	4	5	18	15	—	45 000
6	2 900	70	200	500	4	5	23	17	16 100	25 000
7	7 800	50	20	200	4	3	24	25	16 100	23 000
8	20 000	40	20	200	6	5	21	114	18 200	28 000

Laboratory experiments were conducted on the collembolan *Ony-chiurus armatus* and the earthworm *Dendrobaena rubida* in an attempt to elucidate the reasons for the changes in population structures of invertebrates near to the mill. Specimens of *Onychiurus armatus* were reared on the fungus *Verticillium bulbillosum* which had been grown on a substrate contaminated with copper and lead (Bengtsson *et al.* 1983a, 1985a). Egg production, growth and longevity of the Collembola were severely reduced at concentrations of copper and lead in the food similar to those in fungi 650 m from the mill. It was calculated that this species should theoretically have been absent at this distance but it survived relatively unaffected by the pollution. The survival of *Onychiurus armatus* in sites where laboratory studies have shown it should be absent was due possibly to evolution of tolerance to metals, or the presence in the most contaminated sites of the protein-rich fungus *Paecilomyces farinosus* which alleviated toxicity symptoms in the laboratory (Bengtsson *et al.* 1985b). Similar toxic effects of copper, lead and zinc were

TABLE 5.6

Concentrations of lead and copper in ants, harvestmen (Opiliones), spiders and beetles trapped in August 1978 at three sites at different distances from the Gusum brass mill (see Table 5.5 for concentrations in leaf litter at these sites) (μg g^{-1} dry weight ± standard deviation for samples with at least 3 replicates) (from Bengtsson & Rundgren 1984, by permission of the Royal Swedish Academy of Sciences)

Site:		1	4	7
Distance from mill (m):		175	1000	7800
Ants				
Myrmica ruginodis	Pb	80.0 ± 16.5	49.3 ± 7.0	7.3 ± 1.0
	Cu	285.8 ± 95.4	101.3 ± 10.4	33.8 ± 2.6
Formica polyctena	Pb	57.4 ± 11.3	19.3 ± 8.8	12.2 ± 5.1
	Cu	595.2 ± 128.3	61.6 ± 15.9	50.9 ± 19.8
Harvestmen[a]	Pb	9.3 ± 3.2	4.6 ± 1.9	8.4 ± 3.6
	Cu	137.7 ± 69.6	34.5 ± 5.6	20.1 ± 9.3
Spiders				
Trochosa terricola	Pb	62.2 ± 24.3	27.0 ± 9.4	8.1
	Cu	770.2 ± 68.2	242.1 ± 39.7	133.8
Beetles				
Calathus micropterus	Pb	18.6 ± 9.8	8.4 ± 3.7	10.8
	Cu	179.7 ± 65.8	44.3 ± 27.5	43.7
Pterostichus melanarius	Pb	11.8	3.0 ± 1.4	ND
and *P. niger*	Cu	91.3	28.6 ± 11.1	23.9 ± 0.4

[a] Pooled samples of *Oligolophus palpinalis*, *Opilio parietinus*, *Phalangium opilio*.

ND, not detected.

observed in *Dendrobaena rubida* but these results tallied more closely with field observations on the distribution of earthworms at different distances from the mill than did the similar laboratory/field comparison with Collembola (Bengtsson *et al.* 1983b, 1986; Gunnarsson & Rundgren 1986).

5.3.4 Copper Refinery, Merseyside, N.W. England

Quantification of rates of transfer of metals in food chains is extremely difficult without detailed knowledge of the diets of animals in the field and the extent of variations in the relative importance of different food items throughout the year. Researchers at the University of Liverpool have gone some way towards achieving this goal in their studies on the dynamics of copper and cadmium in grassland ecosystems around a copper refinery in Merseyside, N.W. England (Hunter 1984; Hunter & Johnson 1982; Hunter *et al.* 1984a, 1984b, 1987a, 1987b, 1987c, 1987d).

The copper refinery is situated in an urban/industrial conurbation surrounded by high density housing and amenity grassland on which the studies were conducted. Emissions from the works, which has been in operation for about 50 years, have contaminated soils with copper and cadmium (Table 5.7). Concentrations of these metals in surface soils decrease in an exponential fashion away from the source and decline to background levels at about 3 km from the refinery (Fig. 5.6).

The flora close to the refinery was impoverished and was comprised almost exclusively of metal-tolerant swards of *Agrostis stolonifera* and *Festuca rubra*. Other plant species could not survive the very high levels of soluble copper in the soils (Table 5.7). Experiments on intact turves transported from the refinery site to an uncontaminated area showed that in *Agrostis stolonifera*, 70% of the copper 'in' the plant leaves was present merely as a surface deposit and that only 30% had entered the plants *via* the roots and was bound within the leaves (Hunter *et al.* 1987a). Analysis of the grasses at monthly intervals showed that concentrations of copper in the leaves tended to be greater in the winter months as the surface deposits built up and internally-bound copper was moved to the older leaves prior to their senescence.

Consumers of dead vegetation at the contaminated sites were exposed to much higher levels of copper and cadmium in their diets than animals eating the living leaves. The concentration of copper in senescent leaves of *Agrostis stolonifera*, for example, was more than four times greater than the level in the fresh tissue at the refinery site when averaged out over the year (Table 5.8). In the winter months, the senescent leaves contained more than $1400\ \mu g\ g^{-1}$ of copper while the concentration in the living leaves was less than $400\ \mu g\ g^{-1}$ (Hunter *et al.* 1987a). A thick layer of undecomposed grass litter was present on the soil surface at the heavily contaminated sites, coincident with much reduced numbers of isopods and oligochaetes. Herbivorous and carnivorous invertebrates were apparently able to survive the copper and cadmium pollution, spiders being particularly numerous at the refinery site.

Substantial differences existed in the concentrations of copper and cadmium in different invertebrate groups from the study sites (Table 5.9). Care should be taken in interpreting these results since pooling animals into taxonomic groups disguises species differences. Gut contents may also contribute substantial amounts of metals to the figures

TABLE 5.7

Fractionation of copper and cadmium in surface soils at different distances from the Liverpool copper refinery (mean ± standard error of 30 sequentially extracted samples, $\mu g\,g^{-1}$ dry weight; ranges in parentheses) (from Hunter et al. 1987a, by permission of the British Ecological Society)

Copper

Site	*Total*	*Extractant*		
		Distilled water	*0.5 M CH₃COOH*	*0.05 M EDTA*
Refinery	9 270 ± 2 200 (2 600 – 18 700)	9.3 ± 2.5 (1.8 – 29.0)	2 850 ± 1 066 (325 – 11 580)	5 800 ± 1 670 (750 – 167 000)
1 km	494 ± 105 (129 – 880)	1.7 ± 0.4 (0.2 – 5.6)	4.9 ± 1.2 (2.3 – 15.8)	268 ± 52 (32 – 589)
'Control'	13.3 ± 1.4 (9.8 – 15.4)	0.17 ± 0.01 (0.1 – 0.2)	0.21 ± 0.03 (0.1 – 0.5)	5.3 ± 0.3 (3.9 – 6.4)

Cadmium

Site	*Total*	*Extractant*		
		Distilled water	*0.5 M CH₃COOH*	*0.05 M EDTA*
Refinery	28.3 ± 6.95 (9.6 – 58.3)	0.05 ± 0.01 (<0.01 – 0.18)	12.6 ± 1.43 (4.6 – 27.2)	8.9 ± 1.68 (4.2 – 21.9)
1 km	7.4 ± 0.9 (5.5 – 9.7)	<0.01	0.94 ± 0.18 (0.47 – 1.86)	2.34 ± 0.17 (1.50 – 3.30)
'Control'	0.96 ± 0.1 (0.5 – 1.2)	<0.01	0.07 ± 0.01 (<0.01 – 0.1)	0.24 ± 0.02 (0.11 – 0.27)

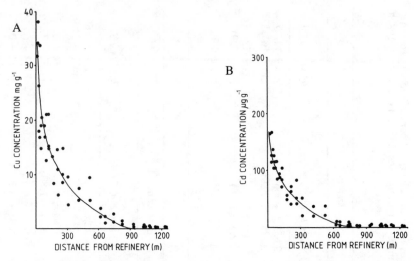

FIG. 5.6 Relationships between (A) surface soil copper and (B) cadmium concentrations (air-dried samples) and distance from a copper refinery near Liverpool, England. (Redrawn from Hunter *et al.* 1987a, by permission of the British Ecological Society.)

TABLE 5.8

Annual mean copper and cadmium concentrations in senescent and live foliar tissue of *Agrostis stolonifera* sampled monthly from three sites at different distances from the Liverpool copper refinery (means of 100 samples, $\mu g\,g^{-1}$ dry weight ± standard error) (from Hunter *et al.* 1987a, by permission of the British Ecological Society)

	Site	*Senescent tissue*	*Live tissue*
Copper	Refinery	680 ± 113	122 ± 31
	1 km	64 ± 7	24.5 ± 2.4
	'Control'	10.9 ± 1.1	9.8 ± 0.7
Cadmium	Refinery	10.4 ± 1.7	3.3 ± 0.4
	1 km	2.06 ± 0.13	1.32 ± 0.07
	'Control'	0.68 ± 0.06	0.63 ± 0.06

(particularly in earthworms). However, there were clearly major differences in, for example, the concentrations of cadmium in isopods and Orthoptera (Table 5.9) which were important to consider when transfer rates of this metal to predators were being calculated (Hunter *et al.* 1987c). Seasonal variations were also evident. The mean annual concentration of cadmium in lycosid spiders was 102 μg g^{-1} (Table 5.9) but the level in the population varied throughout the year from a minimum of 68 μg g^{-1} in December to a maximum of 150 μg g^{-1} in July (Hunter *et al.* 1987b). This fluctuation was probably due to loss of older individuals and recruitment of juveniles to the population at different times of the year. Values substantially higher than these are likely to be reached by individual spiders.

The data on concentrations of metals in invertebrates was used to calculate levels of copper and cadmium in the diets of small mammals at the sites. The Common Shrew *Sorex araneus* which is a carnivore, ingested about three times more copper and 12 times more cadmium than the herbivorous Field Vole *Microtus agrestis* in the same sites because the invertebrates which form its diet were more contaminated than the vegetation (Hunter *et al.* 1987c). Spiders were the most important component of the food of the shrew and contributed more than 50% of the copper and cadmium contained in the diet. However, other invertebrate groups can be greater sources of metals than predation on them would suggest. During December and January, for example, isopods and Collembola comprised only about 8% of the weight of the diet of *Sorex araneus* but contained 28% of the copper and 25% of the cadmium in the food due to their ability to accumulate the metals (Table 5.9). The liver and kidneys of shrews collected from the refinery site contained levels of cadmium which were high enough to cause disruption of cellular ultrastructure (Hunter *et al.* 1984b).

5.4 PREDICTION OF THE EFFECTS OF METAL CONTAMINATION ON TERRESTRIAL ECOSYSTEMS

5.4.1 Introduction
One of the main aims of research on terrestrial ecosystems which are aerially contaminated with metals is to be able to predict the consequences of a new source of pollution on a previously uncontaminated area. Some general principles have emerged from studies such as those conducted at Avonmouth, Gusum and Merseyside which enable us to

TABLE 5.9

Concentrations of copper and cadmium in invertebrates collected by pitfall trapping at different distances from the Liverpool copper refinery ($\mu g\,g^{-1}$ dry weight ± standard error, number of samples in parentheses) (from Hunter et al. 1987b, by permission of the British Ecological Society)

		'Control'		1 km		Refinery	
		Cu	Cd	Cu	Cd	Cu	Cd
Collembola	(30)	49.5 ± 2.2	2.1 ± 0.3	175 ± 27	11.7 ± 2.3	2 370 ± 510	51.7 ± 9.7
Isopoda	(60)	77.8 ± 7.7	14.7 ± 1.0	836 ± 126	130 ± 46	2 390 ± 270	231 ± 131
Diplopoda	(60)	138 ± 6	5.6 ± 0.2	511 ± 40	14.2 ± 0.4	780 ± 126	18.9 ± 2.4
Oligochaeta	(60)	23.7 ± 1.3	4.1 ± 0.3	115 ± 19	34.0 ± 2.8	1 170 ± 300	107 ± 25
Diptera (larvae)	(50)	25.9 ± 2.0	2.2 ± 0.4	85.4 ± 15.7	6.9 ± 0.9	210 ± 26	24.8 ± 3.9
Orthoptera	(500)	37.5 ± 2.1	0.19 ± 0.01	66.4 ± 5.9	0.32 ± 0.04	333 ± 38	1.96 ± 0.10
Formicidae	(50)	32.6 ± 3.1	1.2 ± 0.1	131 ± 4.9	5.4 ± 1.3	731 ± 166	37.7 ± 5.5
Hemiptera	(100)	30.2 ± 2.7	0.8 ± 0.1	61.7 ± 5.3	3.5 ± 1.2	265 ± 98	10.6 ± 3.2
Lepidoptera (larvae)	(30)	13.8 ± 1.1	0.6 ± 0.2	68.7 ± 12.9	7.1 ± 1.3	160 ± 22	22.3 ± 4.3
Coleoptera							
Curculionidae	(250)	29.7 ± 1.6	0.6 ± 0.1	65.6 ± 3.9	3.6 ± 0.4	421 ± 91	15.3 ± 1.4
Staphylinidae	(400)	29.4 ± 1.8	0.6 ± 0.1	57.7 ± 4.1	4.9 ± 0.5	522 ± 100	14.0 ± 1.6
Carabidae	(400)	21.3 ± 0.9	0.7 ± 0.1	45.6 ± 2.9	5.6 ± 0.9	460 ± 96	15.1 ± 1.1
Predatory larvae	(80)	33.2 ± 3.8	2.2 ± 0.3	53.3 ± 14.1	8.1 ± 0.6	298 ± 34	20.9 ± 1.2
Araneidae							
Lycosidae	(400)	57.7 ± 4.2	2.6 ± 0.3	160 ± 15	34.5 ± 5.0	887 ± 171	102 ± 8
Linyphiidae	(200)	89.1 ± 10.7	2.4 ± 0.3	200 ± 13	18.7 ± 1.3	1 020 ± 140	88.6 ± 9.8
Opiliones	(200)	42.2 ± 3.6	2.8 ± 0.3	100 ± 9	25.4 ± 1.8	1 010 ± 110	92.7 ± 10.0

do this to a limited extent.

First, concentrations of metals in air, soil and vegetation decrease in an exponential fashion with distance from the source of contamination (Figs. 5.6, 8.1). Factors such as the emission levels, height of the source above the ground, physical and chemical characteristics of the metals (particle size etc.) and local climatic conditions (wind speed and direction and rainfall patterns) will determine the distance over which the pollutants are carried, but not the exponential shape of the curve which is altered only by the local topography.

Second, metals are relatively immobile after they reach the soil/litter layers. Concentrations are greatest in the leaf litter and top 5 to 10 cm of soil declining to background levels below 30 cm (Fig. 5.3). The exception to this rule is where precipitation is particularly acidic and metals are transported much more rapidly through the soil (Fig. 5.4).

Third, leaf litter which is heavily contaminated with metals accumulates on the surface of the soil in an undecomposed or partially decomposed state. The metals appear to interfere with decomposition directly, by adversely affecting microorganisms which are responsible mainly for chemical decomposition and indirectly, by poisoning detrivorous invertebrates which stimulate microbial activity. Much more research is required on the effects of metals on microbial ecology.

Fourth, there are considerable differences in the mobility of metals in terrestrial food chains. Cadmium gives most cause for concern because of its extreme toxicity and the fact that animal:diet concentration factors for this metal often exceed 5:1. Lead, in contrast, is a relatively immobile metal and is rarely concentrated by animals to levels greater than those in the diet.

Fifth, the metal content of species of invertebrate in the same site may differ substantially depending on diet and physiology. For sites contaminated with cadmium, enough information has been gathered for us to reliably predict that isopods will be among the invertebrates which contain the highest concentrations of this metal and that Orthoptera will be among those animals which contain the lowest (Table 5.9). This does not of course indicate the relative sensitivity of the different groups to cadmium poisoning and much more laboratory-based research is required into the ecophysiology of metal dynamics in terrestrial invertebrates before toxicity levels can be reliably predicted.

5.4.2 Levels of Contamination

The ecological impact of pollution of terrestrial habitats by metals, can

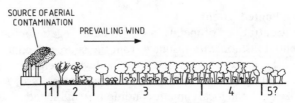

FIG. 5.7 Schematic diagram (not to scale) showing the four 'levels of contamination' of the terrestrial environment away from a source of aerial metal contamination (see text for further details). Beyond level 4, a fifth area can be envisaged where some trace elements are deficient. In such circumstances, slight metal pollution may be beneficial if levels of an essential element such as copper are increased.

be classified broadly into four 'levels of contamination' based on effects on normal ecosystem function (Fig. 5.7).

Level 1 areas are heavily contaminated and few organisms manage to survive. Those which do may contain exceptionally high concentrations of metals. Such areas are found typically on mine waste or close to major sources of aerial metal contamination.

Level 2 areas are not so severely affected but the environmental concentrations of metals are sufficient to disrupt the ecological balance of the habitat in question. Species sensitive to elevated levels of metals in the diet are absent from such areas. One typical feature of level 2 environments is the reduction in decomposition rates of dead vegetation coincident with the absence of species of detrivorous invertebrates such as earthworms and millipedes which would normally be common (Hopkin *et al.* 1985a).

Level 3 areas are not heavily polluted and ecological disturbance may be difficult or impossible to demonstrate. Plants and animals from these regions of low metal contamination contain concentrations which are significantly greater than those in the same species in sites more distant from the source of pollution. Roadside verges, for example, are typical level 3 habitats. There is no evidence that lead pollution from car exhausts has prevented any species from colonising the margins of busy roads, even though most organisms in such sites contain higher concentrations of lead than the same species from uncontaminated areas (Davies & Holmes 1972; Quarles *et al.* 1974).

The boundary dividing a level 3 area from an uncontaminated *level 4* area is difficult to define due to the considerable variation in background levels of metals in air and soil in regions away from sources of pollution

(Mitchell & Burridge 1979; Salomons & Forstner 1984; Sposito & Page 1984; Salomons 1986; Wiersma 1986). This variation in background levels is due partly to natural geological differences and partly to deposition of metals transported in fine particles from distant pollution sources by the wind. Indeed, the presence of anthropogenically-derived elements in polar snow and ice indicates that even the most remote terrestrial habitats have been contaminated to some extent by aerial deposition of metals released into the atmosphere by the industrial activities of man (Hughes *et al.* 1980; Wolff & Peel 1985; Boutron 1986; Legge & Krupa 1986; Nriagu & Davidson 1986). The level 3/level 4 boundary lies approximately 3 km from the Gusum brass mill (Table 5.5) and the Liverpool copper refinery (Fig. 5.6) but at Avonmouth, it may be as much as 25 km from the smelting works (Figs. 8.1, 8.3).

The area affected by a particular level of contamination may of course change depending on whether the amounts of metals released by the pollution source increase or decrease. Even if the output of pollutants remains constant, the boundaries between levels may extend further away from the source due to the build up of metals in the soil with time.

In areas where the soil is deficient in an essential metal, contamination by that element may actually prove to be beneficial. However, an apparent stimulation of a population by low levels of metal enrichment does not necessarily indicate that a deficiency is being rectified (see also the discussion of the phenomenon of hormesis in Section 2.2). The rate of production of offspring by aphids living on beans, for example, increases following application of low doses of zinc, cadmium, lead, copper and mercury to the plants but this stimulation of the population is interpreted as a response to increased levels of nitrogen in the phloem sap following metal-induced damage to the growing shoots (Walther *et al.* 1984).

5.5 RECOVERY OF TERRESTRIAL ECOSYSTEMS FROM METAL POLLUTION

The overriding problem of metal pollution is long-term residence in soils (Martin & Coughtrey 1981). There are numerous examples of mine sites which have been disused for several hundreds of years where concentrations of metals in soils have remained so high that plants have been unable to colonise (see Section 3.3.2).

Recovery of soils contaminated by aerial deposition following closure

of a source of metal pollution is not likely to be any more rapid
(Roberts & Goodman 1973). However, since a large proportion of metals
associated with plants in such sites is present as a surface deposit, levels
on vegetation may return to normal within a year or so, especially in
leaves of deciduous trees where the roots penetrate below the contaminated surface soil (Fig. 5.1). When these leaves become senescent
and fall to the ground, they provide a source of relatively uncontaminated
food for detrivores. Thus, species which were previously excluded due
to the toxic levels of metals in the food, may be able to recolonise
habitats as metal levels in the diet decline.

CHAPTER 6

Factors Controlling Uptake, Storage and Excretion of Metals by Terrestrial Invertebrates

6.1 INTRODUCTION

In Section 1.2, it was argued that the fundamental biochemical systems for regulating the uptake, storage and excretion of metals in the digestive system of marine invertebrates evolved in response to the unpredictable nature of concentrations of essential and non-essential elements in the food. In aquatic invertebrates, exchange of metals is possible with the external medium but in terrestrial invertebrates, the epithelium of the gut has assumed almost total responsibility for exchange of metals with the environment.

Invertebrates which colonised the land from the sea, such as isopods, had to adapt their digestive and excretory systems for a terrestrial existence. The strategies which they were able to adopt for metal regulation were constrained by a biochemistry and internal anatomy which had evolved originally in response to selective pressures in the marine environment. Groups such as insects, myriapods and arachnids, have undergone much greater evolutionary change on land. However, the strategies for metal regulation which they have been able to adopt have also been affected by the nature of their food (Section 6.2), the structure and function of the digestive system (Section 6.3) and a number of other factors such as the necessity to moult their exoskeleton (Section 6.4).

Specific examples, arranged taxonomically, of metal dynamics at the species and organ level are reviewed in Chapter 7, and at the ultra-structural level in Chapter 9.

6.2 FEEDING STRATEGIES

6.2.1 Introduction

The main selective pressures which have acted on the evolution of the digestive and excretory systems of terrestrial invertebrates are the maximisation of nutrient assimilation from a particular food source together with minimisation of water loss and energy expenditure (Sibly & Calow 1986). While an individual animal can not alter digestive processes by changing its internal anatomy, net assimilation of components of the food can be affected by choice of diet and rate of feeding.

6.2.2 Diet

Terrestrial invertebrates show considerable discrimination in the types of food they will eat. Isopods (Neuhauser & Hartenstein 1978; Beck & Brestowsky 1980; Hassall & Rushton 1984; Gunnarsson 1987a, 1987b), millipedes (Kheirallah 1979) and molluscs (Dirzo 1980; Richardson & Whittaker 1982), for example, are able to detect quite subtle differences in the chemical composition of plant material.

Difficulties in determining the exact nature of the diets of terrestrial invertebrates in the field has complicated attempts to construct pathways of metal transfer between trophic levels in ecosystems. Different species of earthworm living in the same site, for example, may consume completely different food (Piearce 1978; Sims & Gerard 1985). Whereas *Lumbricus terrestris* lives in permanent burrows and eats leaf material at the surface, *Allolobophora rosea* ingests soil and remains permanently below ground level (Bolton & Phillipson 1976). The situation is made more problematic by the tendency of some terrestrial invertebrates to indulge in coprophagy. This behaviour has been shown to be essential for the survival of some beetles and primitive cockroaches (Burnett *et al.* 1969; Mason & Odum 1969), and at least one species of millipede (McBrayer 1973). The faeces are reingested to facilitate extraction of nutrients not made available on the initial passage of food through the gut, or to re-inoculate the digestive system with symbiotic micro-organisms after ecdysis of the cuticular lining. Terrestrial isopods may be coprophagous, but faeces constitutes less than 8% of the diet in the field and although its consumption may improve performance slightly, it is not essential for normal growth and reproduction (Hassall & Rushton 1982, 1985).

The choice to ingest or reject food is governed by stimuli from receptors sensitive to taste, smell, moisture content, texture and/or visual appearance. The animal will only eat the food if the sensory information from the receptors is in the correct combination. Some organisms have exploited this situation by producing defensive chemicals that invoke rejection by the potential consumer. Plants may store substances such as tannins and polyphenols in their leaves, which herbivores find difficult to digest, or powerful poisons which kill animals attempting to eat them. Many invertebrates also produce defensive secretions to deter predators (Hopkin *et al.* in press, a).

An animal does not need to know the concentrations of all substances in a potential item of food before it decides whether or not to eat it. Selective pressures have ensured that food which an animal finds acceptable, contains the right balance of essential nutrients for normal growth and reproduction, and a minimum of non-essential substances which it has to detoxify and/or excrete. In most sites, terrestrial invertebrates do not need the ability to detect the levels of metals in their food. Physiological mechanisms have evolved which are able to regulate the concentrations of essential and non-essential metals in the tissues over the range of dietary levels the species has experienced during its evolution. Problems arise when the concentrations of metals in the food fall outside the normal range. The most obvious example is of sites polluted with metals where the physiological mechanisms of an animal can not cope with dietary levels above those they have evolved to control. A less obvious example is of areas which are deficient in an essential metal. In sites where the levels of copper, for example, are particularly low in the soil, plants may be able to grow but the concentration they contain is too low to supply the dietary requirement of livestock. Sheep and cattle become sick and eventually die unless the deficiency is rectified. It would be extremely interesting to investigate whether herbivorous invertebrates in such sites suffered from copper deficiency also.

Some authors (e.g. Dallinger 1977) have suggested that terrestrial invertebrates may be able to select food containing optimum amounts of essential metals. This would require the presence of positive and negative feedback mechanisms in the animal which would control feeding behaviour. An internal deficiency of an essential metal would result in the animal selecting food with high levels of the metal whereas an excess in the body would stimulate feeding on food containing low levels. Such a concept is difficult to accept because there is no physiological or biochemical evidence for the presence of such feedback mechanisms controlling metal regulation in any organism. Furthermore, chemo-receptors which respond to specific metals (other than the major ions), have not been discovered on the mouthparts or in the gut of any animal.

Dallinger (1977) showed that the isopod *Porcellio scaber* was able to discriminate between three batches of leaves which differed, apparently, only in their copper concentrations. 'Copper deficient' isopods ate pro-portionately more leaf material with high levels of copper than 'copper enriched' isopods which preferred the leaves with low levels of copper. However, extreme care should be taken in interpreting the results of such experiments. It is possible that the isopods were not discriminating on the basis of copper levels in the food but that their 'choice' was in response to different levels of the anion (in this case nitrate) which they could probably taste. Application of metals as salts to leaf material may also change the texture and taste of the food by killing surface microorganisms. For further discussion of Dallinger's experiments, see Section 7.4.2.

Discrimination against food contaminated heavily with metals is easier to explain. Van Capelleveen *et al.* (1986) demonstrated that *Porcellio scaber* preferred uncontaminated oak leaves (*Quercus robur*) to leaves to which lead salts had been added, but this does not prove that isopods possess 'lead receptors' on their mouthparts. They are merely rejecting food which tastes unpleasant. Humans are able to avoid food containing exceptionally high levels of lead because of its unpleasant metallic taste but this information is supplied by stimuli to the brain from the sweet, sour, bitter and salt receptors on the tongue, not by a chemoreceptor sensitive specifically to lead.

Some invertebrates in metal-polluted sites may not be exposed to higher concentrations of metals because their diet is not contaminated. Insect larvae living deep within a dead tree in an aerially polluted site, for example, do not become contaminated because metals can not penetrate below the surface layers of the wood. Similarly, spiders may survive in heavily contaminated sites if they feed on flying insects which have migrated in from an adjacent uncontaminated area.

6.2.3 Rates of Food Consumption and Assimilation
The rate at which a terrestrial invertebrate needs to consume food is dictated by the concentration of the most limiting essential nutrient in the diet. If such a substance is at a very low concentration, the animal has to eat a large amount of food in order to satisfy its needs. As a consequence, many other substances will be ingested in amounts far in

excess of requirements. These may include essential and non-essential metals, hence the need for systems to regulate their uptake, transport and excretion. Aphids, for example, need to pass a massive amount of phloem sap through their gut in order to extract enough nitrogen for protein synthesis. The excess sugar solution is voided from the anus as 'honeydew' and provides a source of carbohydrates for many species of ants. Similarly, dead wood is a poor food source for invertebrates until the ratio of carbon to nitrogen compounds is reduced below about 20:1 by the degradation of cellulose by fungal hyphae and bacteria. If the C/N ratio is greater than 20:1, animals can not pass the wood through the gut at a fast enough rate to supply nitrogen needs for growth (Swift *et al.* 1979).

When food of 'high quality' is plentiful, herbivorous and detrivorous invertebrates adopt a strategy of feeding as fast as possible because relatively little energy is required to solubilise the initial fractions of essential nutrients which are released from the food (Hassall & Rushton 1984). When food of poor quality predominates, feeding rates drop because less energy is required to retain food in the gut and allow digestive enzymes to work on it for longer, than to go in search of food in which the nutrients are more available (Hubbell *et al.* 1965).

Such differences in feeding rates will have a profound effect on assimilation of metals by terrestrial invertebrates. For example, Dallinger & Wieser (1977) showed that terrestrial isopods assimilate more than 90% of copper from leaves when this is present as a surface deposit of copper sulphate. If the same high assimilation rate of copper is maintained irrespective of feeding rates then isopods consuming large amounts of surface-contaminated food of high quality, will accumulate much greater amounts of copper than the same animals feeding at lower rates on low quality food which is retained in their guts for longer (see also Section 6.3.3). Conversely, metals which are bound firmly within plant tissue may be released only if food is retained in the gut for a long enough period for digestive enzymes to act. Similar principles apply also to carnivorous invertebrates although they usually spend some time digesting prey because it is nutrient-rich and energy has been expended in searching for and/or catching it.

6.2.4 Concentrations and Chemical Forms of Metals in the Food

It is well-established that rates of assimilation and toxicity of metals in animals are strongly dependent on the ligands to which the elements are bound in the food. The differences relate to the ease with which metals are converted to an 'available' form in the lumen of the gut. Net assimilation of methylated mercury by blowfly larvae, for example, is ten times greater than inorganic mercury at the same concentration (Nuorteva & Nuorteva 1982) and is eight times more toxic to larvae of *Drosophila melanogaster* (Sorsa & Pfeifer 1973a). Larvae of the silkworm *Bombyx mori* die within a few days when cadmium is added to their diet at a concentration of 200 $\mu g\, g^{-1}$ (wet weight), but survive if the metal is chelated in the food with 0.04M EDTA (Nakayama & Matsubara 1981).

The ratios of different metals in the food are also important. Two metals which compete for the same routes of uptake, or sites of biochemical activity, are said to be *antagonistic*. One of the best documented examples in vertebrates is the copper/molybdenum interaction. High concentrations of either essential metal in the diet of cows can inhibit uptake of the other causing deficiency diseases (see Alary *et al*. 1983 for references). In invertebrates, the toxic effects of high doses of copper to lepidopteran larvae may be due in part to the inhibition of the activity of the molybdenum-containing enzyme xanthine dehydrogenase (Marešova *et al* 1973).

Antagonism between the uptake of the class B metals zinc, cadmium and copper has been documented in a wide range of terrestrial invertebrates including beetles (Medici & Taylor 1967), flies (Aoki & Suzuki 1984) and earthworms (Beyer *et al*. 1982). This may indicate the presence of uptake sites which are shared by all three metals in the cell membrane. Zinc, cadmium and copper are certainly bound to metallothionein proteins on entering the cell (Section 9.3.2). The uptake sites can be saturated by a single metal if it is present at very high concentrations in the diet. For example, the proportion of cadmium assimilated during feeding experiments on the snail *Helix aspersa* was 59% at 25 $\mu g\, g^{-1}$ in the diet (2.5 mg ingested, 1.5 mg retained) but only 7% at 1000 $\mu g\, g^{-1}$ (37.3 mg ingested, 2.6 mg retained)(Russell *et al*. 1981).

Two metals are said to be *synergistic* if the presence of one stimulates the uptake or effects of the other. In an interesting paper on ovarian symbionts in cockroaches, Brooks (1960) demonstrated that supplementation of the diet of *Blatella germanica* with manganese, improved the transmission of bacteria to succeeding generations. The passage of symbionts from mother to offspring was further improved by the addition of zinc to the diet which acted synergistically with manganese to improve reproductive performance of the insect. Transmission of symbionts was not stimulated to the same extent if manganese or zinc were added alone.

6.3 STRUCTURE AND FUNCTION OF THE DIGESTIVE SYSTEM

6.3.1 Introduction

The epithelium of the digestive system of terrestrial invertebrates is usually only one cell in thickness and acts as a barrier between the internal environment of the animal (i.e. the blood bathing the organs) and the food in the lumen. For example, the slug *Agriolimax reticulatus* survives an injection of 100 µg of copper into its stomach but dies rapidly when only 20 µg are injected into the blood (Henderson 1965). The gut is thus involved intimately in the uptake, transport, storage and excretion of metals. The structure and function of the gut of each group of terrestrial invertebrates has evolved to maximise assimilation from the food of essential major nutrients such as fats, carbohydrates and proteins, and to minimise net loss of water. The need to regulate essential and non-essential metals is unlikely to have exerted a strong selective pressure on the evolution of the gross morphology and histology of the digestive organs. The strategies which terrestrial invertebrates have been able to adopt for such regulation have, therefore, been constrained by their internal anatomy.

6.3.2 Anatomy of the Gut

Simplified diagrams of the gross morphology of the digestive and excretory systems of the main groups of terrestrial arthropods are presented in Fig. 6.1. More detailed descriptions of the structure and function of the gut of molluscs, earthworms and isopods are given in Chapter 7. Three main regions of the gut can be recognised in all groups.

The *foregut* receives food *via* an oesophagus. Some discrimination as to which components of the food are ingested may take place at the mouthparts. Spiders, for example, have filters on their mandibles which prevent the passage of coarse particles into the stomach (Foelix 1982). Secretions may be added to the food in the foregut from salivary glands (molluscs, centipedes) or midgut diverticulae (isopods). The foregut may be expanded considerably to form a crop for the temporary storage of ingested material (dipteran and lepidopteran larvae), or be elaborated with structures for grinding and/or sorting components of the food (molluscs, isopods).

The *midgut* is the main site of secretion of enzymes and absorption of products of digestion of the food. In some groups it is a simple tube (centipedes, millipedes, lepidopteran larvae) whereas in others, longi-

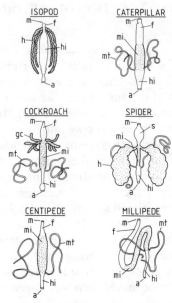

FIG. 6.1 Schematic diagrams of the digestive and excretory organs of some terrestrial arthropod groups (for molluscs, earthworms and isopods in greater detail, see Figs. 7.1, 7.3 and 7.6). These organs may be folded extensively to increase the internal surface area. The 'true' midgut (derived from endoderm) is shaded in each case. a, anus; f, foregut; gc, gastric caeca; h, hepatopancreas; hi, hindgut; m, mouth; mi, midgut; mt, Malpighian tubules; s, sucking stomach.

tudinal folds (such as the typhlosole of earthworms) increase the surface area in contact with the contents of the lumen (Perel 1977). In isopods, the midgut comprises four blind-ending tubules whereas in spiders and molluscs it is elaborated into many hundreds of diverticulae forming the so-called 'midgut gland' or hepatopancreas.

All arthropods, with the exception of isopods, possess blind-ending Malpighian tubules which open into the gut at the midgut/hindgut junction. These tubules are involved in excretion of nitrogenous wastes, ionic balance and in some cases, excretion of metal-containing granules (Chapter 9). Some insects secrete a peritrophic membrane from the foregut/midgut junction which surrounds the food in the lumen of the digestive tract (Chapman 1985).

Material from the midgut, together with secretions from the Malpighian tubules (if present), pass into the *hindgut*. In most groups this is a simple tube where waste matter is compacted and water is absorbed from faeces before they are voided. However in termites, the hindgut is formed into a large pouch where symbiotic microorganisms and Protozoa are mixed with food to aid digestion.

In arthropods, the foregut and hindgut are lined with a layer of cuticle across which only water, ions and small molecules (<1000 Daltons) can pass (Madrell & Gardiner 1980). Metals derived from the food which are deposited in the cells of the hindgut (e.g. the ileum of cockroaches, Ballan DuFrançais *et al*. 1979, 1980; Jeantet *et al*. 1980) may not necessarily have passed through cuticle to get there. The elements could have entered the cells *via* the blood across the basement membrane of the hindgut after passing through the midgut. The midgut has no such cuticular lining to restrict the passage of food components. The apical membranes of the cells are formed into microvilli which increase the absorptive surface area. Food particles of up to 100 nm in diameter are assimilated directly by phagocytosis in the hepatopancreas of molluscs (Walker 1970).

If we accept that terrestrial invertebrates do not possess negative feedback mechanisms to restrict the transport of class B and borderline metals into the cells of the midgut when sufficient amounts are present within the body to satisfy their immediate needs, then on occasions when essential metals are assimilated in excess of requirements, the animals must have methods for detoxifying the metals in a biologically inert form to prevent them from interfering with biochemical reactions in the tissues. Non-essential metals which enter the cells of the midgut must be detoxified also. In terrestrial invertebrates, metals in midgut cells which are surplus to requirements are stored in metal-binding proteins and three types of membrane-bound granules (see Chapter 9 for a full description of these systems). The granules are insoluble and once formed, do not appear to be broken down intracellularly. They form, therefore, a true storage-detoxification system.

Selective pressures have ensured that terrestrial invertebrates have adopted strategies for the regulation of metals which involve expenditure of the least amount of energy. Permanent storage of the insoluble granules would be the least expensive option in energy terms so long as the inclusions did not occupy such a large volume of the cells in which they were deposited that they interfered with normal cytological functions. This appears to be the strategy adopted for copper and cadmium in terrestrial isopods (Hopkin & Martin 1984a). In uncontaminated sites, there is enough 'room' in the S cells of the hepatopancreas to store all unwanted copper and cadmium until the animals die. Retention of the granules for such long periods also indicates that there is very little turnover of S cells in this organ. Adults of some Diptera are also able to adopt this strategy for copper and iron because they live only for a few weeks after emerging (Tapp 1975). Metals may be stored in other tissues which do not have a direct connection with the gut such as the sub-cuticular tissues of centipedes where substantial amounts of zinc may be deposited (Hopkin & Martin 1984b).

Metal-containing granules can be excreted if they are stored in cells which breakdown as part of normal digestive processes and spill their contents into the lumen of the gut. This may occur at the end of each digestive cycle, as in spiders and centipedes, or during moulting as in Collembola (Joosse & Buker 1979) and larvae of the clothes moth *Tineola biselliella* when the whole lining of the midgut is replaced (Waterhouse 1952). Metal-containing granules stored in the hindgut of arthropods can be voided only by breakdown of the epithelium during moulting due to the presence of the layer of cuticle at the surface of the cells. In earthworms, metals are stored in granules in the chloragogenous tissue surrounding the gut and may be excreted *via* the nephridia when the chloragocytes degenerate and discharge their contents into the coelomic fluid (Morgan 1982).

Internal anatomy has, therefore, had a major influence on the strategies which have evolved for the regulation of metals in terrestrial invertebrates and has dictated which animals are best able to survive in metal-contaminated ecosystems. However, the chemical environment of the lumen of the gut has also been important in affecting the concentrations and forms of metals present at the sites of uptake. This factor is considered in the following section.

6.3.3 Chemical Environment in the Lumen of the Gut
The chemical composition of the material in the digestive tract of terrestrial invertebrates depends on the nature of the food and its reaction with secretions from salivary glands (if present) and digestive enzymes produced by cells of the midgut, or symbiotic microorganisms. Some digestion may take place externally if digestive enzymes are regurgitated onto the food as in spiders and some adult Diptera. Once food enters the foregut it may be filtered and the size fractions directed to different parts of the digestive tract. In isopods, for example, only

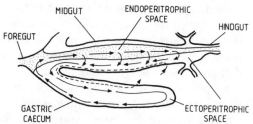

FIG. 6.2 Diagrammatic representation of the endo–ectoperitrophic circulation of digestive enzymes in the midgut of larvae of *Rhynchosciara americana* (Diptera). Endogenous fluid (dotted arrows) is transported into the posterior region of the midgut and moves towards the caeca where it is absorbed. Digestive enzymes (solid arrows) enter into the endoperitrophic space anteriorly, from which they are recovered stepwise as the polymeric molecules they hydrolyse become sufficiently small to accompany them through the peritrophic membrane. The enzymes and oligomeric molecules are then displaced towards the caeca where terminal digestion and absorption occur. (Redrawn from Terra & Ferreira 1981, by permission of Pergamon Journals Ltd.)

fluids and particles with a diameter of less than 40 nm pass through the filters in the foregut into the lumen of the hepatopancreas (Hames & Hopkin, in press).

The cuticular lining in the anterior and posterior parts of the digestive tract of arthropods, prevents large molecules from entering the cells of the foregut and hindgut. In the midgut, the composition of the material to which the microvillus borders of the cells are exposed may be controlled also by the presence of a peritrophic membrane surrounding the food. In larvae of the sciarid fly *Rhynchosciara americana*, the peritrophic membrane separates two extracellular sites as an adaptation to conserve secreted enzymes (Terra & Ferreira 1981; Fig. 6.2). The endoperitrophic compartment is where the initial digestion of food takes place. Products of digestion pass through the peritrophic membrane into the ectoperitrophic space where a different set of enzymes completes the final stages of digestion. The ectoperitrophic fluid passes into the gastric caeca where absorption of nutrients takes place. Such a system would affect the assimilation of metals if, for example, the pH was higher in the endoperitrophic space than in the gastric caeca. In these circumstances, metals would remain insoluble within the peritrophic membrane and be voided in the faeces. Without this restriction, the metals would be soluble in the more acidic pH of the fluid in the gastric caeca and be available for uptake by the cells.

Movement of fluids within the digestive system to facilitate digestion has evolved in other groups also. In terrestrial isopods, digestive enzymes secreted from the hepatopancreas, pass through the food in the hindgut towards the posterior, but are then returned in the reverse direction *via* a pair of dorsal channels formed from longitudinal folds of the wall of the hindgut and a typhlosole (Fig. 7.6). These fluids, containing soluble products of digestion, re-enter the hepatopancreas where absorption by the cells takes place. Thus, only components of the food, including metals which are soluble in the digestive fluids, are exposed to the main absorptive surface of the gut (Hames & Hopkin, in press). Lack of such basic knowledge has led to confusion in the literature as to the role of the digestive system in affecting uptake of metals by terrestrial isopods (see Section 7.4.2.2 for further discussion of this point).

Terrestrial invertebrates possess a wide range of digestive enzymes which break down food into components which can be absorbed by the digestive epithelia. These enzymes are most active under specific chemical conditions, the most important of which is pH. Consequently, the digestive fluids are buffered to maintain the pH within a fairly narrow range. In isopods (Hartenstein 1964), Collembola (Humbert 1974b), earthworms (Wallwork 1983) and most other detrivores, the pH of the gut contents is within one unit of neutrality. In herbivorous Lepidoptera, pH 8 to 9 is more usual. In some species, however, the pH of particular regions of the gut may be strongly acidic or alkaline to promote degradation of particular fractions of food. The pH of the midgut of larvae of the clothes moth *Tineola biselliella* is 10 (Waterhouse 1952). This extreme alkalinity (together with a strongly negative redox potential of -200 mV) enables the insect to split the disulphide bridges of the keratin in the wool which forms its diet. Metals introduced into the lumen of the gut of *Tineola* form colloids of insoluble sulphides which are accumulated in the cavities of the goblet cells. In larvae of the dipteran fly *Lucilia cuprina*, the anterior and posterior parts of the midgut are weakly alkaline while the mid region is strongly acid (Waterhouse & Stay 1955). Copper and iron in the diet are much more soluble at this low pH and are absorbed by the midgut cells (Fig. 6.3).

Many detrivorous and herbivorous invertebrates maintain large populations of microorganisms in their digestive tracts which may be essential for their survival (Anderson & Ineson 1984; Bignell 1984). The list includes millipedes (Crawford *et al.* 1983; Ineson & Anderson 1985), molluscs (Simkiss 1985), isopods (Donadey & Besse 1972; Reyes & Tiedje 1976a, 1976b; Kaplan & Hartenstein 1978; Coughtrey *et al.* 1980;

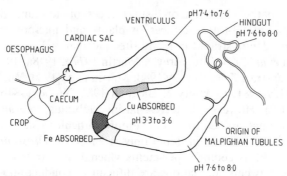

FIG. 6.3 Diagram of the gut of *Lucilia cuprina* larva (Diptera) showing pH of gut contents and regions of iron and copper absorption. (Redrawn from R. F. Chapman, *The Insects: Structure and Function*, London, English Universities Press, 1971, by permission of Hodder & Stoughton Ltd.)

Griffiths & Wood 1985; Ineson & Anderson 1985; Hames & Hopkin, in press), cockroaches (Brooks 1960), beetles (Kiffer & Benest 1981), termites (Bignell *et al.* 1980) and aphids (Auclair & Srivastava 1972; Ehrhardt 1968). Indeed, in the latter group, it is thought that the minimum levels of copper and zinc in the diet are dictated by the requirements of the intestinal symbionts rather than the needs of the aphids

One of the most important functions of these symbionts is to produce enzymes such as cellulase which the invertebrates are not able to synthesize themselves (Martin 1983; Kukor & Martin 1986a, 1986b). The microorganisms have a major influence on metal uptake by their host if they accumulate metals (e.g. in isopods — Hopkin & Martin 1982b), or secrete substances into the digestive fluids which alter metal availability. In the snail *Helix aspersa*, for example, sulphate-reducing bacteria in the crop appear to facilitate copper absorption by incorporating the metal into siderochromes (Simkiss 1985).

6.4 OTHER FACTORS

6.4.1 Tolerance

When a population of organisms has lived in a metal-polluted environment for a number of years, natural selection may take place for those individuals which are better able to grow in contaminated conditions.

This leads to the evolution of distinct races of 'tolerant' organisms. Evolution of metal-tolerant races of plants and microorganisms has been clearly demonstrated in many sites (Jordan & Lechevalier 1975; Tatsuyama *et al.* 1975; Coughtrey & Martin 1977b, 1978a, 1978b; Doelman & Haanstra 1979c; Bewley 1980; Martin *et al.* 1980; Wigham *et al.* 1980; Bewley 1981; Duxbury & Bicknell 1983; Wood 1984b; see also Section 5.1.2). However, in terrestrial invertebrates, tolerance to pollutant metals has proved to be much more difficult to substantiate and has begun to be studied in detail only recently (in *Drosophila* — see Section 9.5.4.9). Breeding experiments which demonstrate clear genetic differences between populations are difficult to conduct and tolerance is invariably acclimatory rather than inheritable.

Some of the clearest cases of tolerance have been demonstrated in South Africa where strains of the tick *Boophilus decoloratus* and the dipteran fly *Lucilia cuprina* have evolved resistance to arsenic (Whitehead 1961; Blackman & Bakker 1975). This metal was used as a pesticide at high concentrations in sheep and cattle dips. Tolerant populations survive immersion in solutions containing four times the concentration of arsenic that is lethal to non-tolerant ticks and fly larvae. The biochemical basis for this tolerance has not been determined.

There are several ways in which a terrestrial invertebrate could reduce the effects of metal contaminants in its diet. The availability of metals in the lumen of the gut could be reduced by raising the pH of the digestive fluids or secreting chelating agents. Such agents could also be produced by symbiotic microorganisms in the digestive tract which can evolve tolerance to metals much more rapidly than their host. Net uptake of metals into the cells could also be reduced by limiting uptake at transport sites, or by increasing the rate of excretion into insoluble intracellular deposits which may be subsequently voided. The few experiments which have been conducted on terrestrial invertebrates suggest that tolerance does not provide a selective advantage except in fairly specialised and subtle ways such as an improved ability to synthesize metal-binding proteins (Tyler *et al.* 1984; Maroni *et al.* 1987). However, further research may demonstrate that tolerance is more widespread than is currently thought as there are several marine groups which exhibit the phenomenon (Bryan 1984).

6.4.2 Moulting
Deposition of elements in the exoskeleton has been suggested as one way in which terrestrial arthropods could excrete unwanted metals (Giles

et al. 1973). However, where this route of loss has been suspected, metals apparently 'in' the cuticle have been shown to be either present as a surface deposit as in spiders (Clausen 1984a), or be associated with other tissues as in Collembola which void the midgut epithelium at each moult (Joosse & Buker 1979). In the centipede *Lithobius variegatus*, the body wall which remains after all the organs and tissues adhering to the internal surface of the exoskeleton have been removed is rich in zinc (Hopkin & Martin 1983). However, moulted exoskeletons of this species do not contain zinc so the metal in the body wall must be associated with tissues bound firmly to the cuticle which are not removed by scraping (Hopkin & Martin 1984b).

6.4.3 Life Histories
The evolution of tolerant races within species which respond to metals in different ways has already been discussed (Section 6.4.1). However, intraspecific differences in metal accumulation may also occur within a population due to factors other than tolerance. This may be due to natural genetic variation, or differences in feeding habits between adults and juveniles. For example, juvenile earthworms feed more on soil organic matter than adults (Satchell & Lowe 1967). Therefore, the young worms may be exposed to greater concentrations in their diet at an early age because metals tend to be associated with the organic rather than inorganic components of the soil. Such differences between adults and juveniles will be even more pronounced if the species has a larval stage that lives in an entirely different environment to the adults. Larvae of chironomid flies, for example, are aquatic and metals accumulated from contaminated freshwater may be retained in the adult stage (Dodge & Theis 1979; Leonhard *et al.* 1980; Rossaro *et al.* 1986).

Castes of the same species of social insect may contain different concentrations of metals due to their divergent lifestyles. Worker termites which collect food growing on serpentine deposits in Zimbabwe, contain much higher concentrations of chromium and nickel than soldiers or the queen which do not venture out of the nest (Wild 1975a). These metals are assimilated from the food by the workers before it is regurgitated and fed to the soldiers and the queen. This 'filtering' of environmental contaminants by the workers is probably an accidental rather than a deliberate phenomenon.

6.4.4 Experimental Design
All the factors discussed so far in this section affect the assimilation of

metals by terrestrial invertebrates in the field. However, when uptake is being studied in the laboratory, the way in which an experiment is conducted has a profound effect on metal accumulation. It is impossible to recreate field conditions exactly in the laboratory and this should always be borne in mind when results are extrapolated to the situation in the wild.

The main abiotic factor which affects metal assimilation is temperature. In general, the higher the temperature, the greater the rate of food consumption. However, Dallinger & Wieser (1977) have shown that isopods extract a larger percentage of the copper in leaf material at low rather than high temperatures because the food is retained in the gut for longer. At higher temperatures, food is eaten at a faster rate and the time during which digestive enzymes can act is much shorter (Hames & Hopkin, in press). Terrestrial invertebrates in the field are also exposed to rapid temperature fluctuations. In marine organisms, fluctuating temperatures can have a considerable effect on the rates of metal accumulation (Watkins & Simkiss 1988a, 1988b) but this phenomenon has not been examined in terrestrial invertebrates.

Daylength may be important also. Populations of the isopod *Oniscus asellus* moult regularly in a 16 hours light/ 8 hours dark cycle but cease to moult if the cycle is changed to an 8 hours light/16 hours dark regime (Steel 1982). Moulting is accompanied by profound physiological changes which may have a considerable effect on metal metabolism in isopods (Wieser 1965b).

The amounts of metals assimilated by terrestrial invertebrates depend on availability in the food. Animals extract a much greater proportion of the total metal when it is present as a deposit of a salt on the surface of the leaf than if it is bound within the leaf tissue (Williamson 1975). The nature of the anion may also affect uptake. Metals present as sulphates, nitrates or chlorides will have different rates of assimilation if the anion alters the chemical environment in the gut.

Many artificial diets have been formulated for rearing insects in the laboratory (Singh 1977). While these have the advantage of standardisation, the results obtained are difficult to extrapolate to the field where animals have a much more heterogeneous diet. Even when invertebrates are provided with a 'natural diet', they often assimilate metals at different rates to the same species in the field. A number of factors may be responsible. With detrivores, the dead leaves supplied may not be the food of choice and the animals will be forced to ingest material they would normally avoid in the field. Furthermore, many detrivores graze fungal hyphae from the surface of dead leaves in preference to ingesting the plant material (Swift *et al.* 1979). The concentrations and chemical form of metals in fungi are likely to be very different to those in vegetation. In the laboratory, the growth of hyphae may differ from that in the wild leading to changes in the amounts consumed by the invertebrates. This will alter rates of metal accumulation. Similar arguments apply to carnivorous invertebrates if they are presented with prey in the laboratory which they would not normally choose in the field.

In feeding experiments, incorrect estimates of rates of assimilation of metals can be made unless precautions are taken to minimise sources of error at each stage of experimentation and analysis. In their experiments on assimilation efficiency of food by the black cutworm *Agrotis ipsilon*, Schmidt & Reese (1986) calculated that cumulative errors became unacceptable unless the caterpillars consumed at least 80% of the food offered.

The density of animals may affect rates of turnover of metals. After 25 days, solitary specimens of the beetle *Popilius disjunctus* labelled with ^{65}Zn, had excreted 90% of the radioactive marker ingested with the food. However, ten beetles crowded together in a small chamber had excreted only 80% of the ^{65}Zn during the same period (Wit *et al.* 1984). Crowding may also lead to an increase in the incidence of parasitism which affects metal loss. Schowalter & Crossley (1982) demonstrated that the cockroach *Gromphadorhina portentosa* infected with the mite *Gromphadorholaelaps schaeferi*, cleared ^{51}Cr and ^{85}Sr from the gut at a faster rate than uninfested insects.

Some authors have included a period of starvation before conducting analyses of concentrations of metals in terrestrial invertebrates in an attempt to remove the contribution due to gut contents. However, in earthworms, a small amount of material is always retained in the digestive tract. In sites where the soil is heavily contaminated, the amounts of metals contained in such material may be considerably greater than are in the tissues of the worm, masking true differences in assimilation by the cells (Walton 1986). The possible contribution of gut contents to concentrations of metals in whole animals should also be borne in mind in laboratory feeding experiments.

Starvation may alter the physiology of invertebrates to such an extent that they do not respond as they would in the field to the ingestion of metal-rich food. Hryniewiecka-Szyfter & Storch (1986) starved isopods for up to six weeks before feeding the animals on a range of diets. It is questionable whether cells of the digestive system responded normally in their experiments after such a long time without food.

CHAPTER 7

Metals in Terrestrial Invertebrates at the Species, Organism and Organ Levels

7.1 INTRODUCTION

Several hundred papers have been published on the concentrations of metals in terrestrial invertebrates. Most of the studies on molluscs, earthworms and arthropods other than insects have reported levels in animals collected from the field whereas in insects, most work has been conducted in the laboratory. In recent years, however, an increasing number of authors have examined assimilation and excretion of metals by all groups of terrestrial invertebrates in laboratory experiments, many with a view to extrapolating their findings to the field. All these papers are cited in this section but only a small, hopefully representative, proportion of the literature is examined in detail.

7.2 MOLLUSCS

7.2.1 Introduction
The main aim of early research on the effects of metals on terrestrial molluscs was to develop agents for the control of slug and snail pests on agricultural crops. In more recent times, it has been shown that slugs and snails may accumulate potentially toxic metals and be important in transferring metals from vegetation to their predators, particularly birds. The widespread occurrence of slugs and snails in a wide range of terrestrial ecosystems has also led to the suggestion that they could be used as biological indicators and monitors of metal pollution (this topic is covered fully in Chapter 8). The large size and ease with which snails

can be kept in the laboratory, has made them ideal model animals for the study of biochemical mechanisms for metal detoxification (Section 9.5.1). All research to date has been conducted on pulmonate gastropods. The dynamics of metals in operculate snails (Helicinidae, Cyclophoridae and Pomatiasidae) has yet to be studied.

Most slugs and snails eat living or recently dead vegetation and play an important role in decomposition processes. In some woodlands, for example, slugs may consume nearly 10% of the annual fall of leaf litter (Jennings & Barkham 1979). A few species of slugs prey on earthworms (e.g. *Testacella* sp.) but no research on metals in this group has been carried out.

7.2.2 Routes of Uptake and Loss of Metals in Molluscs

During feeding, fragments of food are removed by the rasping action of the radula, lubricated with secretions from the salivary glands, and passed along the oesophagus to the crop where they are mixed with digestive enzymes produced by the hepatopancreas (Fig. 7.1). Products of this digestion are sorted in a complex of ciliated channels and only liquids and particles with a diameter of less than about 1 μm pass into the lumen of the hepatopancreas where most absorption takes place (Walker 1972). Larger material is formed into a 'faecal string' which also receives waste products from the hepatopancreas at the end of each digestive cycle. These 'hepatic faeces' are an important route for loss of metals by snails. In *Cepaea hortensis* collected from a contaminated roadside site and fed on uncontaminated carrot for 48 hours, faecal material derived from the hepatopancreas constituted only 13% of total faecal mass but was responsible for 56% and 29% of the losses of zinc and lead respectively (Williamson 1980). The amounts of metals excreted *via* the urine and foot mucus in terrestrial molluscs are insignificant in comparison to losses from the digestive tract.

Studies on metals in molluscs can be divided into three main areas. First, the development of metal-based molluscicides, second, analysis of concentrations of metals in slugs and snails collected from field sites and third, experiments on metal uptake from food contaminated in the laboratory. The ecophysiology of copper in terrestrial molluscs has been reviewed by Wieser (1979).

7.2.3 Molluscicides

The development, use and effectiveness of molluscicides for the control of slugs and snails has a fascinating history which has been reviewed

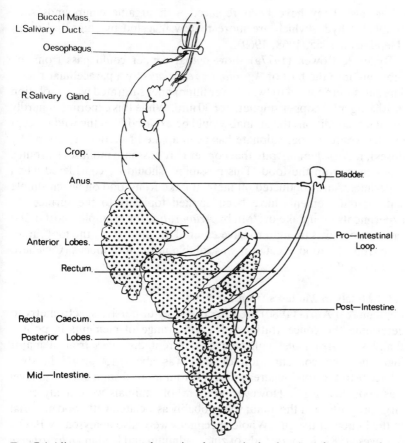

FIG. 7.1 Alimentary tract and associated organs in the slug *Agriolimax reticulatus*. The stomach is occluded by the lobes of the hepatopancreas (shaded). (From Walker 1972, by permission of the Malacological Society of London.)

recently by Godan (1983). The early molluscicides were based on copper sulphate, sodium arsenate, and organic tin and zinc compounds which were sprayed onto crops at application rates of up to 3 g m^{-2} (Gould 1962). The main effect of the metals was, however, not as contact or stomach poisons but as feeding deterrents. Slugs and snails starved rather than eat the crops which remained protected for a considerable time due to the persistence of the chemicals (Strufe 1968). The use of metal-based molluscicides is now discouraged because of their toxicity to non-target organisms. They have been replaced with organic compounds such as metaldehyde, which are more readily ingested by slugs and snails (Henderson 1965, 1968, 1969).

Ryder & Bowen (1977a) showed that copper could pass from the substrate into the foot of *Agriolimax reticulatus* by a paracellular route. The slugs were forced to 'walk' over filter paper saturated with a solution of 1000 µg ml^{-1} copper sulphate for 30 min. While this experiment hardly simulates conditions the animals would be exposed to in the wild (except in sites where copper sulphate has been applied to crops in very heavy doses), it does demonstrate that copper may pass into molluscs by routes other than from the food. This possibility should be considered when experiments are conducted on metal uptake from food on which simple salts of the elements have been applied topically to the surface. In experiments on uptake of ^{54}Mn by *Arion rufus*, for example, most of the radioisotope was thought to have entered the animal *via* the foot rather than from the food because it was soluble in the foot mucus (Cavellero & Ravera 1966).

7.2.4 Metals in Molluscs from the Field

Popham & d'Auria (1980) used X-ray energy-dispersive spectrometry to determine the concentrations of a wide range of elements in pooled samples of *Arion ater* collected at different distances from a major highway. The concentration of lead was about 150 µg g^{-1} in slugs adjacent to the road whereas in slugs from more distant sites, the level was about 20 µg g^{-1}. However, since whole animals were analysed, a large proportion of the metal was probably associated with food material in the lumen of the gut. Whole specimens were also analysed by Beeby & Eaves (1983) in their study of zinc, cadmium and lead in *Helix aspersa* but the snails were starved for 24 hours to allow gut contents to be evacuated before analysis. Analyses of concentrations of metals in whole snails from contaminated sites have also been conducted by Coughtrey & Martin (1977a) and Williamson (1979a).

The hepatopancreas contains the highest concentrations of zinc, cadmium and lead in all species of molluscs in which the tissue distribution of metals has been determined (Coughtrey & Martin 1976a; Cooke *et al.* 1979; Ireland 1979b, 1984a; Williamson 1980). Copper is more evenly distributed in the body because it is associated with haemocyanin and is stored in cells surrounding the blood vessels, rather than in the hepatopancreas (Mason *et al.* 1984).

Concentrations of zinc, cadmium, lead and copper in *Helix aspersa* from an isolated rural area, and a site close to a major aerial source of the metals, are presented in Table 7.1. Concentrations in snails from the uncontaminated site are in good agreement with the levels found in other molluscs from uncontaminated sites and are representative of natural background levels (Avery *et al.* 1983; Carter 1983). The snails from the contaminated site are clearly able to store zinc, cadmium and lead in the hepatopancreas but copper concentrations are elevated in all the body tissues. This may explain the sensitivity of molluscs to copper compounds since they are not able to 'buffer' copper levels in the rest of the tissues by accumulating the metal in the hepatopancreas in the same way as terrestrial isopods. For example, Marigomez *et al.* (1986c) measured a reduction in food consumption of *Arion ater* above a dietary copper level of only 25 μg g^{-1} (fresh weight). The slugs were not deterred from feeding at zinc, mercury and lead concentrations of 300, 100 and 1000 μg g^{-1} in the food respectively.

The concentrations of zinc, cadmium, lead and copper in the hepatopancreas of *Helix aspersa* from Avonmouth (Table 7.1) are the highest reported for snails from any site. However, much higher concentrations of zinc have been recorded in the hepatopancreas of slugs from disused mine sites. Ireland (1979b), for example, reported a mean concentration of zinc of 7130 μg g^{-1} in the hepatopancreas of *Arion ater* collected from the Cwmystwyth lead and zinc mine in the Ystwyth Valley, Wales.

Care should be taken when acid digests of snail shells are analysed by flame atomic absorption spectrometry. The very high concentrations of calcium in the samples may cause severe interference effects leading to erroneous readings, particularly for lead (Coughtrey & Martin 1976b). Where the effects of such interference have been accounted for, levels of zinc, cadmium, and lead have proved to be very low. Therefore, the shell does not appear to be an important site for storage-detoxification of unwanted metals in molluscs (Cooke *et al.* 1979; Williamson 1980).

Concentration factors are difficult to calculate for molluscs because their diet in the field is so varied. Williamson (1979b) showed that concentration factors for lead in *Cepaea hortensis* varied by a factor of 50 from 0.09 to 4.9 depending on the year class of the snails chosen for the calculation and the component of living or dead vegetation which was taken as being representative of their diet. Concentration factors for zinc ranged from 0.14 to 11.8 and for cadmium from 4.2 to 21.2. Consequently, accurate determination of assimilation rates of metals must be measured in the laboratory where food availability can be closely controlled and consumption and defaecation monitored.

TABLE 7.1

Concentrations of metals in *Helix aspersa* from an uncontaminated site (Kynance Cove, Cornwall, England) and a site 1 km from the Avonmouth smelting works (see Section 5.3.2) (mean ± standard error, $n = 6$; original data collected by the author, $\mu g\, g^{-1}$ dry weight—compare with values for the isopod *Oniscus asellus* from the same sites in Table 7.9)

Kynance Cove

	Dry weight (mg)	Zn	Cd	Pb	Cu
Hepatopancreas	0.120 ± 0.021	429 ± 71	20.2 ± 5.2	31.8 ± 4.8	118 ± 41
Gut	0.130 ± 0.041	67.4 ± 9.6	9.51 ± 4.16	5.56 ± 0.89	94.8 ± 33.1
Reproductive tissue	0.181 ± 0.082	51.5 ± 4.5	5.52 ± 4.23	0.75 ± 1.04	55.8 ± 15.9
Rest	0.389 ± 0.110	49.2 ± 2.1	1.12 ± 0.16	0.63 ± 0.66	138 ± 36
Total	0.820 ± 0.202	103 ± 14	4.88 ± 0.62	5.79 ± 1.06	104 ± 28

Avonmouth

	Dry weight (mg)	Zn	Cd	Pb	Cu
Hepatopancreas	0.136 ± 0.033	1780 ± 300	271 ± 17	490 ± 96	186 ± 38
Gut	0.112 ± 0.032	204 ± 17	30.1 ± 14.1	143 ± 46	181 ± 39
Reproductive tissue	0.346 ± 0.129	58.2 ± 3.6	7.46 ± 1.81	16.4 ± 3.3	135 ± 27
Rest	0.396 ± 0.093	95.6 ± 21.7	16.5 ± 2.5	18.0 ± 6.4	347 ± 60
Total	0.990 ± 0.153	418 ± 71	49.6 ± 10.7	121 ± 24	228 ± 43

TABLE 7.2

Concentrations of metals in uncontaminated and metal-enriched lettuce and definition of stock solutions in which the lettuce had been soaked for 1 hour ($\mu g\,g^{-1}$ dry weight ± standard deviation, $n = 12$). This material was used in metal uptake experiments on *Helix pomatia* (from Dallinger & Wieser 1984a, by permission of Pergamon Journals Ltd)

	Uncontaminated	Enriched	Stock solution
Zn	205 ± 75	761 ± 330	$ZnCl_2$ 10 µg ml^{-1}
Cd	4.5 ± 2.2	160 ± 87	$CdCl_2$ 1 µg ml^{-1}
Pb	3.1 ± 1.3	936 ± 320	$Pb(NO_3)_2$ 1 µg ml^{-1}
Cu	28.0 ± 12.4	534 ± 116	$CuCl_2$ 10 µg/ml^{-1}

7.2.5 Experiments on Metals in Molluscs

The dynamics of assimilation and excretion of zinc, cadmium, lead and copper in the internal organs of the Roman snail *Helix pomatia* are fairly well understood following the comprehensive experiments of Dallinger & Wieser (1984a). Snails were fed for 32 days on lettuce which had been soaked for 1 hour in solutions of salts of either zinc, cadmium, lead or copper, followed by a feeding period of 40–50 days on uncontaminated lettuce (Table 7.2). Groups of four snails were dissected at intervals of four days during the first 32-day period and then at days 40, 50 and 80 after which the experiment was terminated. Concentrations of zinc, cadmium, lead and copper were determined in 11 tissues and net changes in metal contents plotted graphically.

It is not possible to determine assimilation rates of metals from Dallinger & Wieser's (1984a) experiments because consumption of treated food was not measured. However, the results confirmed the importance of the hepatopancreas as the main storage site for zinc, cadmium and lead, but not for copper which was always distributed evenly throughout the tissues (Table 7.3). Laboratory experiments on metal uptake by *Arion ater* have confirmed that the hepatopancreas is the most important storage organ for zinc, cadmium and lead in slugs also (Ireland 1981, 1982, 1984b).

In an earlier experiment on copper uptake by *Helix pomatia* (Moser & Wieser 1979), the efficiency of assimilation of the metal was always high. In short term feeding experiments with lettuce containing about

TABLE 7.3

Amounts of metals in the hepatopancreas of *Helix pomatia* (μg) for controls (day 0), end of exposure (32 days) and recovery (70–80 days), and percentage of total snail metal content in hepatopancreas (mean ± standard deviation, n = 48–52) (from Dallinger & Wieser 1984a, by permission of Pergamon Journals Ltd)

Metal	Dry weight of hepatopancreas (g)	Metal content		
		Day 0	*Day 32*	*Day 70–80*
Zn	0.356 ± 0.075	178 ± 74 (54.7%)	413 ± 53 (72.0%)	388 ± 131 (71.9%)
Cd	0.350 ± 0.075	21.9 ± 7.5 (53.5%)	75.0 ± 18.7 (73.8%)	58.3 ± 8.4 (60.2%)
Pb	0.343 ± 0.070	13.1 ± 15.8 (38.3%)	446 ± 144 (92.1%)	413 ± 67 (89.7%)
Cu	0.326 ± 0.056	26.2 ± 12.4 (15.1%)	42.7 ± 6.9 (15.1%)	23.1 ± 5.4 (14.4%)

1400 μg g^{-1} copper, 97% of the ingested metal remained in the snails. The concentration of copper in the hepatopancreas of treated snails increased four-fold during the winter aestivation period due to a combination of a reduction in the weight of the gland and movement of copper into the organ. In sites contaminated with copper, winter mortality may be exacerbated by this effect. Uptake of zinc and lead in *Helix pomatia* has also been studied by Meincke & Schaller (1974).

Dallinger & Wieser (1984a) showed that zinc was accumulated in most organs during the first 32 days and was then re-distributed to the hepatopancreas which held about 70% of this metal in the animals at the end of the experiment (Fig. 7.2A). Cadmium was not assimilated at all by most organs but it accumulated steadily in the anterior gut and hepatopancreas throughout the loading period and was not lost when snails were transferred onto uncontaminated lettuce (Fig. 7.2B). Lead

---

FIG. 7.2 Concentrations of (A) zinc, (B) cadmium, (C) lead and (D) copper in the hepatopancreas of the snail *Helix pomatia* fed on contaminated lettuce for 32 days (see Table 7.2) followed by uncontaminated lettuce (each point is the mean concentration ± standard deviation of 4 snails). (Redrawn from Dallinger & Wieser 1984a, by permission of Pergamon Journals Ltd.)

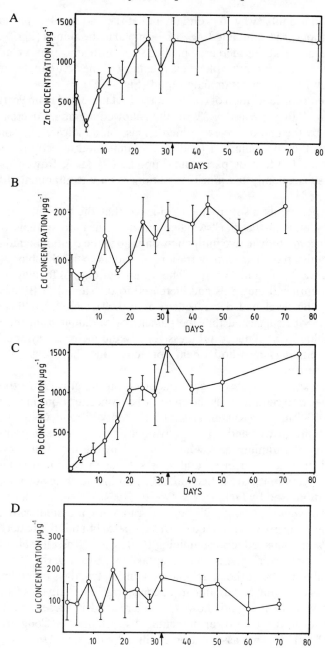

behaved in a similar way to cadmium and at the end of the experiment, the hepatopancreas contained 90% of the metal in the animal (Fig. 7.2C). Copper concentrations increased in all tissues during the experiment but the snails were apparently able to excrete most of the accumulated metal when they fed on uncontaminated food (Fig. 7.2D).

The mean concentration of copper in the foot increased from 96 $\mu g\, g^{-1}$ at day 0 to 150 $\mu g\, g^{-1}$ at day 32 but then showed a gradual decrease to 82 $\mu g\, g^{-1}$ by the end of the experiment. This could indicate that soluble copper on the contaminated food was accumulated directly from the substrate *via* the foot by paracellular uptake (Ryder & Bowen 1977a) and lost subsequently *via* mucus secretion when uncontaminated food was supplied.

The experiments also provided evidence that the presence of high levels of metals in the diet 'switched on' detoxification systems which did not become fully active until a few days after the contaminated food was administered. The concentration of cadmium in the blood, for example, was 0.07 $\mu g\, ml^{-1}$ at day 0 but after four days of feeding on the cadmium-enriched diet, this had increased to 0.29 $\mu g\, ml^{-1}$. By day 28, however, the level had decreased to only 0.09 $\mu g\, ml^{-1}$. Dallinger & Wieser (1984b) suggested that the passage of cadmium from the food into the blood was reduced progressively by the increased synthesis of cadmium-binding metallothionein proteins in the cells of the hepatopancreas.

Similar experiments have been conducted on the effects of exposing *Helix aspersa* to cadmium (as the chloride) at concentrations in the food of up to 1000 $\mu g\, g^{-1}$ (Russell *et al.* 1981). Marked reductions in food consumption, growth and reproductive activity were observed at concentrations of cadmium as low as 25 $\mu g\, g^{-1}$ in the diet. This level is exceeded in vegetation in several sites subject to cadmium pollution (e.g. the Avonmouth area, Section 5.3.2) and is much lower than the concentration used by Dallinger & Wieser (1984a) in their experiments on *Helix pomatia*. About 60% of the cadmium ingested with food containing 25 $\mu g\, g^{-1}$ was retained by *Helix aspersa* but this figure declined to only 7% in snails fed food containing 1000 $\mu g\, g^{-1}$ (Russell *et al.* 1981).

In short-term experiments on lead uptake by *Helix aspersa*, the snails retained about 50% of the metal ingested during two days of feeding on agar blocks containing lead sulphate at concentrations of up to 500 $\mu g\, g^{-1}$ (Beeby 1985). In the field, however, where lead is present in a less available form, a much lower percentage is assimilated (Coughtrey & Martin 1976a). The net assimilation of silver by the slug *Arion ater* and the Giant African Snail *Achatina fulica* has been compared recently by Ireland (1988).

7.3 EARTHWORMS

7.3.1 Introduction
The importance of earthworms in promoting the fertility of soils was recognised originally by Darwin. He calculated that in a typical English pasture, the earthworms deposited a layer of soil of about 5 mm in thickness at the surface each year. Since concentrations of metals may reach very high levels in soils contaminated by mining waste, aerially-deposited metal-containing particles from smelting activity, agricultural chemicals or sewage sludges, earthworms are often exposed to severe stress.

Some species are also important in mixing soil layers for as well as bringing material from depth to the surface, worms may also deposit surface material deep within their burrows. In an experiment on the role of worms in such mixing, *Lumbricus terrestris* were shown to promote the incorporation of ^{109}Cd into the soil by pulling leaves contaminated with this isotope into their burrows (Van Hook *et al.* 1976).

Worms are important prey items for many vertebrates and provide a route through which pollutants are transferred to higher levels in food chains. Particular problems arise with metals which are concentrated by worms to levels which are much greater than those in the soil. Worm/soil concentration factors may exceed ten for such elements as mercury (Siegel *et al.* 1975; Bull *et al.* 1977; Beyer *et al.* 1985a), selenium (Nielsen & Gissel-Nielsen 1975) and cadmium (Ma *et al.* 1983).

Recognition of the important role of earthworms in the transport of metals in terrestrial ecosystems has resulted in a large number of studies on the uptake, tissue distribution and effects of metals on these extremely common soil organisms. Indeed, more than 70 papers have been published on such topics, not including publications concerned exclusively with intracellular storage (which are considered in Section 9.5.2). The subject has been reviewed recently by Beyer (1981) and Ireland (1983).

The ability of earthworms to accumulate metals was recognised at the end of the last century. In what was probably the earliest paper on metals in a terrestrial invertebrate, Hogg (1895) reported that earthworms living in mining spoil accumulated lead in their intestinal tissue. However, despite the intensive research conducted since Hogg's paper, most with modern analytical techniques, there is still confusion in the literature with regard to lethal levels and concentration factors for metals in different species. Contradictions can be explained in part by differences in the availability of metals in soils which depend to a large extent on the amounts of organic matter present (Streit 1985). The toxicity of copper to *Octolasium cyaneaum*, for example, is directly related to the percentage of carbon in the experimental soil (Jaggy & Streit 1982).

Some of these contradictions can also be explained by determining the 'ecophysiological group' to which species belong (Piearce 1972, 1978). Two species living in the same soil may have different diets and be exposed to metals in different forms and concentrations. *Allolobophora caliginosa*, for example, consumes only organic matter in an advanced state of decomposition whereas *Lumbricus rubellus* eats recently-fallen leaf litter. However, even when species are reared under identical conditions, differences in metal assimilation still occur which must be due to differences in the structure and physiology of the digestive and excretory systems. These factors are examined in the next section.

7.3.2 Routes of Uptake and Loss of Metals in Earthworms

The digestive tract of earthworms is a straight tube running from the mouth to the anus which does not possess any large diverticulae (Fig. 7.3). The internal surface area may be increased by the presence of a longitudinal typhlosole which projects into the lumen of the gut. A gizzard for grinding food material, and pouches containing deposits of calcium carbonate (the calciferous glands), are also present in some species (Edwards & Lofty 1977; Wallwork 1983).

The gut is surrounded by a loose assemblage of cells, known as the chloragogenous tissue, which accumulate waste products of digestion. These waste products may include metals such as lead and cadmium which are stored in granules in the chloragocytes (Ireland & Richards 1981). The granules or 'chloragosomes', are discharged into the coelomic fluid and may be excreted either *via* the nephridia in each segment, or stored permanently in large amorphous 'waste nodules', along with other excretory material in the posterior coelomic sacs. In *Lumbricus terrestris* and *Allolobophora longa*, these sacs may become completely packed with nodules which can be lost only by autotomy of the posterior segments of the animal (Andersen & Laursen 1982).

Thus, lead assimilated by the intestinal cells of *Lumbricus terrestris* can be excreted in several ways (Fig. 7.4).

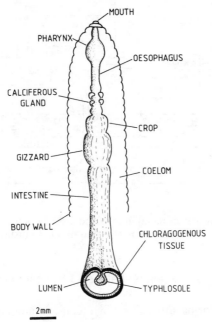

FIG. 7.3 Schematic cut-away diagram to show the internal structure of the anterior alimentary tract of *Lumbricus terrestris*.

(1) Breakdown of the intestinal cells into the lumen of the gut.
(2) Permanent storage-excretion in the chloragocytes.
(3) Temporary storage in the chloragocytes followed by release into the coelomic fluid from which the metal could be excreted by three routes;
    (a) *via* the nephridia through the body wall,
    (b) back into the gut *via* the calciferous glands, or
    (c) by permanent storage-excretion in waste nodules.

The relative importance of each of these pathways depends on the species in question. For example, *Lumbricus terrestris* assimilates large amounts of calcium from the food which it excretes *via* its very active calciferous glands. Since lead is thought to follow similar biochemical pathways to calcium, *Lumbricus terrestris* tends to accumulate less lead than species with less active calciferous glands such as *Allolobophora longa*. The latter species relies more on waste nodule formation for the excretion of lead and accumulates greater concentrations of the metal

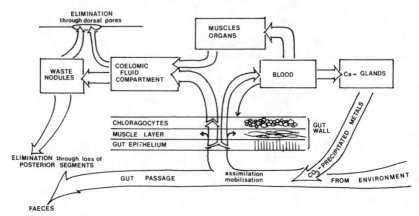

FIG. 7.4 Possible routes of metal uptake and loss in *Lumbricus terrestris*. (From Andersen & Laursen 1982, by permission of VEB Gustav Fischer Verlag.)

(Andersen 1979; Andersen & Laursen 1982; Morgan 1982; Morris & Morgan 1986).

When earthworms are fed on a metal-contaminated diet and are transferred subsequently to uncontaminated food, the curve describing rate of loss of the metal with time can be split into two components (Crossley *et al.* 1971; Neuhauser *et al.* 1985). An initial rapid loss represents voiding of gut contents but this is followed by a much slower loss resulting from turnover of metal assimilated by the tissues of the worm. The duration of this second component depends on the lability of the metal in question i.e. to what extent it has been stored in the sites of permanent storage detoxification described above. The half life of ^{60}Co in *Eisenia foetida*, for example, is more than one year (Neuhauser *et al.* 1984b) whereas for more mobile elements such as caesium (a potassium analogue), the half life is only a few days (Crossley *et al.* 1971).

7.3.3 Metals in Earthworms from the Field
Observations on the concentrations of metals in field populations of earthworms have shown consistently that concentration factors for cadmium are at least five, for zinc are between about two and five and for lead and copper, are always less than one (Van Hook 1974; Van Rhee 1977; Czarnowska & Jopkiewicz 1978; Carter *et al.* 1980).

Soil particles are invariably present in the gut contents of earthworms, even in species which feed almost exclusively on vegetation. Contamination of tissues is thus a serious problem and great care should be exercised in removing material from the digestive tract before analysis (Bengtsson *et al*. 1983b; Walton 1986). This can be done by cutting open and flushing out the gut with distilled water, or by feeding the animals on an inert substance such as filter paper until soil particles are no longer present in the faeces (Honda *et al*. 1984). Stafford & McGrath (1986) have suggested a formula which can be used for correcting for the presence of metals in gut contents of earthworms.

In contaminated soils there are several factors other than the concentrations of metals which determine whether particular species of earthworm are able to survive. The most important of these is the pH (Bengtsson *et al*. 1983b; Beyer *et al*. 1987). The toxic effects of acidic soils are due not only to the high levels of hydrogen ions, but to an increase also in the 'availability' of metal ions to the worms. At a soil pH of about 4, for example, zinc, cadmium and lead are assimilated by *Lumbricus rubellus* at a much greater rate than at a neutral pH whereas net uptake of copper is not affected (Ma *et al*. 1983). Species with highly active calciferous glands are better able to survive in acidic soils than species in which these glands are poorly-developed since secretion of calcium into the gut raises the pH of the contents (Morgan 1982). Calcium may also interfere with uptake of pollutant metals. In the field, the concentration of lead in *Dendrobaena rubida* is much less in calcareous than in non-calcareous soils containing the same levels of lead (Ireland 1975a).

Considerable interest has been expressed in using earthworms to 'rehabilitate' contaminated soils (Schreier & Timmenga 1986). Their action would be two-fold. First, to mix polluted and unpolluted soil horizons by burrowing activity and second, by accumulating metals which could be removed from the soil ecosystem by 'harvesting' worms at regular intervals. However, the ability of earthworms to accumulate toxic metals has restricted their potential use as a source of protein for farm livestock, particularly if worms have been reared in soils to which metal-contaminated sewage sludge has been applied (Andersen 1980; Dindal *et al*. 1979; Hartenstein & Hartenstein 1981).

Numerous studies have examined the accumulation of lead in earthworms from contaminated roadside ecosystems (Gish & Christiensen 1973; Goldsmith & Scanlon 1977; Czarnowska & Jopkiewicz 1978; Scanlon 1979; Ash & Lee 1980; Pulliainen *et al*. 1986). However, concentrations greater than 50 μg g^{-1} have not been recorded in whole

TABLE 7.4

Ranges of concentrations of metals in soils and *Lumbricus rubellus* from 31 sites in The Netherlands ($\mu g\ g^{-1}$ dry weight) (from Ma *et al.* 1983, by permission of Springer–Verlag)

Metal	Soil	Lumbricus rubellus
Zn	10–1 220	717–3 500
Cd	0.1–5.7	20–202
Pb	14–430	9–670
Cu	1–130	12–58

animals from which gut contents have definitely been removed, even in sites adjacent to very busy roads. There is no evidence that lead derived from automobile exhausts causes mortality in earthworms but they can provide a route for transport of the metal to predators.

Aerial contamination of soils with metals derived from smelting activity is a much more serious problem and it is clear that some species of oligochaetes are unable to survive close to such industrial activity. Bengtsson & Rundgren (1982) and Bengtsson *et al.* (1983b) were able to show that the abundance and diversity of enchytraeids and lumbricids in coniferous forest soils in Sweden was reduced by metal pollution from a brass mill for a distance of more than 10 km from the works. The density of lumbricids at a particular site could be predicted from a knowledge of the concentration of zinc in the soil. *Lumbricus rubellus* was the only species found in soils of a deciduous woodland contaminated by metals, 3 km downwind of a primary zinc, lead and cadmium smelting works at Avonmouth, South West England, where one would normally expect to find a much greater diversity of oligochaetes (Hopkin *et al.* 1985a).

In the Kempen region of the Netherlands, soils have been polluted with zinc, cadmium, lead and copper from a smelting works. This has led to extensive contamination of the biota (Ma *et al.* 1983). Very high concentrations of zinc were detected in *Lumbricus rubellus*, although the concentration factors (worm/soil) for this metal were lower than those for cadmium which at one site, exceeded 30 (Table 7.4). Concentration factors for lead and copper were about one or less. Concentrations of the same order have been detected in earthworms from contaminated soils in the Avonmouth area (Martin & Coughtrey 1975, 1976; Wright & Stringer 1980).

Despite the apparent inability of earthworms to accumulate copper to the same extent as cadmium, the metal is surprisingly toxic. Indeed, solutions of copper sulphate are often applied to the putting greens of golf courses to inhibit worm activity to prevent their casts from interrupting the run of the ball! Levels of copper as low as 50 µg g^{-1} in the soil are sufficient to interfere with reproductive activity of *Allolobophora caliginosa* (Van Rhee 1969). Earthworms collected close to a copper refinery in Liverpool, England, had poorly developed clitella resulting probably from copper contamination of their diet (Hunter *et al.* 1987b). Copper levels of about 100 µg g^{-1} in soils to which pig slurry had been applied, inhibited reproduction in earthworms in grassland and pastures (Van Rhee 1975; Curry & Cotton 1980). Chromium (IV) is even more toxic to earthworms. A level of only 10 µg g^{-1} in soil contaminated by effluent from a tannery was sufficient to cause mortality in *Pheretima posthuma* and other species (Soni & Abbasi 1981; Abbasi & Soni 1983).

The most contaminated soils occur in mining areas. The concentrations of lead and zinc in earthworms from disused mine sites in Wales have been reported in a number of papers by Ireland (1975a, 1975b, 1975c, 1976, 1979a) and Ireland & Wooton (1976). *Lumbricus rubellus* collected from the spoil tips of the Cwmystwyth mine contained 0.35% lead on a dry weight basis (Table 7.5). Other studies on worms from mine sites have confirmed the high concentration factors for cadmium in comparison to zinc and lead (Roberts & Johnson 1978; Andrews & Cooke 1984).

Copper is an effective fungicide and has been applied in the past to a range of crops at high dose rates. Soils have become contaminated with the metal which tends to remain in the surface layers of the soil. The toxicity of copper to earthworms was recognised first by Nielson (1951) who reported that densities of *Allolobophora caliginosa* in the soil were reduced following accidental spillage of a copper-based fungicide (Table 7.6). The most serious problems seem to have occurred in orchards where, in many places, copper fungicides have completely wiped out earthworm populations (Van Rhee 1967, 1969; Niklas & Kennel 1978). The resultant reduction in earthworm feeding activity has led to an accumulation of undecomposed leaf litter. A detailed description of the effects of the lack of earthworms on the structure of orchard soils has been given by Van de Westeringh (1972).

TABLE 7.5

Concentrations of metals in soil and *Lumbricus rubellus* from three contaminated sites in Wales, Cwmystwyth (lead mine), Dolgellau (copper mine) and Borth (waste ground on which sewage sludge had been deposited) (μg g^{-1} dry weight, mean ± standard error, $n = 6$) (from Ireland 1979a)

	Zn	Cd	Pb	Cu	Mn	Ca
Cwmystwyth						
Soil	138 ± 1	2.0 ± 0.1	1310 ± 40	20 ± 1	1330 ± 30	998 ± 40
L. rubellus	739 ± 231	15 ± 5	3590 ± 900	13 ± 6	82 ± 22	18 100 ± 1800
Dolgellau						
Soil	100 ± 2	4.0 ± 0.3	42 ± 2	335 ± 15	164 ± 3	4920 ± 120
L. rubellus	416 ± 34	25 ± 3	28 ± 8	11 ± 2	27 ± 4	13 200 ± 500
Borth						
Soil	992 ± 53	4.0 ± 0.2	629 ± 80	252 ± 5	226 ± 9	32 100 ± 700
L. rubellus	676 ± 25	4.0 ± 0.1	9 ± 1	11 ± 1	28 ± 1	12 600 ± 1500

TABLE 7.6

Density of earthworms and concentrations of copper in soil ($\mu g \ g^{-1}$ dry weight) at different distances from a leaking bath containing copper sulphate solution for the treatment of foot rot in livestock on a farm in Hamilton, New Zealand (from Nielsen 1951)

	Distance from bath (m)	Number of worms m^{-2} (range)	Soil depth (cm)	Copper in soil
Affected soil	1	0	0–10	360
			10–20	256
	2	20–60	0–10	266
			10–20	176
	3	260–400	0–10	144
			10–20	66
Normal soil	30	800–1 000	0–10	48
			10–20	40

7.3.4 Experiments on Metals in Earthworms

The results of laboratory and field experiments on the uptake of metals by earthworms are difficult to interpret due to the wide range of experimental conditions under which animals have been kept. Standard procedures have been suggested only for tests on the effects of organic chemicals on earthworms (Karnak & Hamelink 1982), not with metals. Streit (1985) has proposed a model for the uptake of copper by earthworms (based on the acetic acid-soluble fraction of copper in soil) which, it is claimed, explains the contradictions in the literature as to the toxicity of the metal.

The rates of uptake of particular metals are affected by the chemical form of the element, temperature and soil type. An extensive series of experiments by Malecki *et al.* (1982), for example, showed that carbonates and oxides of copper, nickel, cadmium, zinc and lead were less toxic to *Eisenia foetida* than acetates, chlorides, nitrates and sulphates of the metals. The effect of soil type on metal uptake was demonstrated clearly by Ma (1982) who showed that the concentration factors for a range of metals in *Allolobophora caliginosa* were considerably higher in a sandy soil of low pH, than in a loamy soil of neutral pH, both of which had been treated with municipal waste compost (Table 7.7). In the same study, cadmium and copper proved to be much more toxic to *Lumbricus rubellus*, than lead or nickel (Table 7.8). Metal uptake differs also between species. Consequently, it is not legitimate to use separate species for control and experimental treatments as was done by Beyer *et al.* (1982).

In a more recent study, Ma (1984) examined the effects of copper on *Lumbricus rubellus* and showed the important influence of pH and temperature on the toxicity of the metal. Significant decreases in cocoon production at two to six weeks (15 °C) were found at concentrations in the range 100 to 150 µg g^{-1} copper, applied either to a sandy or loamy soil. The relative effect of copper was greater generally at lower temperatures.

Studies on *Dendrobaena rubida* have also confirmed the greater toxicity of lead and copper to earthworms at lower pH's (Bengtsson *et al.* 1986; Fig. 7.5). In the most heavily contaminated animals, lead and copper had accumulated in the seminal vesicles and nervous tissues, in addition to the sites of detoxification associated with the gut, and cocoon production and hatching rates were depressed in comparison with worms living in control soils.

The greatest number of experiments on metal uptake by earthworms have been conducted with soils to which sewage sludge has been added (Suzuki *et al.* 1980; Wade *et al.* 1982; Carter *et al.* 1983; Pietz *et al.* 1984; Kruse & Barrett 1985). Sludge derived from sewage works in industrial areas may contain considerable amounts of metals (Table 3.1). However, there is marked disparity between the rates of assimilation by, and toxicity of particular metals to, earthworms, even in the same species. This is due, presumably, to the heterogeneous composition of sewage sludge which contains large amounts of organic matter to which metals become firmly complexed rendering them unavailable for uptake by the worms (Hartenstein *et al.* 1980a, 1980b, 1981).

Despite this disparity, cadmium emerges in all experiments as the metal with by far the highest concentration factor (Andersen 1979; Helmke *et al.* 1979; Beyer *et al.* 1982; Neuhauser *et al.* 1984a). *Eisenia foetida*, grown on composted sewage sludge, assimilated 25% of the cadmium which passed through its gut whereas only about 2% of the zinc, chromium and copper in the food were absorbed (Mori & Kurihara 1979). As in the field, concentration factors for zinc and copper decrease with increasing levels of these metals in the soil due possibly to saturation of uptake sites for the metals, or active regulation of their assimilation (Neuhauser *et al.* 1985).

TABLE 7.7

Chemical properties of two soils subjected to annual application of municipal sewage sludge for a decade, and concentration factors for a range of metals in *Allolobophora caliginosa* (worms/soil concentrations μg g^{-1} dry weight) (from Ma 1982, by permission of VEB Gustav Fischer Verlag)

	Soil properties			Concentration factors for adult Allolobophora caliginosa							
	CEC	OM	pH	Cd	Cr	Cu	Fe	Mn	Ni	Pb	Zn
Loamy soil											
Untreated	26.3	5.8	7.1	12.3	0.01	1.15	0.05	0.04	0.05	<0.01	4.0
20 tonnes ha^{-1} yr^{-1}	24.5	6.7	7.0	10.7	0.03	0.57	0.05	0.04	0.11	0.16	4.4
40 tonnes ha^{-1} yr^{-1}	25.1	8.4	6.9	11.3	0.07	0.34	0.11	0.10	0.12	0.30	3.6
Sandy soil											
Untreated	5.3	2.8	4.8	140.0	<0.01	2.45	0.36	0.32	1.86	2.63	81.9
20 tonnes ha^{-1} yr^{-1}	6.1	3.7	5.5	59.4	0.11	1.37	0.33	0.39	0.55	1.24	23.8
40 tonnes ha^{-1} yr^{-1}	7.1	4.3	6.0	40.5	0.13	1.08	0.36	0.32	0.38	0.88	13.3

CEC = cation exchange capacity, meq 100 g^{-1}; OM = organic matter, g 100 g^{-1}.

TABLE 7.8

Toxicity of metals to adult *Lumbricus rubellus* in a sandy loam soil (n = 25–30) (from Ma 1982, by permission of VEB Gustav Fischer Verlag)

Dosage (μg g^{-1} dry wt soil)	% Mortality							
	Cd		Cu		Pb		Ni	
	week 6	week 12	week 6	week 12	week 6	week 12	week 6	week 12
Control[a]	4	12	4	12	3	13	3	13
20	0	12	0	8	7	30	10	10
150	3	3	0	8	0	13	7	23
1 000	100	100	52	100	0	17	13	40
3 000	100	100	100	100	7	20	73	100

[a] The uncontaminated control soil contained 0.5 μg g^{-1} Cd, 12 μg g^{-1} Cu, 26 μg g^{-1} Pb and 17 μg g^{-1} Ni.

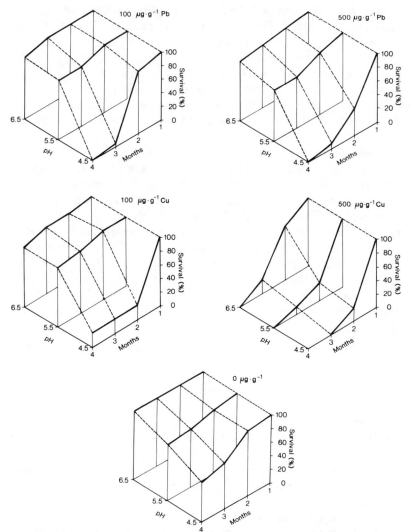

FIG. 7.5 Survival of adult *Dendrobaena rubida* reared in soils of different pH and concentration of lead and copper during a 4-month study. (From Bengtsson *et al.* 1986, reprinted by permission of Kluwer Academic Publishers, ©1986 Reidel Publishing Company.)

Very few experiments have been conducted on the essentiality of metals to earthworms. This is due to the lack of a suitable synthetic diet. However, *Eisenia foetida* grows faster when its diet contains about 25 µg g^{-1} cobalt than on a diet containing 10 µg g^{-1} or 90 µg g^{-1} cobalt (Neuhauser *et al.* 1984b). The biochemical basis for this stimulation of growth is not clear. It could be an indirect effect with higher levels of cobalt stimulating symbiotic gut microorganisms, or inhibiting the development of nematode parasites in the worms (Gunnarsson & Rundgren 1986).

Beyer (1981) invoked the class A/class B metal ion concept (Nieboer & Richardson 1980) to explain the differences in concentration factors for metals in earthworms in field and laboratory experiments. However, his suggestion that class B metals are preferentially accumulated is not supported by the facts. While the class B metals cadmium and mercury are avidly assimilated, copper, another class B metal is not. The low net accumulation of copper by earthworms may be due to restricted uptake mechanisms since the blood pigment of oligochaetes is the iron-based haemoglobin, rather than the copper-based haemocyanin of molluscs and isopods (Ireland & Fischer 1979). The extreme toxicity of copper, and the affinity of earthworms for cadmium, have, however, still to be explained.

7.4 CRUSTACEA

7.4.1 Introduction
Of the many Crustacea which have invaded terrestrial environments, representatives of only two orders, the isopods and amphipods, do not need to return to the water to breed. They have achieved emancipation from an aquatic existence by developing a watertight brood pouch in which the embryos are protected from desiccation. The isopods (Section 7.4.2) have been by far the most successful colonisers and have invaded some of the most extreme environments on earth including North African deserts where they are the most common invertebrates (Warburg 1987). Amphipods (Section 7.4.3) are more restricted in their distribution. In temperate zones, amphipods are rarely found more than 100 m from the sea, although in tropical regions they can be very common inhabitants of leaf litter.

The dynamics of uptake and loss of metals in marine Crustacea have been studied extensively. Decapods (shrimps, crabs etc.) regulate their net assimilation of zinc and copper whereas barnacles tend to permanently store these metals (White & Rainbow 1985; Bryan *et al.* 1986b). Rainbow (1985a, 1985b) has calculated that the soft tissues of a decapod need to contain about 70 μg g^{-1} of zinc (35 μg g^{-1} for enzymes, 35 μg g^{-1} to stabilise haemocyanin) and about 80 μg g^{-1} of copper (35 μg g^{-1} for enzymes, 45 μg g^{-1} for haemocyanin) to satisfy the requirements of enzymes and respiratory proteins. These levels are similar to those found in crabs collected from the field (and in the soft tissues of isopods other than the hepatopancreas—see below), but barnacles contain very much greater concentrations, particularly of zinc, even in sites which are not contaminated with the metal. In crabs, internal concentrations of zinc and copper are 'buffered' by temporary storage in the hepatopancreas (Hopkin 1980; Hopkin & Nott 1979, 1980) whereas in barnacles, the metals are stored permanently in specialised 'parenchyma' cells which surround the midgut. Both crabs and barnacles do not appear to regulate the uptake of cadmium (White & Rainbow 1985).

Experiments to determine the essentiality of particular metals to marine organisms are technically demanding due to the difficulty of obtaining ultrapure seawater, and the problems associated with administering an artificial diet. In freshwater organisms, however, it has been possible to establish that selenium is an essential element for the cladoceran *Daphnia* (Keating & Dagbusan 1984).

7.4.2 Isopods

7.4.2.1 Introduction

Isopods are an important group to study physiologically because they are the only order of animals which has representatives in all the major ecosystems of the world. Isopods are common in the abyssal depths of the oceans, Antarctic seas, the littoral zone of all continents, and in most terrestrial environments. The members of the Sub-Order Oniscidea are the familiar 'woodlice' which are encountered so frequently in and around human habitations (synanthropic sites). The oft-repeated slogan of entomologists — 'Crustacea are the insects of the sea', should surely be ammended to — 'Insects are the Crustacea of the land'!

Several factors have contributed to the success of isopods as land animals. Aquatic representatives are dorso-ventrally flattened and possess a brood pouch, so isopods were to a certain extent 'pre-adapted' for a terrestrial lifestyle. Several physiological mechanisms have evolved, mainly to do with water conservation, since isopods first moved on to the land more than 400 million years ago. These have included excretion of nitrogenous waste as ammonia gas, the development of a water conducting system over the surface of the body and biphasic moulting where calcium reabsorbed from one half of the cuticle is stored temporarily before being used to calcify the other half (Edney 1968; Steel 1982; Wieser 1984; Eisenbeis & Wichard 1987; Warburg 1987).

Considerable research activity has been directed towards understanding how and why metals are assimilated by terrestrial isopods, particularly copper which is accumulated to a greater extent the more 'terrestrial' is the species of isopod being considered (Wieser 1965c, 1967; Hayes 1970). A discussion of the physiological and ecological factors which have resulted in some isopods containing the highest concentrations of zinc, cadmium, lead and copper so far recorded in a soft tissue of any terrestrial animal (Hopkin & Martin 1982a), forms the remainder of this section.

7.4.2.2 Routes of uptake and loss of metals in isopods

Terrestrial isopods provide a good example of how a lack of understanding of the structure and function of the digestive system of an invertebrate can lead to confusion and disagreement with regard to sites of metal assimilation.

The digestive system is composed of a foregut which masticates, filters and sorts food material, a hindgut which is differentiated into an anterior chamber, 'papillate' region and rectum, and a midgut represented by four blind ending tubules (six in some genera) which collectively form the hepatopancreas (Fig. 7.6). The foregut and hindgut are lined with cuticle whereas the cells of the hepatopancreas have a dense microvillus border (Holdich & Mayes 1975). The dorsal wall of the anterior chamber of the hindgut possesses a typhlosole which, together with lateral infoldings of the wall of the hindgut, form the pair of typholosole channels which run from the foregut to the papillate region. The hepatopancreas is by far the most important storage organ for zinc, cadmium, lead and copper which are accumulated in granules in the S and B cells (see Section 9.5.3.1).

Until recently, it was proposed that the function of the typhlosole channels was to carry digestive enzymes from their site of production in the hepatopancreas to food in the papillate region (Hassall 1977; Hassall & Jennings 1975). Under this scheme, there were two ways in which the considerable amounts of metals contained in the hepatopancreas could

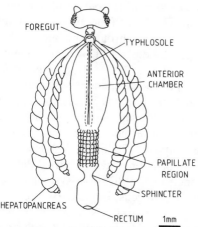

FIG. 7.6 Schematic diagram of the digestive system of *Oniscus asellus*. (Redrawn from Hopkin & Martin 1984a, by permission of the Zoological Society of London.)

have been assimilated. First, from liquids in the lumen which passed through the filters in the foregut as food was ingested. This route was likely to be of limited importance because to enter the hepatopancreas, metals had to be soluble in the liquid component of the food. Second, by transport across the basement membranes of the cells of the hepatopancreas after passage from the lumen of the hindgut across the epithelium of the papillate region and *via* the blood. Lead and cadmium have been detected in the cells of the papillate region so the latter route of movement of metals to the hepatopancreas seemed to be most important (Prosi & Back 1986). This scenario led Wieser (1968, 1979) to suggest that isopods resorted to coprophagy to obtain essential amounts of copper which were not available sufficiently in food after an initial passage through the gut.

Recently, Hames & Hopkin (in press) have re-examined the structure and function of the digestive system of terrestrial isopods. Juvenile isopods, in which the digestive organs could be observed directly through the cuticle of the body wall, were fed a diet labelled with coloured dyes. These observations demonstrated clearly that fluid flow in the typhlosole channels occurs in a direction opposite to that which had been assumed. Digestive enzymes are secreted by the hepatopancreas onto food as it passes through the foregut and into the anterior chamber. Products of digestion are forced *back* along the typhlosole channels from the papillate region to the foregut. This liquid (containing metals released into solution by digestive enzymes) passes through the filters into the hepatopancreas where absorption can take place. This arrangement maximises absorption of nutrients from the food as the internal surface area of the microvillus border of the hepatopancreas is at least 30 times greater than that of the entire hindgut (Hames & Hopkin, in press). It also explains why the assimilation of copper by isopods is in excess of 80% when the metal is administered in lettuce leaves which have been soaked in a solution of copper sulphate (Dallinger & Wieser 1977).

Because the hindgut is lined with cuticle, the only route of loss of metals *via* the lumen of the digestive system is likely to be from the hepatopancreas by lysis of cells. Urine is excreted into the closed-circuit water-conducting system (Eisenbeis & Wichard 1987) so loss of metals by this route is not an available option. A site of permanent storage-excretion, other than the hepatopancreas, has not been discovered in terrestrial isopods.

The diet of isopods in the field consists mainly of leaf litter. However, considerable discrimination for and against particular plant species is exercised. The most important factors controlling this discrimination are texture of the leaf material, concentrations of defensive chemicals and extent of microbial decay (Hassall & Rushton 1984; Eisenbeis & Wichard 1987). Isopods definitely prefer leaves with a rich fungal growth to uninfected food and have been shown to selectively graze on the hyphae (Gunnarsson & Tunlid 1986). When given faeces as a food source, they show a preference for pellets of two to three weeks in age which have the richest fungal growth (Hassall & Rushton 1985).

One way of interpreting coprophagous behaviour in isopods is to say that the animals select the most 'palatable' of all foods present. On some occasions, they may choose to eat their own faeces if the alternative is a recently fallen leaf with high levels of defensive chemicals. Indeed, Hames & Hopkin (in press) observed coprophagy in *Oniscus asellus* only when there was no other food available and Hassall & Rushton (1982) showed that the extent of beneficial effects of faecal consumption on growth of *Porcellio scaber* depended on the food with which faeces was compared. Isopods fed on carrot root grew better than a similar group fed on faeces and birch litter.

Most of the fungal hyphae associated with leaves are digested and assimilated on a single passage of food through the gut whereas the numbers of bacteria increase more than 100-fold (Anderson & Ineson

1983; Ineson & Anderson 1985). However, viable fungal spores may pass through the digestive system and have been observed germinating in the lumen of the rectum of *Oniscus asellus* (Hames & Hopkin, in press). Actinomycetes have been detected in the gut also (Coughtrey *et al.* 1980) and bacteria in which deposits of lead and zinc have been observed may be present in considerable numbers in the lumen of the hepatopancreas (Hopkin & Martin 1982b; Hames & Hopkin, in press). The role of these microorganisms in digestion and metal assimilation is not clear. Some bacteria may supply vital amino acids if these are not present in the diet (Carefoot 1984a, 1984b). However, Kukor & Martin (1986a, 1986b) have pointed out that the presence of a microbially-derived enzyme in the gut of an isopod does not prove that the animal is utilising it in digestion.

Isopods may alter their rate of feeding as well as the specific food materials they ingest. In general, foods which contain a high proportion of easily assimilatable nutrients are ingested rapidly whereas foods in which nutrients are less available are retained in the digestive system for a longer period (Wieser 1984). Thus, the degree of assimilation of a metal by an isopod from a food source will depend on the solubility of the element in the digestive fluids (e.g. whether the metal is present as a surface deposit or bound within the leaf, fungal material etc.), the rate at which the food is ingested by the animal (controlled by the nutrient status, texture, and level of defensive chemicals), and the relative net uptake in the hepatopancreas and hindgut. All these variables should be carefully considered when one is attempting to interpret field data, and design experiments on metal assimilation in isopods.

7.4.2.3 Metals in isopods from the field

The phenomenal ability of terrestrial isopods to accumulate metals was first reported more than 25 years ago (Wieser 1961). The hepatopancreas of specimens of *Porcellio scaber* collected from the spoil tips of disused mining areas in Cornwall, England, contained more than 1.5% copper on a dry weight basis (Wieser & Makart 1961). More recent studies on *Oniscus asellus* from a wider range of sites have shown that the concentrations of zinc, cadmium, lead and copper in the hepatopancreas of individuals of this species from sites contaminated with metals may exceed 1.2%, 0.4%, 2.5% and 3.4% of the dry weight respectively with no apparent ill effects (Hopkin & Martin 1982a; Table 7.9). However, when the concentrations of zinc and cadmium in the hepatopancreas are greater than about 1.5% and 0.5% respectively, the woodlice are always

TABLE 7.9

Concentrations of metals in *Oniscus asellus* from an uncontaminated site (Kynance Cove, Cornwall, England) and a site 1 km from the Avonmouth smelting works (see Section 5.3.2) (mean ± standard error, $n = 6$; original data collected by the author; $\mu g \ g^{-1}$ dry weight—compare with values for the snail *Helix aspersa* from the same sites in Table 7.1)

Kynance Cove

	Dry weight (mg)	Zn	Cd	Pb	Cu
Hepatopancreas	0.998 ± 0.101	652 ± 120	94.2 ± 11.1	31.5 ± 6.3	1050 ± 200
Gut	0.707 ± 0.110	256 ± 83	0.71 ± 0.12	14.3 ± 5.8	62.0 ± 17.4
Rest	14.14 ± 2.13	62.3 ± 6.6	0.09 ± 0.04	1.72 ± 0.44	51.5 ± 9.2
Total	15.84 ± 1.60	108 ± 25	5.93 ± 1.01	4.15 ± 0.91	115 ± 42

Avonmouth

	Dry weight (mg)	Zn	Cd	Pb	Cu
Hepatopancreas	1.245 ± 0.141	6600 ± 1570	2280 ± 580	3100 ± 1080	8090 ± 2260
Gut	0.955 ± 0.094	375 ± 113	64.0 ± 17.2	585 ± 232	138 ± 25
Rest	16.91 ± 1.74	60.6 ± 5.6	5.98 ± 1.37	46.0 ± 13.1	54.3 ± 4.0
Total	19.11 ± 1.93	488 ± 103	152 ± 37	413 ± 150	567 ± 138

moribund and move their limbs only weakly when stimulated (Hopkin & Martin 1984a). The metals are stored intracellularly in granules which are described in detail in Section 9.5.3.1.

The hepatopancreas is by far the most important storage organ of zinc, cadmium, lead and copper. Although this organ constitutes only about 5% of the dry weight of the animal, it may contain more than 75% of the zinc, 95% of the cadmium, 80% of the lead and 85% of the copper in the whole body (Hopkin & Martin 1982a). Results of some of my own unpublished experiments have shown that iron tends to be stored in the hindgut although accumulations of this metal are present in the B cells of the hepatopancreas also (Hopkin & Martin 1982b). In isopods from uncontaminated sites, about 20% of the copper in the hepatopancreas is water soluble, the remainder being bound in insoluble granules (Wieser 1968; Wieser & Klima 1969). Despite the massive concentrations of zinc, cadmium, lead and copper attained in the hepato-pancreas of isopods from metal-contaminated sites, the levels in tissues other than the gut are remarkably constant. Zinc and copper in particular are regulated at around the $50\,\mu g\,g^{-1}$ level which is similar to that calculated by Rainbow (1985a) as being required to supply the needs of enzymes and respiratory proteins in the soft tissues of marine Crustacea.

Zinc and cadmium appear to be toxic to *Oniscus asellus* when the detoxification mechanisms in the hepatopancreas become saturated allowing the metals to pass into the blood to interfere with biochemical reactions with which they are not normally involved (Hopkin & Martin 1984a). At Avonmouth, this occurs when *Oniscus asellus* reaches a dry weight of about 30 mg (Hopkin & Martin 1982b) putting a limit on the maximum size of adults in the population. However, weight may not be synonymous with age because high concentrations of zinc in the diet may reduce the growth rate of isopods (Joosse *et al.* 1983; Van Capel-leveen 1985).

Considerable overlap of age classes occurs when the weight of some isopods born in one year exceeds the weight of the heaviest individuals born in the previous year (Fig. 7.7). Wieser *et al.* (1977) reported that there were seasonal differences in the levels of copper in the tissues of *Porcellio scaber* collected at monthly intervals from a garden in Austria. However, the ages of the isopods sampled was not determined and it is probable that the apparent post-winter decline in copper levels in individuals was due actually to mortality of older adults which contained higher levels of copper. Studies on the population dynamics of isopods in metal-contaminated sites are required to determine whether longevity

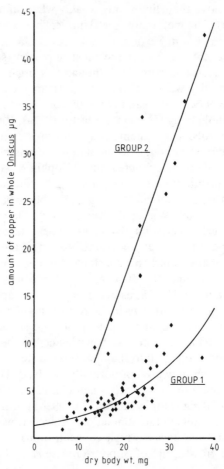

FIG. 7.7 Plot of total content of copper against dry body weight of 60 specimens of *Oniscus asellus* collected from a disused copper mine at Caradon, Cornwall, England. The animals can be split into two groups which may be different year classes. (From Hopkin & Martin 1982a, by permission of Springer–Verlag.)

is affected, as well as growth rates.

Concentrations of pollutant metals have been measured in isopods collected from roadside sites (Williamson & Evans 1972; Williamson 1979b), spoil tips of metal mines (Avery *et al.* 1983) areas adjacent to smelting works (Martin & Coughtrey 1976; Martin *et al.* 1976; Coughtrey *et al.* 1977; Hunter & Johnson 1982; Joosse & Van Vleit 1984; Hunter *et al.* 1987b) and sites subject to contamination with metal-containing fungicides (Wieser *et al.* 1976, 1977). The wealth of information provided by these studies allows certain conclusions to be drawn.

First, concentration factors (isopod/leaf litter) for cadmium and copper are always much greater than for zinc, lead, cobalt, nickel, chromium, iron and manganese at the same site. In aerially contaminated sites, cadmium, copper and zinc are more 'available' in leaf litter than lead (Martin *et al.* 1976) but the differences are not sufficient to account for the differences in net metal uptake in isopods.

Second, concentration factors for zinc and copper are much greater in sites where levels of the metals are low in the vegetation relative to contaminated sites (Fig. 7.8). This limitation of net assimilation of these essential metals may be due to saturation of cellular uptake mechanisms, or an active increase in excretion rates. It could be due also to the evolution of tolerance to metals in the diet but such races of isopods have yet to be identified categorically (Hopkin & Martin 1984a). Higher concentration factors at low levels in the diet for the non-essential metal lead have not been demonstrated clearly in isopods although there is a suggestion that this phenomenon might be occurring with cadmium (Fig. 7.8).

The aspect of isopod physiology which is most difficult to explain is the difference in net accumulation of metals observed in different species collected from the same microhabitat (Table 7.10). This phenomenon has been noted by other authors (Williamson 1979b; Joosse & Van Vleit 1984) and has been observed for other pollutants such as fluoride (Walton 1987). Some of my previously unpublished laboratory experiments have shown that the differences in accumulation of zinc, cadmium and lead between *Porcellio scaber* and *Oniscus asellus* indicated in Table 7.10, occur also when the two species are reared from birth on identical diets. There are no differences in the structure of the digestive systems of the two species (Hames & Hopkin, in press) so the differences must be due to the relative efficiency of uptake and excretory mechanisms for each metal at the gut/lumen interface.

Edney (1968) made the following statement regarding individual differences within species of isopods.

'If 60% of a population of *Oniscus asellus* moves to the warm end of a temperature gradient and 40% settles for the other, are we justified in saying that *Oniscus* shows a slight preference for the higher temperature? If, as it seems, 60% show a complete preference for one

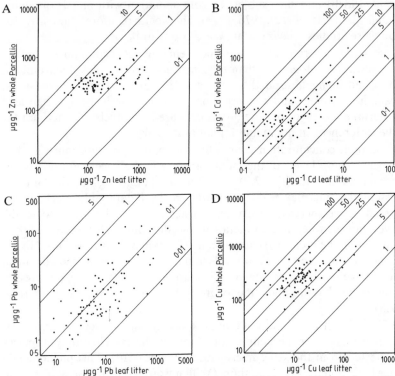

FIG. 7.8 Scatter diagrams relating concentrations of (A) zinc, (B) cadmium, (C) lead and (D) copper in 86 pooled samples of 12 *Porcellio scaber* from Avon and Somerset, England, against concentrations in leaf litter at the same sites. The lines on the graphs join points with the same concentration factor (concentration of metal in whole woodlice/concentration of metal in leaf litter). In general, cadmium and copper are accumulated by the isopods to a greater relative extent than zinc, and to a much greater relative extent than lead. (From Hopkin *et al.* 1986.)

temperature and 40% show a complete preference for the other then surely the question as to how these two kinds of individuals differ is the important one'.

Individual differences in concentrations of metals in the hepato-pancreas of terrestrial isopods are extremely pronounced. For example, a group of 12 *Ligia oceanica* with a mean concentration of 82.8 μg g^{-1} of cadmium in the hepatopancreas contained individuals with a concentration ranging from 23 to 211 µg g^{-1} of cadmium (Table 7.10). Elucidating the physiological basis for these differences in isopods, and other terrestrial invertebrates, provides the most exciting challenge for 'metalworkers'.

7.4.2.4 Experiments on metals in isopods

Attempting to interpret the results of experiments on the dynamics of metals in terrestrial isopods is as difficult as in earthworms. Not only have experiments been conducted at different temperatures on a range of species (which the previous section has shown may accumulate metals at different rates), but a variety of foods have been used which differ in their palatability and moisture content. In addition, some workers have administered metals to animals by providing naturally-contaminated leaf litter (Hopkin & Martin 1984a) whereas others have soaked leaves in solutions of metal salts (Dallinger & Wieser 1977).

The apparent rates of food consumption of isopods will depend on the way in which results are expressed. For example, fresh carrot root has a moisture content of about 80% whereas leaves of field maple (*Acer campestre*) collected from the litter layer, dried and rehydrated for 24 hours at 100% relative humidity, contain about 20% water (Hames & Hopkin, in press). If consumption rates of these diets are expressed in terms of dry weight of food ingested per unit wet weight of isopod (used by most authors), then if one animal (wet weight 100 mg) eats 10 mg (wet weight) of carrot and another animal (wet weight 100 mg) eats 10 mg (wet weight) of field maple, the ingestion rates on a dry weight of food basis will be radically different. The first isopod has consumed 2 mg (dry weight) of carrot whereas the second has eaten 8 mg (dry weight) of field maple, an apparent four-fold difference in ingestion rates! This disparity persists when consumption rates are expressed as dry weight of food consumed per unit *dry* weight of isopod. A woodlouse of 100 mg has a dry weight of about 30 mg. Thus, the first isopod has consumed 6.7% (2/30 mg) of its dry body weight in dry carrot whereas the second animal has ingested 26.7% (8/30 mg) of its dry body weight in field maple leaf. A large proportion of the variability in consumption rates of isopods reported by Wieser (1979) can be explained therefore by differences in moisture contents of foods since it is the bulk fresh volume which sets the maximum rate at which the animals can pass food through their guts.

Some of my own unpublished observations have shown that adult *Oniscus asellus* and *Porcellio scaber* fed on field maple leaves (air-dried

TABLE 7.10

Concentrations of metals in three species of isopod from a supra-littoral site near Portishead on the Severn Estuary, England ($\mu g\ g^{-1}$ dry weight, ± standard error) (from Hopkin et al. 1985b)

Ligia oceanica (n = 12)

	Dry weight (mg)	Zn	Cd	Pb	Cu
Hepatopancreas	8.941 ± 1.024	1 420 ± 190	82.8 ± 16.4	3.16 ± 0.76	1 850 ± 270
(hepatopancreas range)		(673–2 740)	(23–211)	(0.90–11.0)	(1 020–4 600)
Gut	1.676 ± 0.188	77.0 ± 10.8	34.9 ± 3.4	<3	37.1 ± 6.3
Rest	84.43 ± 6.66	56.8 ± 1.6	1.01 ± 0.13	0.73 ± 0.11	53.9 ± 1.5
Total	94.98 ± 7.33	173 ± 9	8.20 ± 0.80	0.90 ± 0.11	203 ± 6

Oniscus asellus (n = 6)

	Dry weight (mg)	Zn	Cd	Pb	Cu
Hepatopancreas	1.463 ± 0.294	963 ± 174	508 ± 96	370 ± 82	2 140 ± 250
(hepatopancreas range)		(460–1 600)	(158–865)	(118–678)	(1 590–3 330)
Gut	0.651 ± 0.160	167 ± 13	12.2 ± 2.2	23.1 ± 4.9	42.8 ± 5.8
Rest	15.90 ± 1.59	52.5 ± 1.4	0.579 ± 0.078	11.3 ± 1.1	57.2 ± 4.1
Total	18.02 ± 1.96	124 ± 11	36.3 ± 6.0	37.8 ± 6.2	196 ± 18

Porcellio scaber (n = 6)

	Dry weight (mg)	Zn	Cd	Pb	Cu
Hepatopancreas	1.622 ± 0.104	3 870 ± 590	128 ± 27	83.1 ± 16.1	2 570 ± 500
(hepatopancreas range)		(957–5 500)	(71–265)	(23–148)	(933–4 470)
Gut	0.713–0.058	116 ± 13	14.9 ± 8.1	73.9 ± 18.6	30.6 ± 4.9
Rest	16.95 ± 0.97	68.0 ± 2.7	0.675 ± 0.099	10.5 ± 0.4	54.8 ± 2.6
Total	19.28 ± 1.09	367 ± 43	11.5 ± 2.1	18.8 ± 1.8	243 ± 32

and rehydrated to a moisture content of 20%) consume a mean of 7% of their dry body weight per day of dry leaf over a five month experimental period (temperature constant 15 °C, 16 hour light/8 hour dark cycle). However, during single 24 hour periods, the food consumption rate of individual isopods can exceed 30% of their dry body weight. These individuals must have filled and emptied the hindgut at least three times to account for all the faecal pellets produced. Isopods stop feeding for about five days when they moult.

Hopkin & Martin (1984a) calculated that even if all the cadmium and copper was extracted from the food of *Oniscus asellus* as it passed through the digestive system, the animals would have to have eaten about 5% of their dry body weight per day in dry leaf litter to account for the total amounts of these metals contained in the body. Thus, net assimilation rates of copper from leaf litter in the field are probably in excess of 50%. In a subsequent paper, Hopkin *et al.* 1985b calculated that in *Oniscus asellus*, net assimilation rates relative to copper (at '100%') were 7% for zinc, 62% for cadmium and only 4% for lead. Why is copper retained to a much greater extent than zinc? The answer may lie in the physiological adaptations isopods had to make when they colonised the land.

Terrestrial isopods must have evolved efficient ways of assimilating essential elements from the food because, unlike their marine ancestors, they could no longer obtain them directly from the external medium across the respiratory surfaces (Edney 1968; Wieser 1968). Copper is present in very low concentrations in leaf litter (only about 10 to 20 µg g^{-1}; Guha & Mitchell 1966; Hopkin & Martin 1982a). Since copper is required as an essential part of the oxygen-carrying blood protein haemocyanin (Bonaventura & Bonaventura 1980), assimilation from the lumen of the digestive system must be very efficient. Zinc, in contrast,

---

FIG. 7.9 Concentrations of zinc, cadmium, lead, copper and iron in the hepatopancreas of *Oniscus asellus* from an uncontaminated site (Wetmoor Wood) and a metal-contaminated site (Haw Wood—see Section 5.3.2), maintained for 5 months on leaf litter (field maple, *Acer campestre*) from the two sites in factorial combination. The isopods were held in four plastic tanks at a constant 16°C in an 8 hour dark/16 hour light regime. 100 isopods were placed in each tank at the start. Each point represents the mean of the 12 animals (± standard error) removed from each tank, dissected and analysed every month. The mean concentrations of metals (µg g^{-1} dry weight) in the leaf litter at the start of the experiment were: Wetmoor, Zn 96.4, Cd 0.72, Pb 42.1, Cu 11.8, Fe 211; Haw, Zn 1430, Cd 26.0, Pb 908, Cu 51.9, Fe 286. (Previously unpublished data.)

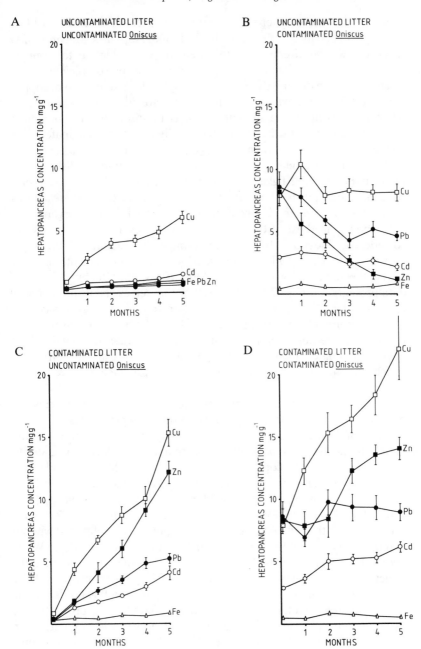

is present in much higher concentrations in uncontaminated leaf litter (about 50 to 300 μg g^{-1}, Guha & Mitchell 1966; Hopkin & Martin 1982a) so even if the requirement for this element was as high as for copper, the mechanisms for its uptake would not need to be so efficient.

The evolution of a highly efficient system for the assimilation of copper may have two disadvantages. First, there are likely to be occasions when more copper is assimilated than is required by the immediate physiological needs of the animal (especially in sites contaminated with copper). Second, cadmium may be taken up along the same biochemical pathway. In a site heavily contaminated with cadmium (a disused zinc mine at Shipham, England — Hopkin & Martin 1982a), the molar concentration of cadmium in the hepatopancreas of *Oniscus asellus* was twice that of copper. Concentrations of copper in this organ were only about one third of those expected when levels in leaf litter were considered, due possibly to competition with cadmium at the same uptake sites. The small proportion of lead accumulated may initially follow the same pathway from the food as calcium (Beeby 1978). The data presented in Figs. 7.9 and 7.10 show that terrestrial isopods appear to have adopted a 'permanent storage-excretion' strategy for copper (and cadmium which may be 'accidental' or 'deliberate'), but that zinc is lost from the hepatopancreas, albeit at a slow rate, if animals which have accumulated high concentrations of the metal are fed on a diet which contains low concentrations of zinc (Lauhachinder & Mason 1979).

These observations and deductions are at variance with conclusions of other authors who have suggested that the main problem isopods face in terms of their copper balance is preventing excessive excretion of the metal. Some workers have proposed that isopods have to resort to coprophagy to obtain copper which they were not able to assimilate on an initial passage of food through the gut (Dallinger & Wieser 1977; Debry & Lebrun 1979; Debry & Muyango 1979; Wieser 1965a, 1966, 1967, 1968, 1978, 1979, Wieser & Wiest 1968, Wieser *et al.* 1977) while others believe that adequate amounts of copper can be obtained without eating faeces (Coughtrey *et al.* 1980; Hassall & Rushton 1982; White 1968). However, Hames & Hopkin (in press) have shown that the structure of the digestive system of terrestrial isopods enables all soluble components of the food to be exposed to the microvillus border of the hepatopancreas. Unless microbial activity increases the proportion of essential nutrients which are soluble in the faeces to a level greater than that in available food, there can be no selective advantage in consuming

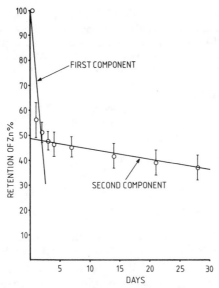

FIG. 7.10 Elimination of ingested ^{65}Zn by *Armadillidium vulgare* held at 15°C for 28 days ($n = 25$, mean ± standard error). The isopods were fed labelled food (lettuce) for 24 hours, then given free access to uncontaminated lettuce. Whole body counts were made subsequently on days 1, 2, 3, 4, 7, 14, 21 and 28. The first component probably represents loss of unassimilated ^{65}Zn from the lumen of the gut of the isopods whereas the second, slower component represents ^{65}Zn which had been assimilated. (Redrawn from Lauhachinda & Mason 1979, by permission of Auburn University Press, Alabama.)

faecal pellets.

In his most recent review, Wieser (1984) has concluded that coprophagy may improve the recycling of copper only if the rate of ingestion of food increases beyond 60-80 mg (dry weight) g^{-1} (wet weight) of isopod per day. This represents a consumption rate of fresh material which, at the very least, would constitute 10% of the fresh weight of an isopod per day and for food with a high moisture content such as lettuce (c. 80% water), a consumption rate of at least 30%. While these rates of consumption may pertain for limited periods in the laboratory at relatively high temperatures, they are unlikely to occur in the wild in temperate climates.

Metal balance experiments are difficult to conduct with small invertebrates and experimental error can be large unless analytical techniques are very precise, large numbers of replicate animals are used and most

TABLE 7.11

Food selection by 'copper-enriched' and 'copper-deficient' *Oniscus asellus*. Ingestion is given in μg dry birch litter mg^{-1} fresh body weight in experiments lasting 2 days (3 replicates, each with 3 isopods; mean ± standard deviation) (from Dallinger 1977, by permission of Springer–Verlag)

Concentration of copper in birch litter (μg g^{-1} dry weight)	'Copper-enriched' Oniscus asellus	'Copper-deficient' Oniscus asellus
20	154.0 ± 60.7	166.6 ± 25.8
340	17.0 ± 9.5	211.6 ± 89.7
5 200	6.3 ± 3.2	11.0 ± 4.2

of the food (>80%) presented is consumed (Schmidt & Reese 1986). The major criticism which can be levelled at the studies by Dallinger (1977), Dallinger & Wieser (1977) and Wieser *et al.* (1977) on copper balance in isopods is that the copper status of the experimental animals was not measured directly, but was calculated from levels measured in leaf material and faeces by the 'zinc dibenzyl dithiocarbamate in CCl$_4$' method. Where levels of copper were analysed in *Porcellio scaber* (Table 4 in Wieser *et al.* 1977) the figures reported are difficult to interpret since the concentrations in the total animals are considerably greater than those in the constituent tissues.

Dallinger (1977) reported that 'copper enriched' specimens of *Porcellio laevis*, *Porcellio scaber* and *Oniscus asellus* 'chose' to eat birch leaves containing low copper concentrations in preference to leaves in which the levels of copper were higher. Conversely, 'copper-deficient' individuals 'chose' to eat leaves which had been enriched with copper. These results (shown for *Oniscus asellus* in Table 7.11) have been quoted in support of the suggestion that the selection of food in isopods is influenced by the status of copper reserves in their bodies (Wieser 1979). Notwithstanding the fact that to do this, isopods would have to possess highly sensitive 'copper receptors' on the mouthparts and in the body tissues, certain aspects of the experimental procedures used by Dallinger (1977) are open to criticism. First, the choice experiment was only conducted for 'two days' (48 hours? — the exact time is not stated). Second, the mean values for food consumption reported for each species were calculated from three replicates, each comprising only three specimens. Third, the 'copper-enriched' isopods were kept for one month prior to the experiment in leaf litter which had been soaked in copper sulphate solution to give a concentration of copper in their food of 340 μg g^{-1}. This is an extremely high level of soluble copper with which to cope in the diet. If 80% of the copper ingested with the food was assimilated by the isopods (as reported to be the case in other experiments by Dallinger & Wieser 1977) then the animals would probably be suffering from toxicity symptoms and would be unlikely to respond in a normal fashion. Dallinger's experiments were valuable but they need to be repeated with a larger number of isopods for a longer period.

Completely artificial diets made from 'off the shelf' chemicals were used by Carefoot (1984a, 1984b) to determine the requirements for major nutrients of the littoral isopod *Ligia pallasii*. Experiments with diets 'deficient' in a range of metals were also conducted. Carefoot reported that the isopods were able to survive for over a year, during which time they doubled in weight, on a 'copper-free' diet. However, since the concentrations of copper were not determined in the food, impurities in the chemicals used to make up the diet may have supplied essential minerals to the animals. Furthermore, the concentrations of metals were not determined in the isopods so it is not legitimate to draw conclusions as to their copper status.

Inheritable tolerance to metals has not been demonstrated conclusively in terrestrial isopods. Hopkin & Martin (1984a) showed that the concentrations of zinc, cadmium, lead and copper in juvenile *Oniscus asellus* fed on uncontaminated or contaminated leaf litter for six weeks, were similar whether or not their mothers had been collected from close to a zinc, lead and cadmium smelting works, or a clean site. However, isopods may reduce their intake of metals in polluted sites by avoiding food which contains very high concentrations (Joosse *et al.* 1981; Van Capelleveen 1983, 1985, 1987). Presumably such food tastes 'unpleasant'.

Most studies on metals in terrestrial isopods (as in other groups) have dealt with zinc, cadmium, lead and copper. However, other metals may be emitted from industrial concerns and Joosse & Van Vleit (1984) have made a study of the uptake and effects of manganese from a blast furnace on *Philoscia muscorum*, *Porcellio scaber* and *Oniscus asellus*. These species normally contain about 10 μg g^{-1} of manganese but this increases to about 500 μg g^{-1} after three weeks of feeding on a diet containing 1000 μg g^{-1} of the metal. It is not known where manganese is accumulated but being a class A metal, one would expect it to follow similar pathways to calcium. If manganese were stored in the hepatopancreas, it would be interesting to know where since type A granules have not been detected in the cells (see Section 9.4).

Experiments on the effects of metals on terrestrial isopods in the laboratory have tended to confirm conclusions drawn from field data. Beyer & Anderson (1985) showed that a level of 1500 μg g^{-1} zinc in leaf litter was sufficient to cause significant mortality in *Porcellio scaber* but that the same effect with cadmium and lead was not observed unless the concentrations exceeded 500 and 12 000 μg g^{-1} respectively. This level of zinc is similar to that in the leaf litter of woodlands near to the Avonmouth smelting works where moribund *Oniscus asellus* are occasionally found (Hopkin & Martin 1982b). The concentrations of metals in the isopods, and food consumption, were not measured in Beyer & Anderson's experiments so the effects they observed may have been due to starvation.

Detoxification and excretion of metals by isopods in polluted sites must involve the expenditure of considerable amounts of energy. Increased rates of oxygen consumption have been measured in isopods exposed to high levels of iron and manganese sulphates (Joosse *et al.* 1981, 1983; Joosse & Van Vleit 1984) and a reduction in growth rates has been detected in isopods exposed to sub-lethal levels of zinc and manganese in the diet (Joosse *et al.* 1983). However, the energy expended in permanently storing the majority of the copper and cadmium which enters the cells of the hepatopancreas is probably less than would be required to excrete the metals continuously by breakdown and replacement of the cells.

7.4.3 Amphipods

Amphipods have not been as successful as isopods in invading the terrestrial environment although in the tropics, they can reach very high population densities (several thousand m^{-2}) in leaf litter (Hurley 1968; Friend & Richardson 1986).

Moore & Rainbow (1987) examined concentrations of copper and zinc in several species of talitroidean amphipods which exhibit increasing 'terrestriality'. The order of species (assessed at a mean standard dry body weight of 10 mg) when ranked according to an ascending series of copper concentrations, reflected their ecological zonation from sea to land. Species living proximal to the sea like *Talitrus saltator* and *Hyale nilssoni* had the lowest concentrations of copper in their bodies (25 to 40 μg g^{-1}) while increasingly supra-littoral species like *Talorchestia deshayesii* and *Orchestia mediterranea* had intermediate values (50 to

55 µg g^{-1}). *Arcitalitrus dorrieni* (euterrestrial), *Orchestia gammarellus* (supra-littoral and semi-terrestrial) and *Orchestia cavimana* (freshwater and semi-terrestrial) had the highest values for copper (60 to 90 µg g^{-1}). No clear relationship was observed between levels of zinc and extent of terrestriality in the amphipods, all species containing a concentration of about 100 to 200 µg g^{-1}.

Semi-terrestrial amphipods do not contain more of the copper-containing respiratory protein haemocyanin than their aquatic relatives (Wieser 1965c). Thus, it is likely that most of the 'extra' copper is stored in insoluble granules in the digestive caeca (Icely & Nott 1980; see also Section 9.5.3.2). The tendency for more terrestrial amphipod species to store copper may reflect a permanent storage-excretion strategy similar to that proposed to have developed in terrestrial isopods as an energy-saving measure (Section 7.4.2.4).

Concentrations of zinc, copper, manganese, nickel, iron, cadmium and silver were determined in beach amphipods by Bender (1975) but since analytical results were not corrected for background interference, the levels reported may be subject to some inaccuracy.

7.5 INSECTS

7.5.1 Introduction

Many papers have been published on the role of metals in insect nutrition and growth but early reviews did not refer to mineral requirements (Dadd 1973; Singh 1977). The topic has been covered adequately only recently by Dadd (1985) and Reinecke (1985). In contrast to other soil animals where the majority of studies have been concerned with metals as pollutants, in insects, most work has been laboratory-based and has concentrated on the effects of metal-based pesticides on growth and reproduction and the essentiality and toxicity of elements in the diet. Ecological studies on accumulation of lead by roadside invertebrates, for example, have tended to lump insects together (Rolfe & Haney 1975; Goldsmith & Scanlon 1977; Udevitz *et al.* 1980), or have subdivided them into 'herbivores' and 'carnivores' (Price *et al.* 1974). Such an approach is of little help in trying to interpret routes of metal transport in contaminated ecosystems since the feeding habits of insects are so diverse.

The presence of copper in several species of insects was reported early in this century by Muttkowski (1921) and Melvin (1931). More recent studies on a wider range of metals and species have shown that there are few examples of insects which accumulate metals to the same extent as isopods, molluscs or earthworms (Levy & Cromroy 1973; Van Rinsvelt *et al.* 1973; Levy *et al.* 1974b), even in sites which are heavily contaminated. Some detrivorous insects contain relatively high concentrations of zinc but much of this is located in the mandibles where it strengthens the cutting edges (Hillerton & Vincent 1982).

Several factors may contribute to the tendency of the majority of insects not to adopt a permanent storage-excretion strategy for metal regulation. Unlike isopods, molluscs and arachnids, insects do not possess large midgut diverticulae in which large amounts of metals can be stored (Chapman 1985). Furthermore, the blood of most species does not contain the oxygen-carrying pigments haemocyanin or haemoglobin so the need for insects to transport copper or iron across the digestive epithelia is not as great as in other invertebrate groups which do not possess a tracheal system. The short life cycle of many species gives little time also for metals to accumulate. However, this may allow adult Diptera to adopt a permanent storage-excretion strategy since they die before metals reach toxic levels in the midgut cells (see Section 9.5.4.9).

About 25 orders of insects are recognised but representatives of only nine of these have been studied in detail with regard to metals. Some authors have investigated metal toxicity in aquatic larval stages of insects which are terrestrial as adults (e.g. Leonhard *et al.* 1980). However, a review of this work is outside the scope of this book.

The structure and function of the digestive system of insects will not be covered in the same amount of detail as for molluscs, earthworms and isopods since the topic has been recently reviewed extensively by Chapman (1985). Simplified diagrams of the digestive organs of representatives of several insect orders are illustrated in Figs. 6.1, 6.2 and 6.3.

7.5.2 Collembola (Springtails)

7.5.2.1 Introduction

Collembola are small (usually <5 mm in length), primitive, wingless insects, which are extremely common in leaf litter and surface soil in temperate ecosystems. They promote decomposition by consuming microorganisms which reside on leaf material and stimulate the growth of fungal hyphae by their feeding activity (Hanlon & Anderson 1979; Hanlon 1981). In coniferous forests where earthworms are absent due to the low pH of the litter, Collembola (together with the mites) are the dominant decomposers and can reach densities in excess of 10,000 m^{-2}. The major role of Collembola in relation to metals is to increase the availability of elements which would otherwise remain immobilised in fungal hyphae. Transfer to predators is thought to be far less important. For example, food chain transfer of lead by *Orchesella cincta* accounts for only about 11 times the population standing pool per year whereas the flux of lead through consumption and defaecation is about 10,000 times the standing pool per year (Van Straalen *et al.* 1985).

Until recently, the dynamics of metals in individual Collembola had not been studied because of the difficulty of measuring concentrations in such small animals. The development of the graphite furnace atomic absorption spectrophotometer (Section 4.4.2.2) in the late 1970s enabled these analytical difficulties to be overcome (Bengtsson & Gunnarsson 1984).

Apart from the reports by Carter (1983) and Hunter *et al.* (1987b) who give concentrations of zinc, cadmium and copper in pooled samples of Collembola, the only detailed studies at the species level have been conducted on the algal-feeding *Orchesella cincta* by a research group based in The Netherlands (reviewed by Joosse & Verhoef 1987) and the fungal-feeding *Onychiurus armatus* by a Swedish team (reviewed by Bengtsson 1986). Considerable differences exist between species in the extent of metal accumulation (Table 7.12) so the pooling of springtails before analysis is not to be encouraged (Van Straalen & Van Wensem 1986).

The only other member of the subclass Apterygota in which metal concentrations have been determined is the dipluran *Campodea staphylinus* in which levels of zinc, cadmium and lead have been measured. The specimens were collected from close to a zinc smelting works (Table 7.12).

7.5.2.2 *Dutch research on* Orchesella cincta

Laboratory experiments on metal uptake and loss in *Orchesella cincta* have used the alga *Pleurococcus* sp. soaked in solutions of metal salts as a food source (Van Straalen & Van Meerendonk 1987). Lead is assimilated very inefficiently by this species. Furthermore, of the few percent of the lead in the food which is retained by the animals, most is stored in the midgut epithelium and is lost during the next moult when the entire lining of the gut is shed (Joosse & Buker 1979). A fast and slow component to the loss of lead has been recognised (Table 7.13; Fig. 7.11).

TABLE 7.12

Concentrations of metals in arthropods collected from pine forest litter 1 km north-east of a zinc factory near Budel, The Netherlands (μg g^{-1} dry weight $\pm$ standard error) (from Van Straalen and Van Wensem 1986, converted from nmol g^{-1} in the original paper to ease comparison with other tables in this book)

	Mean individual dry weight (mg)	n	Pb	Cd	Zn
Beetles					
Calathus melanocephalus	6.574	7	2.54 ± 0.66	1.18 ± 0.37	70.6 ± 5.9
Notiophilus biguttatus	2.966	8	1.10 ± 0.15	2.04 ± 0.29	81.7 ± 9.8
Notiophilus rufipes	2.837	8	1.97 ± 0.14	2.37 ± 0.81	56.2 ± 3.9
Lathrobium brunnipes	1.248	7	1.47 ± 0.41	4.98 ± 0.70	235 ± 16
Centipedes					
Lithobius forficatus	5.082	2	6.61 ± 2.48	2.23 ± 0.71	186 ± 12
Schendyla nemorensis	0.275	7	0.54 ± 0.31	16.8 ± 3.5	395 ± 45

Spiders				
Centromerus sylvaticus	0.456	10	1.04 ± 0.35	19.9 ± 2.7	286 ± 43
Pseudoscorpions				
Neobisium muscorum	0.401	8	ND	17.4 ± 2.5	319 ± 23
Oribatid mites				
Chamobates cuspidatus	0.0034	10[a]	ND	3.10 ± 0.81	365 ± 62
Collembola				
Orchesella cincta	0.148	10	1.12 ± 0.64	1.36 ± 0.09	71.9 ± 8.5
Lepidocyrtus cyaneus	0.0627	18	ND	2.78 ± 0.40	45.7 ± 7.2
Isotoma notabilis	0.0048	9[a]	ND	7.33 ± 3.24	54.9 ± 30.7
Diplura				
Campodea staphylinus	0.0390	8	ND	15.9 ± 2.3	205 ± 22
A Horizon forest litter			663 ± 97	11.0 ± 1.8	1 670 ± 240

[a] Each individual sample consisted of 10–20 pooled individuals.
ND, not detected.

TABLE 7.13
Estimates for rate constants and half lives for lead in *Orchesella cincta* (95%
confidence limits in parentheses) (from Van Straalen & Van Meerendonk 1987,
by permission of Springer–Verlag)

Compartment	Rate constant (day^{-1})	Half life *(days)*	% of total burden
Gut	2.02 (0.68–3.35)	0.34 (0.21–1.02)	48%
Fast body burden	0.094 (0.058–0.130)	7.37 (5.33–11.95)	36%
Slow body burden	0.032 (−0.021–0.085)	21.66 (8.15–∞)	16%

In subsequent experiments, Joosse & Verhoef (1983) showed that
Orchesella could reduce the amount of lead which entered the lumen of
the gut by 'choosing' to eat food which contained lower levels of lead.
The Collembola could discriminate between control algae, and algae
which had been soaked in a solution of 25 µg ml^{-1} of lead as the nitrate.
However, the concentration of soluble lead in the food would have been
very high and the animals may have been discriminating on the basis of
nitrate rather than metal (see also Section 7.4.2.4 for discussion of similar
experiments on isopods).

Recent work has examined whether populations of *Orchesella* which
have resided in metal-contaminated sites for long periods of time respond
differently to metal stress than Collembola from uncontaminated sites
(Van Straalen *et al.* 1986a, 1986b; Nottrot *et al.* 1987). Excretion
efficiency calculated from the percentage of total body lead or cadmium
which populations of *Orchesella cincta* could lose in gut pellets when
they moulted, was greater in individuals from a disused Belgian lead
mine than in other sites. However the effect, although statistically
significant, was very slight. Median efficiency of excretion of metals by
Collembola from the lead mine was 57.3% for lead (compared with a
mean of 49.7% at the other eight sites) and 36.4% for cadmium (compared with a mean of 27.6% at the other sites).

In a further study on cadmium, Van Straalen & De Goede (1986)
showed that the growth of female *Orchesella* was inhibited at concentrations in the food as low as 5 µg g^{-1}. This level is considerably less

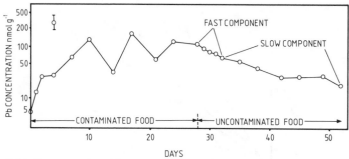

Fig. 7.11 Concentration of lead in *Orchesella cincta* (Collembola) fed on a suspension of the green alga *Pleurococcus* containing 2000 µg g^{-1} Pb for 28 days followed by uncontaminated food. Note the logarithmic scale and the molar units. Each mean is based on 5 replicate observations. Each observation was based on analysis of 2 pooled individuals. Each mean has a standard error as shown in the left upper corner, which was derived from the analysis of variance error mean square. Lead in gut contents was excluded from the concentration determinations. There are two components to loss of assimilated lead from the tissues. The fast phase probably represents loss via exfoliation of the intestinal epithelium whereas the slow component represents lead which has accumulated in the rest of the body tissues and is more difficult to excrete. (Redrawn from Van Straalen & Van Meerendonk 1987, by permission of Springer–Verlag.)

than that found in field sites where *Orchesella* is common. The authors concluded that the sublethal effects of cadmium on growth are obscured in the wild because there is such a high level of natural mortality from predation.

7.5.2.3 Swedish research on Onychiurus armatus

Onychiurus armatus is exposed to high concentrations of metals in contaminated sites because its diet of fungal hyphae accumulates metals so extensively. The fungus *Verticillium bulbillosum* grown on substrates with 300 µg g^{-1} of copper and lead, can contain as much as 3500 µg g^{-1} of these metals (Bengtsson *et al.* 1983a).

Laboratory experiments conducted by Bengtsson *et al.* (1983a) showed that higher rates of growth and average maximum lengths relative to controls were attained by Collembola subjected to relatively low levels of copper (45 µg g^{-1} in the substrate — Fig. 7.12). This effect may be due to the elimination of a copper-sensitive gut parasite (Bengtsson *et al.* 1985a). However, concentrations of copper *and* lead in the substrate of 45 µg g^{-1} (corresponding to levels of more than 1000 µg g^{-1} of these

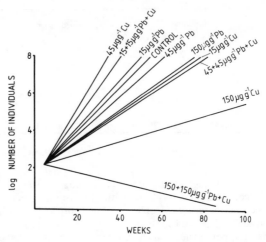

FIG. 7.12 Predicted population development of *Onychiurus armatus* (Collembola) reared on laboratory cultures of fungal hyphae growing on substrates containing different concentrations of lead and copper. The predictions are based on a Leslie matrix calculation. A straight line with a positive slope of ln Nt versus time indicates exponential growth. (Redrawn from Bengtsson *et al.* 1985a, by permission of the British Ecological Society.)

metals in the fungi), were sufficient to reduce population growth rates (Fig. 7.12). The threshold whole body concentrations of copper and lead in *Onychiurus* at which significant reductions in growth rates could be detected was 200 $\mu g\,g^{-1}$, similar to levels found in a population of the same species 650 m from a brass mill at Gusum, S.E. Sweden (see also Section 5.3.3).

In a subsequent paper (Bengtsson *et al.* 1985b), these effects were examined in greater detail by rearing *Onychiurus* on five species of fungi isolated from metal-polluted soils close to the Gusum mill. The effects of metal pollution were exacerbated by limiting the availability of food, and alleviated by supplying the protein-rich fungus *Paecilomyces farinosus* which they preferred to eat given the choice of a range of species.

Thus, it is crucial to understand the mechanisms by which growth and survival of different species of Collembola are affected by food availability if the effects of metal contamination on this diverse group of invertebrates are to be fully understood.

7.5.3 Orthoptera (Grasshoppers, Locusts, Crickets)

7.5.3.1 Introduction

Orthoptera in contaminated sites do not accumulate metals to the same extent as other invertebrates (Table 5.9). This is due, in part, to the short life span of most species and the fact that their diet consists of living vegetation in which levels of pollutant metals are lower. Considering the economic importance of locusts as crop pests, it is surprising that relatively little work has been conducted on metals in the group. However, the research which has been done is of considerable interest since it covers elements which are rarely studied (e.g. arsenic), interactions of metals with other dietary components, and essentiality.

7.5.3.2 Metals in Orthoptera from the field

In one of the earliest studies on cadmium in terrestrial food chains, Munshower (1972) showed that invariably, grasshoppers contained a similar, or lower, concentration of the metal than was in their presumed diet. This is in marked contrast to isopods and earthworms in which concentration factors are usually five or more (Sections 7.3.3, 7.4.2.3). Levels of iron, manganese and zinc were reported in samples of grasshoppers (unidentified) from two sites in The Netherlands by Joosse & Van Vliet (1982). They showed that the insects contained naturally high levels of zinc (c. 200 µg g^{-1}), moderate levels of iron (c. 100 µg g^{-1}) and low levels of manganese (10 to 15 µg g^{-1}). The grasshoppers appeared to be able to regulate their net uptake of iron as animals collected from a site near to a blast furnace where levels of iron were elevated considerably in vegetation contained concentrations which were only slightly higher than in grasshoppers from an uncontaminated site. Much of the zinc in orthopterans is located in the mandibles where it helps to strengthen the cutting edges (Hillerton & Vincent 1982).

Three species of grasshopper (*Chorthippus parallelus, C. brunneus* and *Omocestus viridulus*) were analysed by Avery *et al.* (1983) as part of a study on the diet of lizards in sites adjacent to a busy road and a disused lead mine. Even in the most contaminated site where vegetation contained 150 µg g^{-1} of lead, the grasshoppers never had more than 100 µg g^{-1} of lead in their bodies. In contrast, concentrations of zinc and copper were about five times greater in the grasshoppers (250 and 50 µg g^{-1} respectively) than in the plants (50 and 10 µg g^{-1} respectively). *Chorthippus brunneus* has also been studied in metal-contaminated grasslands by Hunter *et al.* (1987d). Goldsmith & Scanlon (1977) found that the concentration of lead in grasshoppers was less than $4 \mu g \, g^{-1}$, even in specimens collected from the edge of very busy roads.

7.5.3.3 Experiments on metals in Orthoptera

The importance of the intestinal epithelium in acting as a barrier to the passage of toxic metals into the blood was demonstrated by Martoja *et al.* (1983) in their experiments on *Locusta migratoria*. Solutions of cadmium or mercury salts were injected into the locusts and it was shown that a concentration of only $5 \mu g \, g^{-1}$ of either metal in the blood was sufficient to cause fatal damage to the fat body. The same amount of cadmium and mercury administered in the food would not cause cellular damage because detoxification systems in the cells of the gut prevent the metals from passing into the blood.

Experiments using [109]Cd have confirmed that concentration factors for cadmium in Orthoptera are low in comparison with most other invertebrates (Van Hook & Yates 1975). The concentration of the isotope in crickets stabilised after nine days at a level of about 60% of that in the diet. In an earlier study, Crossley & Van Hook (1970) used ^{51}Cr as a marker to quantify energy assimilation rates in the house cricket *Acheta domesticus*.

7.5.3.4 Essentiality of metals to Orthoptera

The importance of copper as an essential micronutrient for insects was recognised 40 years ago by Bodine & Fitzgerald (1948). Their experiments demonstrated that females of the grasshopper *Melanoplus differentialis* provided copper in the yolk of their eggs for development of the embryo (Fig. 7.13).

More recent work has examined the dietary requirements of copper and zinc in the house cricket *Acheta domesticus* (McFarlane 1974, 1976). The addition of copper to an artificial diet improved the growth and survival of the crickets, the optimum concentration being $2 \mu g \, g^{-1}$ for females and $10 \mu g \, g^{-1}$ for males (on a fresh weight basis). The sex difference in requirement for the metal was attributed in part to the accumulation of copper by the larval testes. Adult males showed a loss in pigmentation, the so-called 'albino male' effect, when the level of copper in the diet was less than $10 \mu g \, g^{-1}$. The requirement for copper was increased by the addition of Vitamin E to the food which stimulated growth of the testes. Additions of selenium in place of Vitamin E did not increase the copper needs of *Acheta* (McFarlane 1972). Adding zinc to the diet improved the growth of the crickets in the presence of added

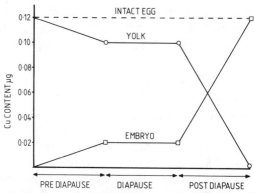

FIG. 7.13 Schematic graph to show changes in the distribution of copper between the yolk and embryo in eggs of the grasshopper *Melanoplus differentialis* during development (dry weight of individual egg = 1.5 mg). (Based on data in Bodine & Fitzgerald 1948.)

copper but not in its absence indicating synergism between the two trace elements. Presumably, sufficient zinc was already present in the artificial diet. McFarlane suggested that the minimum requirements of zinc and copper for normal growth of crickets were $23 \mu g\, g^{-1}$ and $14 \mu g\, g^{-1}$ of fresh diet respectively.

7.5.3.5 Effects of metal-based pesticides on Orthoptera
One of the best papers on metals in terrestrial invertebrates concerns the uptake and effects of compounds of arsenic on fourth instar nymphs of the meadow katydid *Conocephalus fasciatus* (Watson *et al.* 1976). Four arsenicals including organic and inorganic forms of both the trivalent and pentavalent states were applied to the food supplied to the animals. These were cacodylic acid, cacodylic acid plus sodium cacodylate, arsenic trioxide, and arsenic pentoxide. At the end of the experiment, tissue concentrations of elemental arsenic were higher for organic forms than for inorganic forms at similar exposure dosages. Life expectancies were reduced to less than 10% of unexposed populations by levels above $5 \mu g\, g^{-1}$ arsenic (dry weight) in dosing formulations. Katydids exposed for 14 days to a chronic dose of $1500 \mu g\, g^{-1}$ arsenic in the food contained $715 \mu g\, g^{-1}$ of the metal. These very high levels are likely to have occurred in the field since the non-selective herbicide Phytar 560-G (which was being sold in the USA in the 1970s for use in non crop growing areas) contained cacodylic acid and its sodium salt in amounts which produced

a total arsenic concentration of 12.7%. The normal level of arsenic in the food of katydids in uncontaminated sites is only about 0.05 µg g^{-1}. The paper by Watson *et al.* (1976) is thoroughly recommended.

7.5.4 Dictyoptera (Cockroaches)

7.5.4.1 Introduction

The presence of copper in cockroaches has been known for many years (Giunti 1879; Haber 1926) but it was not until the paper by Cunningham (1931) that a value for the concentration was reported (32 µg g^{-1} of copper in whole *Blatta orientalis*). Most recent research on metal dynamics in cockroaches has used laboratory populations of exotic species which have become pests of human habitations in temperate climates.

7.5.4.2 Routes of uptake and loss of metals in cockroaches

The roles played in metal regulation by the different regions of the digestive and excretory systems of cockroaches (Fig. 6.1) are far from clear. Semi-quantitative X-ray microanalysis of the internal distribution of elements in three species in the genera *Blabera*, *Blatella* and *Periplaneta* have shown that the ileum and Malpighian tubules contain the most calcium, phosphorus, zinc and iron (Ballan-Dufrançais 1974). The midgut is apparently of little importance in metal storage. However, while subsequent studies by electron microscopy have demonstrated that numerous intracellular metal-containing granules are present in the ileum of cockroaches (Jeantet *et al.* 1980; Fig. 9.15), there do not appear to have been any detailed studies on the midgut epithelium.

The ileum is lined with cuticle which in *Gromphadorhina portentosa*, is permeable to small molecules and ions (Madrell & Gardiner 1980). However, the metals stored in the ileal cells could have been transported there from the midgut *via* the blood. Jeantet *et al.* (1980) could find no evidence that intracellular granules could be voided into the lumen of the gut by lysis of the cells so this organ may represent a site of permanent storage-excretion. The Malpighian tubules, on the other hand, may be able to excrete the type A calcium phosphate granules containing zinc (see Section 9.4), but more research is required before the relative importance of the different digestive organs in metal assimilation and excretion can be elucidated. It is interesting to note that some assimilated [85]Sr in *Gromphadorhina portentosa* was voided in the faeces *before* strontium which passed straight through the gut without being assimilated was lost (Schowalter & Crossley 1982). The assimilated strontium may have been lost in granules *via* the Malpighian tubules.

7.5.4.3 Experiments on metals in cockroaches

Blatella germanica is very resistant to zinc and can survive on an artificial diet containing 1600 µg g^{-1} of the metal (Gordon 1959). However, this species is killed rapidly if it feeds on the same food containing either 1600 µg g^{-1} copper, 1500 µg g^{-1} manganese, 200 µg g^{-1} strontium or 300 µg g^{-1} cadmium. These deleterious effects did not occur if the diet was supplemented with calcium at a concentration in the food of 6000 µg g^{-1}. Thus, calcium at this high level counteracted the toxic effects of copper, manganese, strontium and cadmium, but since concentrations in the tissues were not measured it is not known by how much assimilation of these metals by the digestive organs was reduced. Omission of zinc, copper and manganese from the artificial diet in Gordon's experiments reduced the growth rate and reproductive success of the cockroaches but the levels in the food were not recorded so the essential requirements have still to be determined.

Many species of cockroaches have a symbiotic relationship with intra-luminal bacteria which aid digestion. The primitive species *Cryptocercus punctalatus* continually re-inoculates the gut with microorganisms by ingesting its own faeces. Coprophagy is particularly important after a moult when the lining of the hindgut together with the resident bacteria are lost. Experiments using ^{65}Zn have shown that metals may be recycled through the gut several times by this coprophagous behaviour (Burnett *et al.* 1969).

The ovaries of female cockroaches pass symbiotic microorganisms into the eggs before they are laid. Brooks (1960) has shown that in *Blatella germanica*, zinc and manganese must be added to certain artificial diet formulations if this process is to be successful in laboratory-reared populations.

7.5.5 Isoptera (Termites)

Termites are the dominant herbivorous and detrivorous soil invertebrates in the tropics and must be of profound importance in the environmental cycling of metals. However, this aspect of their biology has been studied only in relation to uptake of nickel and chromium by different castes on serpentine deposits in Zimbabwe where the soil contains up to 10% and 0.6% of the metals respectively (Wild 1975a, 1975b).

Wild's work provides a classic example of how the feeding behaviour of an animal influences its exposure to metals. Worker termites contained 5000 µg g^{-1} nickel and 1500 µg g^{-1} chromium because they feed directly on plant material and soil. The soldiers had only 100 µg g^{-1} nickel and 300 µg g^{-1} chromium, and the queens only 20 µg g^{-1} of both metals because these castes are fed on saliva produced by the workers from which most of the metals have been removed. The offspring produced by the queen contain no detectable nickel or chromium until they start to feed.

Control of wood-eating termites in buildings is achieved by soaking the timbers in saturated solutions of chromium, arsenic and copper (Usher & Ocloo 1975). The toxic effects of these metals may however be exerted on the symbiotic Protozoa and bacteria in the guts of termites rather than on the epithelial cells of the digestive tract.

Zootermopsis angusticollis, contains about 1 µg g^{-1} of Vitamin B$_{12}$ which is produced entirely by symbionts in the digestive tract (Wakayama *et al.* 1984). Since this essential vitamin contains 4% cobalt, the metal must be present in the diet if termites are to grow and reproduce normally.

7.5.6 Hemiptera (Aphids, Scale Bugs)

7.5.6.1 Introduction

Hemipteran insects possess piercing and sucking mouthparts which are specialised for feeding on liquid diets. The only studies of which I am aware where concentrations of metals have been measured in identified species of Hemiptera from the field were those of Bowden *et al.* (1985a, 1985b) and Cohen *et al.* (1985). These authors attempted to use the elemental composition of several species of aphids as a biological marker of the geographical source of the insects. However natural variation was so great that the technique was unsuccessful. Hunter *et al.* (1987b) reported levels of copper and cadmium in Hemiptera from near to a copper refinery (Table 5.9), but the species were not given.

Among the Hemiptera, aphids are the most serious pests of crops for as well as damaging plants directly by removing material from the phloem vessels, they act also as agents in the transmission of viral diseases (Crawford 1983). The economic importance of aphids has stimulated several researchers to study their mineral requirements in the search for an optimal artificial diet on which the insects can be reared in the laboratory.

7.5.6.2 Essentiality of metals to aphids

The major problem faced when conducting experiments on essential levels of metals in animals using artificial diets is obtaining components of sufficient purity. Low levels of iron and zinc may be present in 'purified' sucrose, vitamins etc. which are sufficient to supply the physiological requirements of aphids. In these circumstances, the essential levels of iron and zinc which are needed for normal metabolism can not be quantified accurately (Dadd & Mittler 1965; Cress & Chada 1971; Srivastava & Auclair 1971b).

In a series of experiments, Dadd and co-workers (Dadd & Mittler 1966; Dadd 1967; Dadd & Krieger 1967) working on *Myzus persicae* and *Aphis fabae*, managed to develop a synthetic diet on which they reared more than 30 generations. Growth of the aphids depended crucially on the presence in the diet of the correct concentrations of iron, manganese, zinc and copper (Fig. 7.14). The optimum levels of the four metals in the artificial diet (on a fresh weight basis) which provided for the needs of all possible growth stages were about $2 \mu g\, g^{-1}$ iron, and $1 \mu g\, g^{-1}$ zinc, manganese and copper. Iron and zinc deficiency symptoms in aphids fed on 'iron-free' and 'zinc-free' diets were observed in the first generation but aphids fed on 'manganese-free' and 'copper-free' diets were not affected until the second and third generations. Evidently, sufficient quantities of these metals were transported to the offspring in the eggs but the amounts were not enough to provide for a second generation.

Akey & Beck (1971, 1972) managed to rear 46 generations of *Acyrthosiphon pisum* on a sterile synthetic diet and found that optimal levels of iron, zinc, manganese and copper in the food were of the same order of magnitude as found by Dadd (1967) in *Myzus persicae*. The ratios of these optimal levels of the four metals to each other in the synthetic diet were very similar to those in the sap of broad bean plants on which the aphids normally feed. Similar experiments on *Acyrthosiphon* have suggested that molybdenum and boron may be essential also but the elements are required more by intestinal symbionts than directly by the aphids (Auclair & Srivastava 1972; Srivastava & Auclair 1971a). A similar conclusion was reached by Ehrhardt (1968) in his studies on iron, zinc, manganese and copper in *Neomyzus circumflexus*.

7.5.6.3 Effects of metal-based pesticides on Hemiptera

Metal-based pesticides are much less effective than organic insecticides

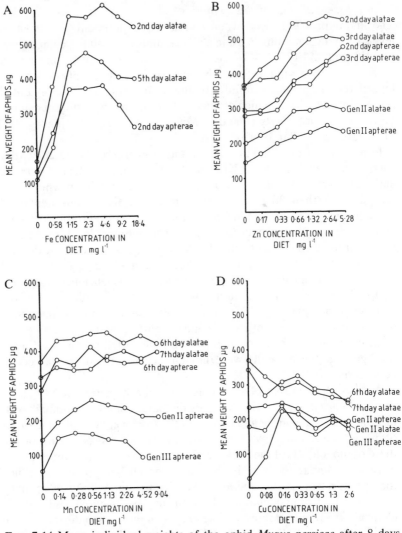

FIG. 7.14 Mean individual weights of the aphid *Myzus persicae* after 8 days growth on a synthetic diet containing a range of concentrations of (A) iron, (B) zinc, (C) manganese and (D) copper. 2nd, 3rd, 6th and 7th day apterae and alatae are larvae of the first generation on a synthetic diet weighed on the 8th day of growth. Gen II and Gen III are larvae of the second and third generations weighed when apterae had become adult on the best diets. Note the classic 'bell-shaped' dose–response curve for iron (A) and compare this with Fig. 2.1.
    (Redrawn from Dadd 1967, by permission of Pergamon Journals Ltd.)

for killing hemipterans (Bartlett 1964a). Sanford (1964) showed that DDT was about 20 times more toxic than lead arsenate to the mirid bug *Atractotomus mali*, a pest of apple trees. However, mercury was still being recommended recently as an effective repellant of the bug *Dysdercus koenigii* in India because it remained active for at least 70 days (Ahmad & Khan 1980). Mercury must be used with extreme care since in its methylated form, it is rapidly accumulated by aphids to concentrations which are toxic to their natural enemies (Haney & Lipsey 1973).

7.5.7 Lepidoptera (Moths, Butterflies)
7.5.7.1 Introduction
The economic importance of moth larvae as pests of crops, and the ease with which they can be reared on artificial diets in the laboratory, has stimulated many researchers to study the toxicity of metals to Lepidoptera. Fewer workers have examined the effects of metal pollutants on field populations of non-pest species or have determined the requirements of essential elements in the diet. However, one of the most interesting suggestions for the role of iron in the biology of terrestrial invertebrates has been made following experiments on orientation in Lepidoptera. Deposits of this metal have been found in a sub-cuticular layer in the abdomen of adult monarch butterflies. It is possible that the iron enables the butterflies to orientate with respect to the earth's magnetic field on their long migratory flights (Jones & MacFadden 1982; Kirschvink *et al.* 1985).

7.5.7.2 Routes of uptake and loss of metals in Lepidoptera
The diet of caterpillars which feed on living vegetation is rich in available nutrients which are easily assimilated. The feeding strategy which most species adopt is to pass food through their guts as rapidly as possible because it is not 'worthwhile' (in energy terms) to solubilise the more refractile components of the diet (Sibly & Calow 1986). The large throughput of food in the digestive tract exposes the larvae to considerable quantities of essential and non-essential metals, particularly in aerially contaminated sites where the elements are present as a surface deposit on leaf surfaces. Caterpillars undoubtedly accumulate cadmium and copper in these circumstances (Table 5.9). The highest concentrations of cadmium occur in the midgut and Malpighian tubules where levels of 1100 and 500 $\mu g\,g^{-1}$ have been recorded respectively (Suzuki *et al.* 1984; Matsubara *et al.* 1986). Histological studies have

shown that metals are stored in granules (Waterhouse 1952), and are lost by lysis of the cells of the midgut and Malpighian tubules (see Section 9.5.4.8).

The different stages in the life cycle in Lepidoptera may have radically different susceptibilities to metals. For example the LD_{50} concentration of copper chloride injected in solution into larvae of *Scotia segetum* is 100 times higher than that for pupae (Škrobák & Weismann 1975, 1979a, 1979b; Škrobák *et al.* 1975, 1976). The larvae are able presumably to excrete the copper before it interferes with biochemical reactions in the tissues whereas the pupae have no route *via* which the metal can be lost.

The mandibles of lepidopteran larvae, like many other herbivorous insects, contain high concentrations of zinc which strengthen the cutting edges (Hillerton & Vincent 1982). Experiments using ^{65}Zn showed that male *Heliothis virescens* transferred 36% of their whole body zinc burden to females at the time of mating (Engebretson & Mason 1980, 1981). Approximately 5% of the male's and 11% of the female's total zinc burden was found in eggs oviposited during the following ten days. Conservation of zinc may represent provision in the eggs for the developing larvae so they can harden their mandibles and begin to feed immediately after hatching. Mason & McGraw (1973) have estimated egg production in *Trichoplusia ni* in the wild by comparing the level of ^{65}Zn in released and recaptured females.

7.5.7.3 Metals in Lepidoptera from the field
Pooling of caterpillars of different species is to be discouraged because this disguises interspecific differences in metal concentrations (Table 7.14). Most species of herbivorous lepidopteran caterpillars feed on specific foodplants. Consequently, the rates of transfer of metals from vegetation to caterpillars can be determined accurately, unlike the situation with detrivores where one is never quite sure what the animal has been feeding on. This enabled Beyer & Moore (1980) to demonstrate clearly that the lead contamination of the plant *Prunus serotina* adjacent to a busy road, was not assimilated to any great extent by eastern tent caterpillars *Malacosoma americanum*. Caterpillars collected 10 m from the road contained $7 \, \mu g \, g^{-1}$ lead whereas their food plant contained $10 \, \mu g \, g^{-1}$ lead.

Analysis of adult Lepidoptera in contaminated sites should be carried out only if the moths or butterflies are seen to emerge. Beyer *et al.* (1985b) managed to collect six species of moths by light-trapping near to a smelting works but the results are difficult to interpret because the

TABLE 7.14

Concentrations of metals in two species of lepidopteran larvae and their food plant (μg g^{-1} dry weight $\pm$ standard deviation) and concentration factors (CF, larvae/plant) in an unpolluted site, and a site contaminated with emissions from a blast furnace in The Netherlands (from Joosse & Van Vliet 1982, by permission of Springer–Verlag)

	Unpolluted site			Blast furnace site		
	Fe	*Mn*	*Zn*	*Fe*	*Mn*	*Zn*
Larvae *Thyria jacobaea*	74.3 ± 3	23.1 ± 0.4	124.8 ± 4.2	341 ± 32	39.6 ± 3.4	128 ± 2
Foodplant *Senecio jacobaea*	225 ± 25	27.6 ± 0.7	61.2 ± 0.7	660 ± 17	78.6 ± 0.8	82.5 ± 3.0
CF	0.3	0.8	2.0	0.5	0.5	1.5
Larvae *Yponomeuta evonymellus*	45 ± 4	7.2 ± 1.7	83.7 ± 9.8	71.4 ± 4.0	8.4 ± 1.7	73.2 ± 1.7
Foodplant *Evonymus europaeus*	174 ± 1	31.8 ± 0.8	75.6 ± 6.8	656 ± 45	60.1 ± 7.3	53.0 ± 1.5
CF	0.3	0.2	1.1	0.1	0.1	1.4

sites where the larvae had been feeding before they pupated were not known.

Attempts to relate the concentrations of a wide range of elements in Lepidoptera to their geographical source (so-called 'chemo-printing') have been largely unsuccessful (Bowden *et al.* 1984; Sherlock *et al.* 1985), except in regions where there are large environmental variations in concentrations of rare elements such as molybdenum (Zhulidov *et al.* 1982).

7.5.7.4 Experiments on metals in Lepidoptera

An extensive study of the toxicity of 12 elements to larvae of the Indian meal moth *Plodia interpunctella* fed on artificial diets was carried out by Perron *et al.* (1966). Considerable differences in the toxicities of the elements were found. Arsenic, which was the most toxic element, had an LD_{50} (on a fresh weight of diet basis) of $60 \,\mu g \, g^{-1}$ whereas the value for zinc was $5000 \,\mu g \, g^{-1}$. The ratios of LD_{50} molar concentrations of metals relative to arsenic (1) were cobalt (3.5), copper (4.8), aluminium (33), zinc (45), iron (48) and molybdenum (139). These differences were attributed to the extent of interference by the metals on enzyme systems in the larvae. Copper, for example, is thought to inhibit xanthine oxidase activity in *Scotia segetum* (Marešova *et al.* 1973).

Sastry *et al.* (1958) examined the toxicity of zinc chloride to larvae of the rice moth *Corcyra cephalonica*. There was no reduction in growth rates of larvae fed on a diet containing $800 \,\mu g \, g^{-1}$ zinc (wet weight). However, at concentrations above $2000 \,\mu g \, g^{-1}$, growth was markedly inhibited and at $4000 \,\mu g \, g^{-1}$ the larvae survived barely a week. Similar figures were obtained for zinc sulphate. Addition of large amounts of RNA and DNA to the diet (at a concentration of 1%) alleviated zinc toxicity, even at the highest levels.

Several researchers have examined the dynamics of metals in larvae of the silkworm *Bombyx mori* since Akao (1939) reported that they contained $10 \,\mu g \, g^{-1}$ copper. More recent work has focussed on whether silk production could be increased by addition of metals to the mulberry leaf diet of the caterpillars.

Whole silk moth larvae reared on mulberry leaves contain about $250 \,\mu g \, g^{-1}$ iron, $30 \,\mu g \, g^{-1}$ copper, $7 \,\mu g \, g^{-1}$ molybdenum, $35 \,\mu g \, g^{-1}$ manganese and $12 \,\mu g \, g^{-1}$ cobalt (Sridhara & Bhat 1966a). Small supplements of all these elements to the diet improved growth and silk production. The concentrations in the food (wet weight) at which particular metals reduced growth rates were $200 \mu g\, g^{-1}$ manganese, $400 \mu g\, g^{-1}$ zinc, $500 \mu g\, g^{-1}$ molybdenum, $600 \mu g\, g^{-1}$ copper and $1000 \mu g\, g^{-1}$ cobalt.

In a subsequent paper, Sridhara & Bhat (1966b) reported that cobalt was particularly effective at promoting silk production and that it exerted a more favourable effect in its free form than when it was incorporated into Vitamin B_{12}. Studies by other authors on *Bombyx* have shown that larvae fed on mulberry leaves which had been soaked in a solution containing $100\ mg\ l^{-1}$ cobalt were 20% heavier at pupation than larvae fed on leaves which had been soaked in water (Chakrabarti & Medda 1978; Young 1979). Supplements of cobalt led to a doubling of the weight of the silk gland relative to untreated larvae, an increase in acid and alkaline phosphatase activity and a reduction in the lipid and glycogen content of the organ relative to controls (Bhattacharyya & Medda 1981a, 1981b, 1983). Cobalt clearly had a beneficial effect on silk production in *Bombyx* larvae but the biochemical basis of this effect has yet to be elucidated.

The effects of other metals on silkworm larvae have been studied in a series of papers by Miyoshi and co-workers. A concentration of $200 \mu g\, g^{-1}$ zinc or $10 \mu g\, g^{-1}$ cadmium (as the sulphates) in the diet led to malformation of the cocoon layer and at higher levels, the larvae failed to pupate (Miyoshi *et al.* 1971). The 'safe' levels of zinc or cadmium in fresh diet were $100 \mu g\, g^{-1}$ and $5 \mu g\, g^{-1}$ respectively. Growth of first instar larvae was inhibited at concentrations of lead, copper and arsenic in the diet of 200, 800 or $40 \mu g\, g^{-1}$ respectively but was not affected at 100, 100 or $5 \mu g\, g^{-1}$. Fifth instar larvae were much more resistant to the toxic effects of the metals (Miyoshi *et al.* 1978a).

Most combinations of these metals acted synergistically when combined in the diet (Miyoshi *et al.* 1978b). In dirty rearing conditions, the threshold toxicity was much lower. The maximum 'safe' levels for metals in the diet (on a dry weight basis) recommended by Miyoshi *et al.* (1978b) (which take into account all possible rearing conditions) were $2.5 \mu g\, g^{-1}$ cadmium, $100 \mu g\, g^{-1}$ zinc, $50 \mu g\, g^{-1}$ lead, $50 \mu g\, g^{-1}$ copper and $1.25 \mu g\, g^{-1}$ arsenic. These levels agree with the 'no effect' dietary concentrations determined for lead in larvae of *Scotia segetum* (Weismann & Svatarakova 1981), for zinc in *Spodoptera littoralis* (Salama 1972; Salama & El-Sharaby 1973), *Ostrinia nubilalis* (Gahukar 1975) and *Heliothis virescens* (Sell & Bodznick 1971) and for copper in *Homona coffearia* (Sivapalan & Gnanapragasam 1980).

Cadmium, copper and zinc exhibited the same toxicity to larvae of *Bombyx mori* whether they were administered as the chloride, oxide, sulphate or acetate (Miyoshi *et al.* 1978c). However, sulphides of cadmium, zinc and copper had very little effect on the larvae because this form of the metals is so insoluble in the digestive juices (Waterhouse 1952). Very high levels of metals in the food may act as feeding deterrents so that the larvae die through starvation rather than from metal toxicity.

Incorporation of the powerful chelating agent EDTA to the diet mitigates the toxic effects of some of these metals (Nakayama & Matsubara 1981; Masui *et al.* 1986). However, high levels of EDTA reduce the availability of essential elements to such an extent that the caterpillars begin to suffer from deficiencies (Sell & Schmidt 1968).

7.5.7.5 Essentiality of metals to Lepidoptera

Plants grown in nutrient solutions from which copper or zinc had been omitted supported normal growth and reproduction of aphids but more than 90% of caterpillars of *Alabama argillacea* died before pupation (Creighton 1938). The few larvae which managed to pupate took at least 24 days to emerge instead of the normal 12 days. Creighton's studies were initiated following the recognition in Florida of several soil-deficiency diseases. Thus the 'bronzing' of tung oil trees and the 'frenching' of citrus trees was due to a deficiency of zinc, the 'everglades chlorosis' of beans to a manganese deficiency and 'salt sick' of cattle to a copper deficiency. Creighton made the interesting observation that populations of aphids and scale insects often increased dramatically after application to crops of Bordeaux mixture, a copper-containing fungicide. It had previously been assumed that this was due to elimination of fungal hyphae which attacked the aphids. However, Creighton put forward the hypothesis that the insect pests may have been suffering from copper deficiency which was rectified following applications of the fungicide.

Ito & Niimura (1966) performed an extensive series of feeding experiments on larvae of *Bombyx mori* fed on diets which lacked mineral elements normally included in the Wesson's salt mixture. As in all such experiments where concentrations of metals in the diet and larvae were not measured directly, it was not possible to attribute apparent deficiency effects to a lack of a specific element because of probable impurities in the dietary components. However, very poor growth was achieved by larvae fed on diets from which iron, copper and manganese were 'absent'.

The optimum concentration of zinc in an artificial diet for silkworm larvae was $20\,\mu g\,g^{-1}$ fresh weight of food (Horie *et al.* 1967) which corresponds with the level found in fresh mulberry leaves (Akao 1939). Concentrations of zinc of less than this value resulted in decreased growth rates.

7.5.7.6 Effects of metal-based pesticides on Lepidoptera

Lead arsenate was used extensively between the wars for control of codling moth larvae on apples (Harman & Moore 1938) but its use on food crops was eventually discontinued because of the danger of poisoning human consumers. More recently, tin and lead-based compounds have been applied to non-food crops such as cotton where their effectiveness as insecticides may have more to do with feeding deterrence than direct toxicity to caterpillars (Wolfenbarger *et al.* 1968; Hueck & La Brijn 1973).

A tin-based compound 'Du-Ter' was ten times more toxic to larvae of *Spodoptera littoralis* when injected into the gut than when applied topically to the dorsal cuticle (Abo-Elghar & Radwan 1971). Histological examination of the tissues following injection showed that the cells of the midgut had completely broken down allowing the pesticide to pass into the blood and disrupt the integrity of the muscles and fat body.

7.5.8 Diptera (Two-Winged Flies)

7.5.8.1 Introduction

The Diptera is a huge order of insects that contains hundreds of thousands of species worldwide. The larvae develop in a diverse range of habitats including Fungi, dead vertebrates and leaf litter. Most species feed on a semi-liquified diet which is pre-digested before consumption by secretion of digestive enzymes onto the food. All adult flies are liquid feeders.

Diptera disperse metals which they have accumulated as larvae in contaminated ecosystems when they metamorphose and fly away as adults. This route of metal transfer exists in freshwater (Dodge & Theis 1979) and littoral ecosystems (Bender 1975) and is thought to be important in agricultural land to which sewage sludge has been applied as a fertiliser (Redborg *et al.* 1983).

More than 50 papers have been published on the uptake and toxic effects of metals in Diptera. However, the requirements for essential elements have not been quantified. It is very difficult to quote accurate figures for toxic levels of specific elements in the diet because of the wide range of experimental procedures which have been adopted. For example, Maroni *et al.* (1986a) found that the LC_{50} for cadmium in

TABLE 7.15
Distribution of $^{109}Cd^{2+}$ between the organs of larvae of *Drosophila* after feeding on labelled medium (from Maroni *et al.* 1986a, by permission of the US Department of Health and Human Services)

Organ	CPM	%
Midgut	$16\,947 \pm 8\,914$	96.0 ± 3.8
Dissection fluids	867 ± 567	2.8 ± 2.2
Fatbody + salivary glands	107 ± 98	0.4 ± 0.6
Body wall, nervous system, imaginal disks	90 ± 88	0.4 ± 0.5
Malpighian tubules	72 ± 114	0.3 ± 0.5

CPM, counts per minute; %, percent of recovered counts in different organs ± 95% confidence interval.

larvae of *Drosophila* was highly dependent on the medium in which the insects had been reared. On standard corn meal, molasses and killed-yeast medium, the larvae could tolerate concentrations of cadmium chloride ten times higher than when they were grown on commercially produced 'Instant Drosophila Medium'. Furthermore, different strains of *Drosophila melanogaster* have differing sensitivities to the same metal (Dugatova & Podstavkova 1978; Jacobson *et al.* 1985) and recent research has shown that races 'tolerant' to metals may evolve rapidly (see Section 9.5.4.9). Authors do not always make clear which strain has been used. The figures quoted in the following sections should be interpreted with these provisos in mind.

7.5.8.2 Routes of uptake and loss of metals in Diptera

Experiments with radioactive isotopes of cadmium (Maroni & Watson 1985; Maroni *et al.* 1986a) and copper (Poulson & Bowen 1951) have provided clear evidence that the middle midgut of Diptera is the most important site for storage of these metals in intracellular granules and metal-binding proteins (Fig. 6.3; Table 7.15, 7.16; Aoki *et al.* 1984). A storage-excretion strategy for detoxification of unwanted iron and copper is possible in Diptera because the adults have a short life span (Sohal & Lamb 1979). Zinc, in contrast, is stored primarily in the Malpighian tubules from which it is excreted probably in type A granules (Table 7.16). Intracellular storage of metals in Diptera is discussed in detail in Section 9.5.4.9.

TABLE 7.16

Distribution of zinc and copper in houseflies of different ages fed on a sucrose diet (control) or a similar diet supplemented with 500 µg g^{-1} wet weight zinc or copper sulphate (from Sohal & Lamb 1979, by permission of Pergamon Journals Ltd)

Diet	Age (days)	Concentration of Zn (µg g^{-1} wet weight)		
		Malpighian tubules	Midgut	Rest of body
Control	1	0	0	4
	10	0	0	7
	17	0	0	11
	25	0	0	7
500 µg g^{-1} ZnSO$_4$	1	40	2	41
	10	77	4	55
	17	105	1	70
	25	182	11	84

Diet	Age (days)	Concentration of Cu (µg g^{-1} wet weight)		
		Malpighian tubules	Midgut	Rest of body
Control	1	4	3	5
	10	4	24	22
	17	5	40	26
500 µg g^{-1} CuSO$_4$	1	5	6	7
	10	8	59	45
	17	14	77	35

Female flies provide iron in the eggs for the needs of the developing embryos (Gautney *et al.* 1981).

7.5.8.3 Metals in Diptera from the field

Contamination of Diptera close to sources of metal pollution has been reported by Andrews & Cooke (1984) and Beyer *et al.* (1985b). However, these results are difficult to interpret because the species of fly were not determined. A concentration of 2000 µg g^{-1} of lead was measured in wild *Drosophila* collected 200 m from a smelting works in southern Missouri, U.S.A. (Koirtyohann *et al.* 1976) but the flies could have been contaminated with a surface deposit of the metal since the method of trapping was not stated.

Hunter *et al.* (1987b) determined the concentrations of cadmium and copper in larvae of the small crane fly *Trichocera annulata* (which feeds on decaying vegetation) collected in the vicinity of a copper refinery. Larvae at the refinery site contained levels of cadmium and copper which were about ten times higher than those in larvae from the control site (Table 5.9).

The degree of assimilation of mercury by dipteran larvae depends very much on the chemical form of the metal (Ramel & Magnusson 1968; Sorsa & Pfeifer 1973a; Nuorteva *et al.* 1978a; Nuorteva & Nuorteva 1982). Larvae feeding on dead fish from a lake subject to mercury pollution concentrated the metal to five times the level in their diet because most of the mercury was in the methylated form which is at least ten times more 'bioavailable' than the inorganic form (Nuorteva *et al.* 1980). Whereas flies normally contained less than 0.05 µg g^{-1} mercury, adults collected on the shore of the mercury-contaminated lake had more than 0.5 µg g^{-1} (Nuorteva & Häsänen 1972). In contrast, larvae of fungus gnats did not accumulate mercury from a contaminated fungal diet because very little of the metal was in the methylated form (Lodenius 1981).

Attempts have been made to use the 'trace element profile' of Diptera to determine the geographical area from which they are derived. Chamberlain (1977) concluded that separate populations of the horn fly *Haematobia irritans* could be identified based on their content of strontium, but not magnesium, manganese or zinc. In a subsequent paper, Chamberlain *et al.* (1977) examined whether flies reared on media containing specific ratios of elements could be distinguished after release and recapture. Silver, cobalt and nickel were not suitable markers because they were toxic when included in the larval rearing medium at levels which could be detected in adults in the field. Rubidium, magnesium, manganese and strontium were not toxic at any level tested but the concentrations in the adults decreased to those of wild flies within three to seven days after release.

Similar studies were carried out by McLean *et al.* (1979) with rubidium in the onion maggot *Hylemya antiqua*. Levels of rubidium remained sufficiently high in the adults for treated flies to be distinguished from wild populations for only about a week. Concentration factors for chromium in Diptera are too low for this metal to be used as a marker (Massie *et al.* 1983a).

7.5.8.4 Experiments on metals in Diptera

Early publications reported the 'presence in trace amounts' of lithium, copper and manganese in adult *Drosophila* (King 1957) and that adult *Musca domestica* contained 78 µg g^{-1} of copper (Cunningham 1931). An excellent series of papers by Waterhouse published during the Second World War described a series of experiments on metal dynamics in the blowfly *Lucilia cuprina* (note the specific name!), an important parasite of sheep in Australia.

Waterhouse (1940) showed that iron was absorbed by the cells of the midgut in the region where the pH of the luminal contents was less than 4 (Fig. 6.3). The concentration of iron in the mid region of the midgut was about 1400 µg g^{-1} compared with 360 µg g^{-1} in the rest of the alimentary canal and only 140 µg g^{-1} in the remaining body tissues. Lennox (1940) measured the solubility of iron in *Lucilia cuprina* and found that 12% was water soluble, 30% was easily dissociable in mild extractants but that 58% was firmly bound in the tissues. Waterhouse made the interesting suggestion that insecticides could be used which were soluble at a pH of less than 4. These would not be available to sheep, but would be assimilated by the larvae.

Copper was also accumulated in the mid-midgut of *Lucilia* to a concentration of 220 µg g^{-1} on the normal larval rearing medium (containing 23 µg copper g^{-1} wet weight of diet), a level some ten times higher than in the other tissues (Waterhouse 1945b). About 60% of the copper in the midgut was bound firmly in intracellular granules. When the level of copper was increased in the diet, the metal was detected in the cells of the Malpighian tubules also (Waterhouse 1945a). Growth rates were retarded at concentrations of copper in the food of more than 70 µg g^{-1} (wet weight). This was similar to the dietary level of copper (1 mmol) which inhibited diuresis in the tsetse fly *Glossina morsitans* (Peacock 1986).

In some parts of New Zealand, selenium was applied to soils at a rate of 10 to 20 g ha^{-1} yr^{-1} to rectify natural deficiencies of this essential element. Watkinson & Dixon (1979) conducted experiments to determine the level at which selenium became toxic to the soldier fly *Inopus rubriceps*. Larvae of this insect feed on ryegrass. Symptoms of toxicity did not appear in the larvae until the rate of application of selenium exceeded 3 kg ha^{-1} yr^{-1}. Ryegrass and *Inopus* larvae normally contained less than 1 µg g^{-1} selenium but in the most heavily treated crops, concentrations exceeded 2000 µg g^{-1} (although most of this was probably present as a deposit on the leaf surfaces). Larvae feeding on these plants

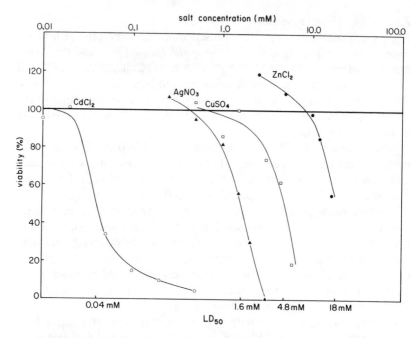

FIG. 7.15 Relative viability of larvae of *Drosophila* in media with one of four metal salts. Each point represents a mean of ratios between the number of larvae that reached pupariation at a given metal salt concentration and the number of larvae that reached pupariation in unsupplemented medium ($n = 3$). (From Maroni & Watson 1985, by permission of Pergamon Journals Ltd.)

accumulated selenium to a maximum concentration of 200 µg g^{-1} before they died.

By far the largest number of laboratory experiments on metal toxicity in Diptera have been conducted on the fruit fly *Drosophila melanogaster*. Williams *et al.* (1982) calculated four day LD$_{50}$'s for one-day-old adult flies fed on a medium containing different concentrations of metal chlorides. The values (l^{-1} of fresh diet) were cadmium 3.6 mmol, mercury 5.7 mmol, nickel 12 mmol, silver 13 mmol, copper 16 mmol, cobalt 16 mmol, chromium 23 mmol and zinc 34 mmol. The differences between the toxicity of metals were even more pronounced when the effects on larval viability were measured (Fig. 7.15).

The extreme toxicity of cadmium and mercury to adult and larval *Drosophila* has been confirmed in several other studies. However, the

TABLE 7.17

Induction of dominant lethals by cadmium chloride in chromosomes of *Drosophila melanogaster* (from Vasudev & Krishnamurthy 1979, by permission of the Current Science Association, Bangalore)

Cd concentration ($\mu g\ g^{-1}$)	Number of eggs counted	Number of eggs unhatched	% dominant lethals
Control	1 076	52	4.83
5	1 244	147	11.8
10	1 375	196	14.3
20	1 390	199	14.3

concentration at which deleterious effects could be detected varied by more than an order of magnitude depending on the experimental techniques used by the different authors (Sorsa & Pfeifer 1973b; Inoue & Watanabe 1978; Cheng 1980; Massie *et al.* 1981, 1983b). Damage to chromosomes may occur at levels of cadmium, arsenic or mercury far below those which disrupt growth and reproduction (Table 7.17; Ramel & Magnusson 1968; Dugatova *et al.* 1978; Kogan *et al.* 1978; Vasudev & Krishnamurthy 1979).

Poulson (1950b) found that larval *Drosophila* grew normally on media containing 160 μg ml^{-1} copper but that they failed to pupate if reared in media containing 320 $\mu g\ g^{-1}$ (approximately 5 mmol — similar to that found by Maroni & Watson 1985). However, Massie *et al.* (1984) discovered that toxic effects of copper could be observed in adult *Drosophila* fed on media containing only 1 mmol of the metal. The copper content of larvae of the fleshfly *Sarcophaga peregrina* was reduced following addition of cadmium to the diet but the level of zinc was not affected (Aoki & Suzuki 1984).

7.5.8.5 Effects of metal-based pesticides on Diptera

Mathew & Al-Doori (1976) tested the mutagenic effect of the mercury-based fungicide Ceresan M on adult *Drosophila* after a mercury poisoning incident in Iraq. A concentration of 400 μg ml^{-1} mercury in their diet was very toxic and resulted in a significant increase in the frequency of sex-limited recessive lethal mutations.

Zinc has been used as an insecticide to control face flies and horn flies around livestock. The metal was included in feed at a concentration that was not toxic to cattle but which was sufficiently high to prevent development of the fly larvae in the faeces of the cows (Bruce 1942; Ode & Matthysse 1964). Insecticides based on triphenyl tin compounds have been used also for the control of the housefly *Musca domestica* (Kenaga 1965; Hays 1968).

Several papers have been published on the use of lead and arsenic-based compounds for the control of dipteran larvae (Maxwell 1961; Maxwell & Parsons 1963; Maxwell *et al.* 1963), but to have any effect, these chemicals have to be applied at rates which result in food concentrations of more than 500 $\mu g\,g^{-1}$ (Pickett & Patterson 1963). These levels are so high as to render products for eventual human consumption unsaleable, even after washing. Beneficial predators of the pests are also killed. Application of metals at very high rates also exerts a strong selective pressure for the evolution of tolerant races of pests. In South Africa, populations of *Lucilia cuprina* became four times more resistant to arsenic in areas where for many years sheep had been dipped in solutions of insecticide containing the metal (Blackman & Bakker 1975).

7.5.9 Hymenoptera (Bees, Ants, Wasps)
7.5.9.1 Introduction
Research on metals in bees, ants and wasps is scattered throughout the literature. A diverse range of topics have been covered including the role of iron in orientation with respect to the earth's magnetic field and effects of metal-containing pesticides on non-target insects. However, I know of no one who has conducted laboratory experiments on the essential requirements or toxicity of metals to Hymenoptera.

7.5.9.2 Bees
Early analyses of the mineral content of *Apis mellifera* detected concentrations of copper of 60 $\mu g\,g^{-1}$ (Aronssohn 1911) and 25 $\mu g\,g^{-1}$ (Cunningham 1931). These values were similar to the more recent determination of 31 $\mu g\,g^{-1}$ in spring bees by Nation & Robinson (1971) who also quoted levels for iron, manganese and zinc of 148, 102 and 130 $\mu g\,g^{-1}$ respectively. Bees collected from gold-mining regions in Russia contained 0.4 $\mu g\,g^{-1}$ (ash weight) which led to the suggestion that they might be useful for 'biogeochemical prospecting' for rare elements (Razin & Rozhkov 1966).

Bees are vulnerable in aerially-polluted sites because pollen becomes heavily contaminated with a surface deposit of metals. In Bulgaria, very high concentrations of about 700 $\mu g\,g^{-1}$ of zinc and copper were detected in pollen brought back to a hive by bees foraging close to a smelting complex (Toshkov *et al.* 1974). The nectar collected by the same bees was not contaminated so the concentrations in the honey were the same as in hives in unpolluted regions. This example demonstrates the importance of knowing the feeding behaviour of invertebrates when one is trying to interpret the effects of metal pollutants on terrestrial ecosystems. The use of honey for indicating environmental metal contamination has been discussed recently by Jones (1987).

Small supplements of some metals may be beneficial. Russian bee-keepers recommend that cobalt should be added to the winter food at a concentration of 1 µg ml^{-1} as this increases brood sizes and honey yields (Young 1979). As in silkworms where the same stimulatory effect of cobalt has been noted (Section 7.5.7.4), the biochemical basis for this phenomenon is not known.

Deposits of iron have been discovered recently in the abdomen of bees. The orientation of their dances is affected by magnets and it has been suggested that the iron is used for orientation with respect to the field lines of the earth's magnetic field (Gould *et al.* 1980). Indeed, the numbers of granules in the fat body of *Apis mellifera* reach a maximum when the bees begin foraging, nine to 12 days post-eclosion (Kuterbach & Walcott 1986a, 1986b). The evidence is, however, contradictory and the review by Kirschvink *et al.* (1985) should be consulted for a recent discussion of the role of magnetic materials in biological tissues.

7.5.9.3 Ants

Ants are of major importance in terrestrial ecosystems as predators and herbivores, particularly in tropical rain forests where they are the dominant ground-dwelling invertebrates. A wide range of elements has been detected in ants but the analytical techniques used in the most comprehensive studies were too inaccurate to allow specific concentrations of metals to be quoted reliably (Ponomarenko *et al.* 1974; Levy *et al.* 1974a, 1979).

Pharoah's Ants (*Monomorium pharaonis*) have become serious pests in buildings where they are difficult to eradicate. However, their foraging behaviour allows highly toxic insecticides such as arsenates to be used in baits which are carried back to the nest by the workers where they kill the larvae and sterilise the queen (Berndt 1974).

In the contaminated grasslands studied by Hunter *et al.* (1987b), ants contained the highest concentrations of cadmium and copper of all the herbivorous insects (Table 5.9). The species analysed by Bengtsson & Rundgren (1984) contained the highest levels of lead of all the insects and spiders collected in pitfall traps close to a brass mill (Table 5.5). Elevated concentrations of mercury were detected in *Formica aquiloma* when fish contaminated with the metal was placed 5 m from the nest (Nuorteva *et al.* 1978b). These examples all show that ants are of considerable importance in the cycling of pollutant metals and would repay more detailed study in the field and in the laboratory.

7.5.9.4 Wasps

Care must be taken to prevent useful predators from being killed when metal-based pesticides are used. Parasitic wasps are of considerable importance in controlling many insect pests and Viggiani *et al.* (1972) have made a detailed study on the toxicity of a wide range of chemicals (including lead arsenate and copper oxychloride) to *Leptomastidea abnormis*, a natural enemy of the hemipteran *Planococcus citri*.

7.5.10 Coleoptera (Beetles)

7.5.10.1 Introduction

It has been estimated that there may be as many as 500,000 species of beetles worldwide which makes them by far the largest group of animals. In temperate ecosystems, the ground-dwelling carabid beetles are often the dominant predatory invertebrates whereas other species are serious pests of crops and stored products. The larvae may have a completely different diet from the adults. This should be borne in mind when one is attempting to determine the routes by which metal pollutants are assimilated by adult beetles.

7.5.10.2 Routes of uptake and loss of metals in Coleoptera

Thousands of papers have been published on the physiology of Coleoptera including more than 40 publications on metals. However, the distribution of metals between the internal organs of beetles has not, to my knowledge, been studied. Indeed, even at the microscopical level, very little is known about the sites of metal storage in Coleoptera (Section 9.5.4.11). This is an area clearly in need of further research.

7.5.10.3 Metals in Coleoptera from the field

The population dynamics of carabid beetles may be disrupted in sites contaminated with metals. Populations in a woodland close to a primary zinc, lead and cadmium smelting works showed an absence of summer diapause and delayed maturation due possibly to scarcity of prey (Read *et al.* 1987). However, accumulation of metals may also have been responsible for the delayed development time since carabid and staphylinid beetles accumulate zinc, copper and cadmium when the metals are present at elevated levels in the environment (Andrews & Cooke 1984; Hunter *et al.* 1984a, 1987b; Van Straalen & Van Wensem 1986; Tables 5.9, 7.13). Accumulation of sub-lethal amounts of mercury leads to reduced growth and reproductive activity in *Tenebrio molitor* (Schütt & Nuorteva 1983).

Lead is not accumulated to any great extent by beetles close to industrial sources of contamination (Table 7.13; Bengtsson & Rundgren 1984; Beyer *et al.* 1985b) or in sites adjacent to busy roads (Giles *et al.* 1973; Maurer 1974; Zhulidov & Emets 1979; Wade *et al.* 1980; Robel *et al.* 1981). Indeed, concentrations of lead of greater than 20 μg g^{-1} have never been recorded in whole beetles. In uncontaminated sites, beetles contain from about 100 to 300 μg g^{-1} of zinc (Carter 1983). Much of this metal in herbivorous species is present in the mandibles which contain more than 1% zinc (Hillerton *et al.* 1984).

Concentrations of copper of 15 to 50 μg g^{-1} were measured in six species of stored product beetles by Levy *et al.* (1973). Gold was detected in cockchafers from an auriferous region of Czechoslovakia at a level of '25 g per ton ash weight of insect' by Babička *et al.* (1945). On serpentine sites in Zimbabwe, beetles of the genus *Catamerus* contained concentrations of nickel and chromium of more than 2000 μg g^{-1} although a large proportion of the metals were probably in the gut contents (Wild 1975b).

7.5.10.4 Experiments on metals in Coleoptera

Interactions between zinc, copper and cadmium have been demonstrated clearly in the confused flour beetle *Tribolium confusum* (Medici & Taylor 1966, 1967). The growth rate of larvae was reduced severely when 50 μg g^{-1} of cadmium were added to a diet containing a minimal level of zinc but normal growth was restored when zinc was supplemented to give a concentration of 50 μg g^{-1} (Table 7.18). Thus, cadmium in the diet raises the requirement for zinc considerably, and the requirement for copper slightly. Levels of metals were not determined in the larvae so it is not known whether the assimilation of zinc, cadmium and copper was affected.

The effects of cadmium on the performance of the alligator weed flea beetle have been studied on host plants grown in nutrient solutions containing 1 μg ml^{-1} of the metal (Quimby *et al.* 1979). The leaves of the alligator weed accumulated cadmium to a concentration of 8.7 μg g^{-1}.

TABLE 7.18

Growth of larvae of *Tribolium confusum* on diets containing different levels of zinc, cadmium and copper ($\mu g\ g^{-1}$ wet weight) (from Medici & Taylor 1967)

Variable cations ($\mu g\ g^{-1}$)		Larval weight at 18 days post-hatching (mg) Level of added Cd			
Zn	Cu	0	50 $\mu g\ g^{-1}$	150 $\mu g\ g^{-1}$	400 $\mu g\ g^{-1}$
50	12.8	2.71	2.63	2.57	1.20
50	0.12	2.70	2.63	2.00	dead
5.0	12.8	2.68	0.95	0.24	dead
5.0	0.12	2.61	0.88	0.15	dead

TABLE 7.19

Bioaccumulation of mercury from contaminated fish flesh through blowfly larvae to predatory staphylinid beetles *Creophilus maxillosus* (from Nuorteva & Nuorteva 1982, by permission of the Royal Swedish Academy of Sciences)

Concentration of mercury ($\mu g\ g^{-1}$ wet weight)		
Fish flesh	Blowfly larvae	Staphylinid beetle
0.7	2.0	6.9
2.7	6.3	17.4
6.9	13.3	33.4

This level was sufficient over a ten-day period to reduce feeding rates in the beetles by a half, decrease the average number of eggs produced by females from 53 to 4, and increase mortality to 35% compared to only 10% in controls.

Davis & Shah (1980) examined the effect of adding zinc to the diet of larvae of *Tenebrio molitor* at levels of 13 to 100 $\mu g\ g^{-1}$. All larvae were able to regulate their zinc concentration to between 80 and 100 $\mu g\ g^{-1}$. However, beetles are not able to regulate internal concentrations of non-essential metals so successfully. For example, Nuorteva & Nuorteva (1982) fed mercury-contaminated blowfly larvae to predatory staphylinid beetles (Table 7.19). In longer term studies, flies containing about

$40 \mu g\, g^{-1}$ mercury were fed to *Tenebrio* larvae which died after four months from 'progressive leg paralysis' (Nuorteva *et al.* 1978a). The dead larvae contained more than $200 \mu g\, g^{-1}$ mercury. It is clear from these experiments that biomagnification of mercury may occur in food chains in the field. During their studies, Nuorteva & Nuorteva (1982) collected specimens of the staphylinid beetle *Aleochara curtula* which had developed as internal parasites in some of the blowfly pupae. These beetles contained $10.6 \mu g\, g^{-1}$ mercury, about the same concentration as their fly hosts.

A research group in the U.S.A. have conducted a series of experiments on rates of energy turnover and the importance of coprophagy in the Passalus beetle *Popilius disjunctus* using radioactive isotopes. Initial studies suggested that the rates of excretion of ^{65}Zn and ^{54}Mn following addition of the markers to the diet, were a good index of metabolic activity (Burnett *et al.* 1969; Mason & Odum 1969) and could be used to assess the effects of pesticides at levels below those which affected growth rates and reproductive activity (Dismukes & Mason 1975). In these early experiments, the metals were treated as inert markers. More recently, it has been recognised that since zinc and manganese are essential elements, their behaviour in the beetles may be more complex than was at first thought (Mason *et al.* 1983b). A higher percentage of ^{65}Zn was retained in beetles fed a low level of zinc in the food than in beetles fed a high level (Fig. 7.16). Thus, the insects were able to regulate the internal concentrations of zinc by a reduction in net assimilation when the metal was present at a high concentration in the diet.

Other factors which affected ^{65}Zn elimination rates were crowding of the beetles (Wit *et al.* 1984) and pretreatment with supplements of zinc chloride which increased rates of ^{65}Zn excretion when the isotope was added subsequently to the diet (Mason *et al.* 1983a).

7.5.10.5 Essentiality of metals to Coleoptera
Fraenkel (1958, 1959) is the only worker to have conducted experiments with the specific aim of quantifying the requirements for a trace metal in beetles. He showed that at least $6 \mu g\, g^{-1}$ zinc (fresh weight) was required in the diet of larvae of *Tenebrio molitor* for normal growth (Fig. 7.17).

Gueldner *et al.* (1975) suspected that the artificial food on which boll weevils (*Anthonomus grandis*) were being reared, was deficient in trace elements because the beetles did not grow as well as on their natural diet of cotton. Analysis of the foods showed that zinc, copper and iron

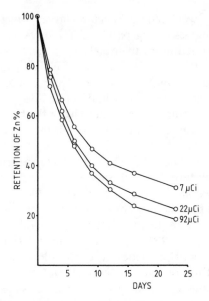

FIG. 7.16 Elimination of ingested ^{65}Zn by *Popilius disjunctus* held at 23°C for 23 days in constant darkness (n = 40 for each treatment). The beetles were fed for 48 hours on wood labelled by topical application of either 7, 22 or 92 μCi ^{65}ZnCl$_2$, then given free access to uncontaminated wood. Coprophagy was prevented by suspending the beetles on a mesh through which faecal pellets could fall. Whole body counts were made subsequently on days 2, 4, 6, 12, 16 and 23. Note that the low-labelled individuals eliminated the radioactive zinc at a relatively slower rate than the beetles in the other two groups. (Redrawn from Mason *et al.* 1983, by permission of Auburn University Press, Alabama.)

were present in similar concentrations in artificial and natural diets. However, levels of calcium and manganese in the artificial diet were only about 25% and 10% respectively, of those in cotton. Thus, deficiencies of calcium and/or manganese could have accounted for the differences in performance of boll weevils on the two diets although this was not confirmed by Gueldner *et al.* (1975).

7.5.10.6 Effects of metal-based pesticides on Coleoptera
As in other insect orders, the effects of particular metals on beetles are related to the chemical form in the diet. For example, there was more than a ten-fold difference in the toxicity of tin to larvae of *Hylotrupes bajulus* which depended on the organic ligand to which the metal was

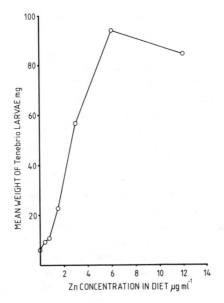

FIG. 7.17 Graph showing the mean individual weight of larvae of *Tenebrio molitor* reared for 11 weeks on a synthetic diet containing different levels of zinc ($n = 20$ for each treatment). The beetles require a concentration of zinc in their food of at least $6\,\mu g\ g^{-1}$ (fresh weight) for normal growth. (Based on data in Fraenkel 1959.)

bound (Becker 1978).

Copper sulphate applied to the surface of beans inhibited oviposition by female bean weevils *Acanthoscelides obtectus* (Muschinek *et al.* 1976). Copper compounds sprayed as fungicides onto potato crops also proved to be toxic to the colorado beetle *Leptinotarsa decemlineata* (Hare 1984; Izhevskiy 1976). However in subsequent years, beetles controlled initially by insecticides may increase in numbers if they develop tolerance or the metals kill off their natural enemies (Viggiani *et al.* 1972). The fungicidal action of mercury applied as a seed dressing may also affect Coleopteran pests. Mercury at a concentration of about 1.25 mmol completely inhibited the activity of amylase in the gut of larvae of *Tenebrio molitor* (Applebaum *et al.* 1961). Ivanova *et al.* (1982) have suggested that the physiological activity of metals can be assessed from their effects on the catalytic activity of immobilised luciferase of the firefly *Luciola mingrelica*.

7.6 MYRIAPODS

7.6.1 Diplopoda (Millipedes)

Considering that millipedes are among the most numerous terrestrial invertebrates in soil/leaf litter ecosystems, it is surprising that so little work on the role of metals in their biology has been carried out. Millipedes feed on living and dead vegetation and may occasionally be serious agricultural pests. Their main role in decomposition is to stimulate microbial activity by fragmentation of leaf litter (Swift *et al.* 1979).

Species such as the pill millipede *Glomeris marginata* consume about 10% of their body weight per day in dead leaves but assimilate only a small fraction of the plant material. However, about 50% of the bacteria and fungal hyphae on the surfaces of the leaves are digested (Anderson & Bignell 1982). Since these microorganisms are known to accumulate metals to levels more than an order of magnitude greater than are in the substrate (Bengtsson 1986), the epithelial cells in the midgut of millipedes may be exposed to much greater concentrations of available metals than would be apparent from considering only total concentrations in the leaves.

In a contaminated grassland, millipedes were only moderate accumulators of copper in comparison to other animals, despite containing the highest concentration of this metal of any of the invertebrates at the control site (Table 5.9). Close to the source of pollution, concentrations of cadmium in millipedes were less than a tenth of those in isopods feeding on a similar diet. Figures for copper and cadmium in millipedes in the uncontaminated site reported by Hunter *et al.* (1987b; Table 5.9) were almost identical to those found in polydesmid millipedes in a control site by Beyer *et al.* (1985b).

When *Glomeris marginata* was fed on leaf litter from metal-contaminated sites, concentrations of cadmium increased in the midgut, but some of the metal passed through the cells to the rest of the body tissues (Fig. 7.18A). Lead was assimilated hardly at all (Fig. 7.18B). In contrast, almost 40% of the zinc was assimilated (Fig. 7.18C).

Millipedes naturally contain high concentrations of zinc of the order of 300 to 500 μg g^{-1} (Carter 1983; Beyer *et al.* 1985b; Read & Martin 1988) and it is clear from Fig. 7.18 that most of the metal is not stored in the digestive organs. Some of my own unpublished observations have shown that zinc is deposited in a sub-cuticular layer which can not be removed from the internal surface of the exoskeleton by scraping with

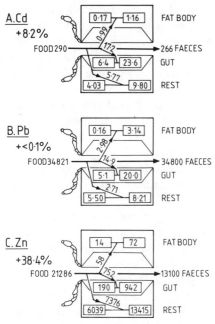

FIG. 7.18 Estimated assimilation (%) and net fluxes in amounts (ng) of metals through *Glomeris marginata*, and between the 'gut', 'fat body' and 'rest' tissue fractions of millipedes from uncontaminated sites fed for 8 weeks on field maple leaf litter (*Acer campestre*) collected from Haw Wood, a site contaminated with cadmium, lead and zinc (see Section 5.3.2). The value given in the left-hand box of each tissue fraction represents the amount of metal estimated to have been present at the start of the experiment. The value in the right-hand box represents the amount present at the end ($n = 12$). (Redrawn from Hopkin *et al.* 1985a, with units converted from moles to ng, by permission of the University of Amsterdam, Commissie voor de Artis-Bibliotheek.)

forceps. However, the metal is not deposited in the cuticle since moulted exoskeletons do not contain any detectable zinc. This is also the case in centipedes (Hopkin & Martin 1983 — see Section 7.6.2).

Myriapods do not have midgut diverticulae or a large organ of storage similar to the hepatopancreas of isopods and molluscs (Fig. 6.1). Therefore, they may have developed a sub-cuticular site for deposition of zinc although why they should accumulate the metal to such high concentrations in uncontaminated sites remains a mystery. Perhaps the zinc is assimilated as a consequence of some, as yet unidentified, biochemical process. Once deposited in the sub-cuticular tissues, the zinc is difficult to excrete because there is no direct route to the lumen of the gut.

Numbers of millipedes were greatly reduced in the litter layer of Hallen Wood close to the Avonmouth smelting works (Table 5.3). In the laboratory, *Glomeris marginata* ate only 3% of its body weight per day when fed on contaminated leaves of *Acer campestre* (field maple) over a two-month period compared to 6.4% of uncontaminated leaves, although there was no difference in survival in the two groups (Hopkin *et al.* 1985a). Read & Martin (1988) showed that immature *Tachypodoiulus niger* and *Glomeris marginata* were much more susceptible to metals in their diet than the adults. Reduced survival rates in juveniles seems therefore the most likely explanation for the lack of millipedes in Hallen Wood.

Considerable differences in the concentrations of mercury have been observed in two species of millipedes in Hawaii (Siegel *et al.* 1975). Decaying vegetation contained 0.056 µg g^{-1} mercury whereas a species of *Trigonialis* contained 0.47 µg g^{-1}. A species of *Oxydus* at the same site contained only 0.082 µg g^{-1} mercury.

Concentrations of copper, cadmium and lead were very similar in two groups of millipedes from an uncontaminated site; however, the level of zinc in polydesmids (980 µg g^{-1}) was almost three times that in spirobilids (370 µg g^{-1})(Beyer *et al.* 1985b). As in other invertebrate groups, elucidation of the underlying reasons for such species differences would be a rewarding area of research.

7.6.2 Chilopoda (Centipedes)
All centipedes are carnivores and possess a powerful pair of poison claws on the head which are used for grasping prey and injecting venom. Centipedes are common in soil and leaf litter, especially in synanthropic sites where they are important in controlling garden pests. In temperate deciduous woodlands, lithobiid centipedes are the most effective invertebrate predators (Wignarajah & Phillipson 1977) and are thus of great potential importance in transferring metals to higher carnivores in the food chain. The general biology of centipedes has been reviewed by Lewis (1981).

The 'presence' of copper in centipedes was reported more than 60 years ago by Muttkowski (1921) who suggested that the metal may be required for the blood pigment haemocyanin (see Section 9.3.4). Concentrations of zinc, cadmium, lead and copper have been measured in whole centipedes from uncontaminated and contaminated sites (Beyer

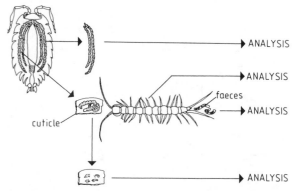

FIG. 7.19 Diagrammatic representation of the experimental procedure used to measure the assimilation of metals by *Lithobius variegatus* fed on the hepatopancreas of *Oniscus asellus*. Comparisons of the levels of metals in the two tubules retained, the remains left by the centipede on a piece of woodlouse cuticle, the faeces, and the organs of dissected animals, enables a 'budget' for each element to be produced (e.g. Fig. 7.20). (From Hopkin *et al.* 1985a, by permission of the University of Amsterdam, Commissie voor de Artis-Bibliotheek.)

et al. 1985b; Van Straalen & Van Wensem 1986; Table 7.12) but Hopkin & Martin (1983) are the only authors to have examined the internal distribution of the metals. Like millipedes, centipedes contain very little lead or cadmium but substantial amounts of zinc which are stored primarily in the sub-cuticular tissues (Table 7.20). The main site of copper storage in centipedes from contaminated sites is the midgut (Table 7.20).

The morphology of the digestive system in centipedes is similar to that of millipedes in that neither group possesses large midgut diverticulae in which metals can be stored (Fig. 6.1). Centipedes consume a diverse range of invertebrates (Sunderland & Sutton 1980) so they have had to evolve detoxification and excretory mechanisms able to cope with unpredictable influxes of metals into the lumen of the gut.

Hopkin & Martin (1984b) conducted an experiment designed to clarify the roles of the different organs of centipedes in metal detoxification. Tubules of the hepatopancreas of the isopod *Oniscus asellus* from uncontaminated and contaminated woodlands were fed in factorial combination to *Lithobius variegatus* collected from the same sites (Fig. 7.19). Net assimilation or loss of zinc, cadmium, lead and copper from the internal organs of the centipedes were calculated.

TABLE 7.20

Concentrations of metals in *Lithobius variegatus* from an uncontaminated site (Midger Wood) and a site 3 km downwind of the Avonmouth zinc, lead and cadmium smelting works (Hallen Wood) ($\mu g\ g^{-1}$ dry wt ± standard error, $n = 7$) (from Hopkin & Martin 1983)

Midger Wood

Tissue sample	Dry weight (mg)	Zn	Cd	Pb	Cu
NC	0.241 ± 0.038	163	<0.6	<3.0	42.0
MT	0.165 ± 0.025	137	<0.9	<4.3	100
OG	0.155 ± 0.013	200	<0.9	<4.6	199
RO	1.723 ± 0.312	93	<0.08	<0.4	9.5
PC	0.614 ± 0.072	84	<0.2	<1.2	21.0
Head	1.735 ± 0.155	155	<0.08	<0.4	11.4
Legs	3.005 ± 0.110	110	<0.05	<0.2	1.77
MG	1.670 ± 0.204	658 ± 35	3.23 ± 1.11	<3.0	61.4 ± 7.2
SCT	0.457 ± 0.090	1 840 ± 430	<1.5	<11	61.4 ± 9.5
EXO	10.212 ± 0.540	439 ± 38	<0.05	<0.5	8.3 ± 1.0
Whole body	19.977 ± 2.720	367	<0.42	<1.0	16.4

Hallen Wood

Tissue sample	Dry weight (mg)	Zn	Cd	Pb	Cu
NC	0.201 ± 0.035	172	<0.7	<3.5	6.41
MT	0.153 ± 0.028	179	<0.9	<4.7	55.3
OG	0.148 ± 0.024	212	<1.0	<4.8	284
RO	1.795 ± 0.311	87.4	<0.08	<0.4	17.5
PC	0.707 ± 0.080	105	1.91	<1.0	15.7
Head	1.922 ± 0.288	210	4.26	<0.4	27.2
Legs	2.114 ± 0.402	144	2.75	<0.3	5.38
MG	1.950 ± 0.341	857 ± 62	85.0 ± 36.0	12.3 ± 6.1	398 ± 95
SCT	1.212 ± 0.174	2 450 ± 750	<0.8	<4.2	82.6 ± 7.8
EXO	7.994 ± 0.631	612 ± 55	1.14 ± 0.22	<0.5	9.4 ± 1.1
Whole body	18.196 ± 2.910	581	<10.5	<2.1	60.9

The nerve cord (NC), Malpighian tubules (MT), oesophageal glands (OG), reproductive organs (RO), poison claws (PC), heads and legs were pooled after weighing. Cadmium and lead were not detected in samples prefixed by <. These figures have been calculated on the assumption that the digests contained concentrations of the metals at just below the detection limit of the analytical equipment. MG, midgut; SCT, sub-cuticular tissues; EXO, exoskeleton.

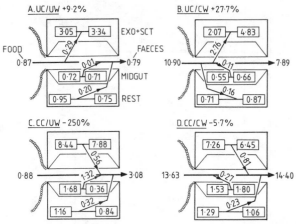

FIG. 7.20 Estimated assimilation (%) and net fluxes in amounts (μg) of zinc through *Lithobius variegatus*, and between the 'midgut', exoskeleton and sub-cuticular tissue fractions (EXO + SCT) and the 'rest' of the body tissues of centipedes from an uncontaminated site (UC) and a contaminated site (CC, Haw Wood—see Section 5.3.2) fed in factorial combination on the hepatopancreas of 6 uncontaminated (UW) or 6 contaminated (CW) specimens of the woodlouse *Oniscus asellus* collected from the same sites (mean of 7 centipedes in each case). The value given in the left-hand box of each tissue fraction represents the amount of zinc estimated to have been present at the start of the experiment. The value in the right-hand box represents the amount present at the end. (Redrawn from Hopkin & Martin 1984b, with units converted from moles to μg, by permission of the British Ecological Society.)

Lead was not assimilated at all by the centipedes but up to 10% of the cadmium and 19% of the copper in the food was retained. Most of this copper and cadmium was in the midgut at the end of the experiment. Thus, the gut appeared to act as a barrier, preventing excessive amounts of the metals from passing into the blood.

Centipedes from the uncontaminated site fed on contaminated wood-louse tissue all died before the end of the experiment. Hopkin & Martin (1984b) suggested that this was due to poisoning of the cells of the midgut by cadmium. Interestingly, centipedes from the contaminated site survived with higher concentrations of cadmium in the midgut and may have evolved tolerance to the metal.

The experiments provided clear evidence of the existence of regulatory mechanisms for zinc in centipedes (Fig. 7.20). Centipedes fed on the hepatopancreas of isopods from their 'own' site assimilated little or none of the zinc taken in with the food (Fig. 7.20A,D). Centipedes from the clean site fed on contaminated isopod tissues, however, assimilated more than 20% of the zinc which was stored mainly in the sub-cuticular tissues (Fig. 7.20B). In marked contrast, centipedes from the contaminated site fed on uncontaminated isopod tissues lost three times more zinc in the faeces than was eaten in the food (Fig. 7.20C). Most of this zinc was derived from the midgut in which concentrations decreased five-fold between the beginning and end of the experiment.

The findings allow a tentative hypothesis for the method of zinc regulation in *Lithobius variegatus* to be put forward. In uncontaminated sites (Fig. 7.20A), centipedes maintain the concentrations of zinc in the tissues by excreting the metal at more or less the same rate as it is assimilated. In contaminated sites (Fig. 7.20D), concentrations of zinc are maintained also, but at somewhat higher levels. When uncontaminated centipedes are fed contaminated food (Fig. 7.20B), almost all the 'extra' zinc in the diet is diverted to the sub-cuticular tissues. Zinc accumulated in the midgut of contaminated centipedes can be lost rapidly from the animals if they are fed on the hepatopancreas of uncontaminated isopods (Fig. 7.20C). However, zinc stored in the sub-cuticular tissues is apparently much less labile and is lost relatively slowly. Thus the midgut prevents excessive passage of zinc into the blood, but some regulation of concentrations in the other organs is maintained by storage in sub-cuticular tissues. It is not known whether the metals in the sub-cuticular tissues perform any essential biochemical functions or whether the tissues are merely used as a convenient place to store zinc and copper which have been assimilated in excess of requirements.

Yates & Crossley (1981) recognised a two-component retention pattern for isotopes of caesium and strontium in *Scolopcryptops nigridia* which probably represented an initial rapid loss from the cells of the midgut, followed by a slower excretion from the other tissues, possibly *via* the Malpighian tubules. The experiments of Hopkin & Martin (1984b) need to be repeated with radioisotopes so that actual movement of metals rather than net fluxes can be determined.

7.7 ARACHNIDS

7.7.1 Aranae (Spiders)

7.7.1.1 Introduction
All spiders are carnivores and are important in controlling pest insects.

Bristow (1958) calculated that during the summer months, fields in southern England (which had not been sprayed with insecticide) contained more than 5×10^6 spiders ha^{-1}. He concluded that 'spiders rank easily first as the enemies of insects, with birds and other insectivorous creatures trailing along far behind'.

Despite their obvious importance as routes for transfer of metals from insects to higher organisms in food chains (many spiders are eaten by birds and lizards), no one has attempted to quantify the extent of such pathways. This is where cooperation between population ecologists and 'metalworkers' would be rewarding.

7.7.1.2 Routes of uptake and loss of metals in spiders

Spiders expend considerable amounts of energy in catching their prey whether this is achieved by building a web, or active hunting. Consequently, spiders have evolved an extremely efficient digestive system which enables them to assimilate a high proportion of the nutrients in the food.

Spiders are liquid feeders and predate almost exclusively on other arthropods. When a prey item has been captured, it is subdued by injection of venom *via* the piercing fangs. Digestive enzymes are poured onto the food, or forced into the unfortunate animal through a small hole made in the cuticle of the prey. Some species mash their food into a pulp but others suck the digested tissues from the prey leaving an empty 'shell' behind. The sucking action is provided by the stomach.

The digestive gland, or hepatopancreas, of spiders is an extensive organ formed from numerous blind-ending diverticulae of the midgut (Fig. 6.1). Branches of this gland ramify throughout the body and extend into the bases of the legs in many species (Foelix 1982; Collatz 1986). This complicated anatomy has prevented the internal distribution of metals in spiders from being determined accurately because the organs are so difficult to separate from one another.

The feeding strategy of spiders exposes the cells of the hepatopancreas to metals which have been solubilised by digestive enzymes outside the animal. Since they only remove the soft tissues from the food where most of the metals are stored, spiders are exposed to much higher concentrations of metals than would be apparent from a consideration of whole body levels in the prey.

Food is retained in the lumen of the hepatopancreas for several hours and this fact, together with the huge internal surface area of the gland, enables spiders to assimilate more than 50% of the food on a weight basis into the body tissues. However, the assimilation of some components of the food is even greater. For example, Nabholz & Crossley (1978) fed a diet labelled with ^{134}Cs (a potassium analogue) to *Pardosa lapidicina*. The isotope was lost subsequently from the spiders at a constant, and very slow rate. Such one-component retention curves represent only biological excretion and suggest complete assimilation of radioisotope. Two-component retention curves (e.g. in centipedes, Yates & Crossley 1981) are characterised by an initial rapid decrease in radioactivity due to the loss of unassimilated tracer from the digestive tract followed by a slower loss of assimilated isotope excreted from the tissues.

The rate of turnover of elements by spiders depends on their rate of feeding. *Steatoda triangulosa* fed initially on houseflies labelled with ^{65}Zn, lost the label subsequently five times faster when fed than when they were starved (Breymeyer & Odum 1969).

Nitrogenous waste products may be stored permanently in sub-cuticular tissues as guanine (visible as white markings on the abdomen of many spiders) or excreted as crystals of uric acid which are formed in the hepatopancreas and voided by lysis of the digestive cells (Seitz 1986). Unwanted metals are formed into intracellular granules which are also voided in the faeces (see Section 9.5.6). Spiders produce liquid faeces which can be stored in a diverticulum of the rectum (stercoral pocket) for sudden release as a defensive measure in situations of stress (Seitz 1986).

7.7.1.3 Metals in spiders from the field

Concentrations of copper and cadmium have been reported in pooled samples of two families of spiders caught in pitfall traps at different distances from a copper refinery (Hunter *et al.* 1987b; Table 5.9). The spiders were among the most contaminated invertebrates close to the refinery and exhibited, apparently, a marked seasonal variation in body levels. However, some of this variation may have been due to inclusion of different species in the samples (which become active at different times of the year), or recruitment of lesser-contaminated juveniles to the population. In an uncontaminated grassland, Carter (1983) found levels of copper, cadmium and zinc in pooled samples of spiders of 53, 2.5 and 253 $\mu g\,g^{-1}$, respectively.

The affinity of spiders for cadmium was also remarked upon by Van Straalen & Van Wensem (1986). They noted that the small linyphiid

Centromerus sylvaticus contained the highest concentration of cadmium of all the arthropods collected in litter from a pine forest 1 km from a zinc smelting works in The Netherlands (Table 7.12). Lead was not accumulated extensively by *Centromerus*.

Specimens of the ground-dwelling lycosid *Trochosa terricola* collected from the most contaminated site studied by Bengtsson & Rundgren (1984) contained only $62 \mu g\, g^{-1}$ of lead (Table 5.6). A more detailed analysis of lead in ten species of spiders was conducted in Denmark by Clausen (1984a, 1984b) to assess their usefulness as indicators of atmospheric lead pollution. Differences in the concentrations of lead between species from the same site were more than five-fold. Such differences may be due to the composition of the diet, or to the possible accumulation of airborne metal particles trapped on the silk of orb-web weaving spiders which eat and re-spin the web every night (Foelix 1982).

Care should be exercised in the storage of spiders before analysis. Clausen (1984a) showed that 25% of the lead 'in' *Araneus umbricatus* was present as a surface deposit which could be washed off with a weak acid.

7.7.1.4 Experiments on metals in spiders

Attempts to formulate an artificial diet for spiders have not met with any success. Peck & Whitcomb (1968) reared *Lycosa gulosa* through two moults on a diet of two parts homogenised milk, ten parts of egg yolk and one part dye but the spiders all died after 90 days. Thus, in order to get spiders to ingest metals, the elements must be assimilated first by an animal on which they would feed in the wild.

Hopkin & Martin (1985b) conducted experiments on the dynamics of zinc, cadmium, lead, copper and iron in *Dysdera crocata*, a spider which feeds exclusively on terrestrial isopods (Pollard 1986). The spiders were collected from waste ground in central Bristol, S.W. England, and were divided into three groups. The first group were fed a specimen of *Porcellio scaber* from their 'own' site at intervals of three days until they had consumed twelve isopods. The second group of spiders were fed in an identical manner but the isopods were from Hallen Wood, a site contaminated with zinc, cadmium, lead and copper by aerial deposition of emissions from a smelting works (see Section 5.3.2). The third group of spiders were starved for the duration of the experiment.

Three days after the last feeding occasion, all the spiders were killed and analysed for their metal content. The amounts of metals contained in the specimens of *Porcellio scaber* before they were fed to the spiders were estimated from plots of wet weight against metal content of whole isopods from the two sites. The amounts of metals removed from the isopods during feeding were calculated by subtracting the amounts in the remains discarded by the spiders from these estimates. In Table 7.21, only the results for starved spiders, and those fed on isopods from Hallen Wood are presented, since there were no significant differences in the concentrations of zinc, cadmium, lead, copper or iron between the three groups of spiders at the end of the experiment.

The hepatopancreas contains more than 50% of the zinc, cadmium, lead and copper in terrestrial isopods (Hopkin & Martin 1982a). Consequently, although the spiders removed only about 15% of the weight of the prey during feeding, they ingested considerably larger proportions of the metals, particularly cadmium and lead (Table 7.21). Spiders which feed on insects remove similar proportions of metals, but a larger proportion of the weight of their prey (Breymeyer & Odum 1969; Lee *et al.* 1978; Van Hook & Yates 1975).

On individual feeding occasions, the combined concentrations of zinc and copper in the tissues removed from the isopods by the spiders exceeded 1% on a dry weight basis. However, *Dysdera* was able to prevent net assimilation of these metals during the experiment. Spiders fed on isopods from Hallen Wood, which consumed 11.2 μg and 18 μg of cadmium and lead respectively, contained only 0.52 μg and 0.27 μg of these metals at the end of the experiment, amounts which were not significantly different from those in starved individuals (Table 7.21). Thus, *Dysdera crocata* from central Bristol would apparently be able to survive in Hallen Wood, even though their food at this site would contain concentrations of zinc, cadmium and lead which were 2, 20 and 10 times greater respectively, than they would experience feeding on isopods from their 'own' site.

The experiment of Hopkin & Martin (1985b) does, of course, provide information only on the *net* assimilation of metals by *Dysdera crocata*. It is probable that much of the metal which is lost eventually in the faeces is assimilated initially by the cells of the hepatopancreas and stored in metal-containing granules. The turnover of cells in this gland is rapid and allows the granules to be voided in the faeces before concentrations of metals reach toxic levels. In *Dysdera*, this system has to be efficient because its prey contains such large amounts of zinc and copper, even in uncontaminated sites. If a permanent storage-excretion

TABLE 7.21

Experiments on assimilation of metals by *Dysdera crocata* (from Hopkin & Martin 1985b, with molar values converted to weight for ease of comparison with other tables in this book, by permission of Springer–Verlag)

A. Consumption of tissues and metals by spiders fed on the isopod *Porcellio scaber* collected from Hallen Wood, 3 km downwind of the Avonmouth zinc, cadmium and lead smelting works ($n = 4$ pooled samples of 12 isopods; SE, standard error).

	d.w.	*Zn*	*Cd*	*Pb*	*Cu*	*Fe*
Wa	87.1	132	15.8	31.5	124	11.4
SE	(20.4)	(33)	(3.9)	(7.9)	(31)	(2.8)
Wb	72.4	76.0	4.6	13.5	66.3	7.4
SE	(18.7)	(18.3)	(1.4)	(3.6)	(20.1)	(1.6)
Wa−Wb	14.7	56.0	11.2	18.0	57.7	4.0
%	17%	42%	71%	57%	46%	35%

Wa = estimated mean dry weight (d.w.mg) and metal content (μg) of isopods fed to *Dysdera crocata*.
Wb = mean dry weight (mg) and metal content (μg) of remains of isopods discarded after feeding by *Dysdera crocata*.
Wa−Wb = amounts of tissue and metals consumed by *Dysdera crocata*.
% = percentage of original amounts of tissue and metals in isopods removed by *Dysdera crocata*.

B. Mean dry weight (d.w.mg) and amounts (μg) and concentrations ($\mu g\ g^{-1}$) of metals in *Dysdera crocata* after starvation for 36 days, or feeding on 12 specimens of *Porcellio scaber* from Hallen Wood (one isopod every 3 days for 36 days) (SE, standard error; $n = 4$).

Starved (d.w. = 12.04 mg; SE = 4.66)

	Zn	*Cd*	*Pb*	*Cu*	*Fe*
μg	13.8	0.54	0.59	12.1	5.4
SE	(4.6)	(0.14)	(0.42)	(4.5)	(1.2)
μg g^{-1}	1 270	54.9	29.4	1 120	450
SE	(200)	(9.0)	(13.3)	(220)	(20)

Fed Porcellio scaber *from Hallen Wood* (d.w. = 11.80 mg; SE = 4.27)

	Zn	*Cd*	*Pb*	*Cu*	*Fe*
μg	17.7	0.52	0.27	8.7	3.6
SE	(6.8)	(0.28)	(0.08)	(1.5)	(0.9)
μg g^{-1}	1 490	37.2	33.8	1 070	420
SE	(150)	(7.2)	(11.8)	(290)	(70)

strategy had been evolved for zinc and copper in *Dysdera*, the hepatopancreas would have a composition similar to that of brass within a year!

7.7.2 Opiliones (Harvestmen 'Spiders')

Harvestmen are easily distinguished from true spiders by their extremely long legs and small body in which the head, thorax and abdomen have fused to form a single unit. Opilionids hunt small, soft-bodied prey which are captured and torn apart by a pair of chelicerae before ingestion (Adams 1984). The digestive system of *Phalangium opilio* has been studied recently by Becker & Peters (1985a, 1985b) but no one has conducted laboratory experiments on the dynamics of metal assimilation in any species.

During the late summer and autumn, harvestmen are very common, especially in synanthropic sites. In a single October night in my modest garden in Reading (10 m by 10 m), 154 specimens of the largest British species *Odiellus spinosus* inadvertently drowned themselves in six yoghurt pots half-filled with brown ale which had been put out as slug traps! At other times of the year in temperate regions, harvestmen are uncommon and this may explain why they feature so rarely in studies on the invertebrate fauna of polluted sites.

In the contaminated grassland studied by Hunter *et al.* (1987b), pooled samples of unidentified 'Opiliones' (incorrectly classified as 'Araneida' in their paper) contained concentrations of cadmium and copper which were very similar to spiders (Table 5.9). However, concentrations of lead and copper in pooled specimens of *Oligolophus palpinalis*, *Opilio parietinus* and *Phalangium opilio* from contaminated sites in Sweden (Bengtsson & Rundgren 1984) were less than a fifth of those in the spider *Trochosa terricola* (Table 5.6). Carter (1983) detected more than 500 μg g^{-1} zinc in 'phalangids' from an uncontaminated site but the species were not given.

More laboratory and field work is required on individual species of harvestmen as they clearly play an important role in the cycling of metals in food chains at particular times of the year.

7.7.3 Acari (Mites and Ticks)

The population density of mites in soil and leaf litter of temperate ecosystems often exceeds 100,000 m^{-2}. Soil mites are a diverse group which includes species that graze fungal hyphae, consume dead vegetation and prey on other soil organisms (Swift *et al.* 1979). However, mites have rarely been studied in detail in sites contaminated with metal pollutants because they are difficult to identify and are too small to analyse on an individual basis (Read 1988). In addition, natural variation in species composition of populations is so large that deleterious effects of metals on the mite fauna of soil and leaf litter is almost impossible to prove. Streit (1985) could only detect a decline in numbers of one out of eight species of orabatid mites when he added copper to the soil to give a concentration of 200 μg g^{-1}. The only soil species in which the concentrations of metals have been determined is *Chamobates cuspidatus* (Van Straalen & Van Wensem 1986; Table 7.12).

Pesticides incorporating lead and arsenic (Bartlett 1964b; Readshaw 1972) or compounds of zinc (Terriere & Rajadhyakasha 1964) are toxic to mites only when applied at levels which are so high as to render the crops unsaleable. Much more success with metal-based acaricides has been achieved in the control of ticks on cattle where rates of application can be strictly controlled. The cattle tick *Boophilus annulatus* was removed by dipping the host in a solution containing 0.22% arsenic (Drummond *et al.* 1964; Gragam *et al.* 1964). However, this has become less effective as the ticks evolved resistance to the chemical (Matthewson & Baker 1975).

In a fascinating historical review on the uses to which arsenic has been put in agriculture, Whitehead (1961) showed that resistant blue ticks were four times more tolerant to arsenic than non-resistant populations. Resistant ticks contained much higher levels of glutathione and free sulphydryl groups which were able to negate the toxic effects of arsenic. The resistant strain of ticks became more susceptable to arsenic poisoning if sulphydryl inhibitors were included in the diet. Interestingly, arsenic resistant ticks were cross-resistant to some other pesticides including triethyl tin hydroxide.

7.7.4 Scorpions

Scorpions are very rare in temperate climates but become more common, the nearer one gets to the equator. Goyffon (1977) determined concentrations of copper in *Androctonus australis* by a colourimetric method but no other researcher has studied metals in any representative of the Scorpiones and only Van Straalen & Van Wensem (1986) have analysed a member of the Pseudoscorpiones (Table 7.12).

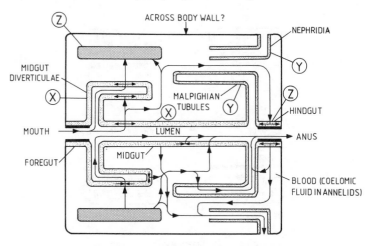

FIG. 7.21 Schematic diagram to show all possible routes of uptake (above), loss (below) and sites of storage of metals in terrestrial invertebrates. Metals are assimilated and lost more easily from type X and Y organs than from type Z organs because they have a direct connection with the lumen of the gut (see text for further explanation). The foregut and hindgut are shown lined with cuticle (solid black), but this would of course be absent from molluscs and annelid worms.

7.8 MODELS OF METAL DYNAMICS IN TERRESTRIAL INVERTEBRATES

7.8.1 Internal Anatomy

The research described in the previous sections of this chapter has highlighted the diversity of responses of the different taxonomic groups of terrestrial invertebrates to metal pollution. At first sight it would not appear to be possible to classify these responses and produce a model into which all examples would fit. However, if one treats the internal organs of the animals as 'black boxes', some general principles emerge.

A hypothetical terrestrial invertebrate, which possesses all the organ types identified to date, is shown in Fig. 7.21 (there is no animal in which this arrangement exists). The internal anatomy of all terrestrial invertebrates can be made to fit one of only five categories (Fig. 7.22) representing (1) molluscs, (2) earthworms, (3) isopods, (4) myriapods

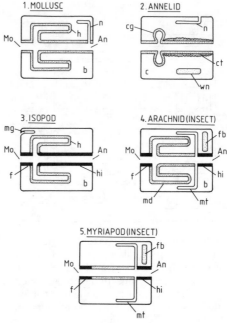

FIG. 7.22 Schematic diagram to show the five main arrangements of digestive and excretory organs in terrestrial invertebrates. Epithelia lined with cuticle are shown in black. Insects can be divided into those groups which possess midgut diverticulae (type 4) and those in which the midgut consists of an unelaborated tube (type 5). An, anus; b, blood space; c, coelom; cg, calciferous glands; ct, chloragogenous tissues; f, foregut; fb, fat body; h, hepatopancreas; hi, hindgut; md, midgut diverticulum; mg, maxillary gland; Mo, mouth; mt, Malpighian tubule; n, nephridia ('kidney'); wn, waste nodules.

and insects without midgut caeca (e.g. Lepidoptera larvae, Fig. 6.1) and (5) arachnids and insects with midgut caeca (e.g. cockroaches, Fig. 6.1). The internal organs of these groups can be classified into three types based on potential routes of assimilation and loss of metals (Fig. 7.21).

Type X organs comprise the cells of the gut which are exposed directly to the food in the digestive tract. In molluscs, isopods, arachnids and some insects, the gut is elaborated to form a number of blind-ending tubules which increase the surface area exposed to the lumen.

Type Y organs are primarily excretory in function and are represented by the Malpighian tubules of insects, myriapods and arachnids, and the nephridia of molluscs and earthworms (the latter being a special case since the nephridia of annelids remove material from the coelomic fluid rather than the blood). Type X and Y organs are separated from the 'external environment' (the lumen of the digestive tract in insects, myriapods and arachnids) only by the plasma membrane of the epithelial cells. Thus, insoluble accumulations of unwanted metals in type X and Y organs can be voided directly from the epithelial cells.

Type Z organs, in contrast, do not have a direct route to the exterior because they are either suspended in the blood (e.g. sub-cuticular tissues of centipedes, fat body of insects and myriapods), or there is a barrier of cuticle which prevents the passage of metal-containing granules (e.g. the hindgut of arthropods). The chloragogenous tissue of earthworms is a special case because the contents of the chloragocytes can be voided *via* the nephridia.

All three types of organ can act as sites of storage-excretion of metal-containing granules. However, the residence time of metals in type Z organs tends to be longer because the rate of breakdown and replacement of cells is slower, and there is no direct route *via* which the granules can be lost without either passing across a layer of cuticle, or across the basement membrane of a type X or Y organ (except in earthworms).

7.8.2 Responses of Internal Organs to Metal Exposure

Dallinger & Wieser (1984a) recognised four types of response of organs of the snail *Helix pomatia* to loading followed by removal of zinc, cadmium, lead and copper from the diet. However, a slightly modified version of their scheme (Fig. 7.23) is applicable to all species in which concentrations have been monitored following exposure to metals in the diet.

Response 1 is characterised by a rapid increase in concentration of the metal during the initial period of exposure followed by a decrease before loading stops. In *Helix pomatia*, this occurred most noticeably with cadmium and lead in the blood. The initial passage of the metals was rapid but decreased substantially when detoxification systems, which take some time to become fully active, were able to trap cadmium and lead in the cells of the digestive organs.

Some organs accumulate a metal in a similar way during loading but behave differently when the animal is transferred onto an uncontaminated diet. In *Response 2*, none of the assimilated metal is subsequently lost but in organs following *Response 3*, there is a rapid decrease in concentrations (Fig. 7.23). Thus, concentrations of cadmium

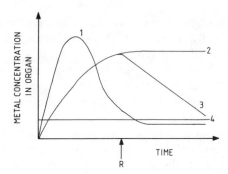

FIG. 7.23 The four main metabolic responses of the organs of terrestrial invertebrates to a diet contaminated with metals. The arrow indicates a return to
an uncontaminated diet. (After Dallinger & Wieser 1984a, by permission of
Pergamon Journals Ltd.)

in the hepatopancreas of *Helix pomatia* (Fig. 7.2), *Oniscus asellus* (Fig.
7.9), and of zinc in the sub-cuticular tissues of *Lithobius variegatus* (Fig.
7.20) remain relatively constant and conform to response 2. In contrast,
concentrations of copper in the hepatopancreas of *Helix pomatia* (Fig.
7.2) and of zinc in the hepatopancreas of *Oniscus asellus* (Fig. 7.9) and
midgut of *Lithobius variegatus* (Fig. 7.20) decrease and conform to
response 3.

Response 4 is exhibited by organs in which the concentration of a
metal is constant, irrespective of changes in the level in the diet (Fig.
7.23). There are several possible reasons why an organ may respond in
this fashion. First, the metal may not be in a form which can be
assimilated by the cells. This may be the case for lead which appears to
be completely unavailable to the midgut of *Lithobius variegatus* (Hopkin
& Martin 1984b). Second, organs in the haemocoel may be 'protected'
if metals are prevented from passing into the blood by detoxification
mechanisms in the midgut. The midgut of centipedes apparently prevents
any cadmium from passing into the blood from contaminated food
(Hopkin & Martin 1984b). Third, the metal may enter the cells of the
organ but it is excreted at the same rate. This possibility should be borne
in mind in all experiments since it is *net* assimilation or loss of metals
which are being measured (except when isotopes are used in which case,
the passage of specific metal ions can be followed).

There are numerous factors which can affect the shape of the curves
and the values of the axes in Fig. 7.23. These include temperature,

length of the experiment, the specific metal being considered and the internal anatomy and physiological status of the animal. All such factors have been discussed at length in Chapter 6.

7.8.3 Experiments on Whole Animals

Experiments have been conducted on the dynamics of assimilation and loss of isotopes of a range of elements in earthworms (Crossley *et al.* 1971), isopods (Lauhachinda & Mason 1979), crickets (Grant *et al.* 1980), cockroaches (Rhodes & Mason 1971), beetles (Gist & Crossley 1975), centipedes (Yates & Crossley 1981) and spiders (Nabholz & Crossley 1978). A number of complex mathematical models have been described also (Goldstein & Elwood 1971; Kowal 1971; Esser & Moser 1982). All these studies have used whole animals treated as 'black boxes' with a single input *via* the mouth and a single route of loss *via* the anus. However, Reichle (1969) has pointed out that total body and faecal analysis can only describe *net* assimilation of substances and can not provide direct information on digestive efficiency or movement of metals between internal compartments.

When an invertebrate is 'dosed' with an isotope, the subsequent rate of loss of radioactivity occurs in two stages (e.g. Fig. 7.10). The first stage is a rapid loss due to voiding of unassimilated label in the gut contents. The second stage is much more gradual and represents loss of isotope which has been assimilated across the plasma membrane of the gut and subsequently excreted. This second component can, of course, contain label derived from several sites within the animal. One could speculate for example that in centipedes, assimilated zinc is initially lost quite rapidly from the midgut, followed by slower excretion from the Malpighian tubules of metal derived from the sub-cuticular tissues (Fig. 7.20).

Such experiments on dynamics of metals in whole animals can provide indirect evidence for the existence of different compartments in invertebrates which are too small to dissect. The presence of two internal compartments in the collembolan *Orchesella cincta*, in which metals could be stored, was proposed by Van Straalen & Van Meerendonk (1987) following feeding experiments with algae contaminated with lead nitrate. The first compartment, from which assimilated lead could be lost most rapidly, comprised the epithelial cells of the gut. Lead was stored in these cells but was lost when the lining of the digestive tract was voided during moulting. The second compartment (possibly the fat body) contained lead which was stored on a more permanent basis. The half life of this component exceeded the life span of Collembola in the field. In contaminated sites, assimilation of lead into the second compartment exceeded excretion resulting in a net accumulation of the metal with age of the insects and a decrease in growth, fecundity and survival.

7.8.4 Conclusions

It is important to recognise that the internal movements of a metal may be entirely different in two species of invertebrate which exhibit identical assimilation rates on a whole organism basis. In food chain studies on the transfer of metals between trophic levels, knowledge of the internal distribution of metals in invertebrates may not at first sight be thought to be important. However, results described in this chapter have shown that a slight increase in concentrations in whole animals may hide a much greater increase in the levels of a metal in a particular organ. Many spiders, for example, ingest only the soft tissues of their food and are exposed to much greater concentrations of metals in their diet than is apparent from a consideration of levels in whole prey. The feeding behaviour of invertebrates in the field is in need of much more research.

The models described in this section are intended to be speculative and provocative! They should be tested rigorously on a wider range of metals and invertebrate groups before being accepted. Particularly rewarding would be research on species which are taxonomically close (in a morphological sense), but which exhibit differences in their physiology, biochemistry and ecology.

CHAPTER 8

Invertebrates as Indicators and Monitors of Metal Pollution in Terrestrial Ecosystems

8.1 INTRODUCTION

To gauge the 'environmental impact' of contamination of terrestrial ecosystems by metals, knowledge is required of the rates at which the pollutants pass through the biota, and their effects on growth and reproductive success. Armed with such information, biologists should, in theory, be able to make informed decisions on the amounts of metals which can be released into the environment without causing undue disruption of natural food webs. In practice, however, the legal limits for emission rates of pollutants from industry are calculated with reference to economic factors and potential effects on the human population only.

The distribution of metals on a geographical basis is assessed invariably by measuring concentrations in abiotic components of ecosystems such as air and soil. If such a study is carried out on one occasion, the results may *indicate* regional variations in the distribution of metals and pinpoint sources of pollution. A good example of such a study in England and Wales is the Wolfson Geochemical Atlas (Webb *et al.* 1978). To produce the Atlas, stream sediments were collected during a ten-week period in 1969 from tributary/road intersections at an average density of one sample per 25 km^2.

Concentrations of a wide range of elements were measured in the sediments and the results presented as attractive coloured maps which highlighted areas of deficiency or enrichment of particular metals. Regions where extensive mining activity had taken place showed up clearly. The study also indicated areas of England and Wales where trace metal deficiency diseases might occur in livestock in, for example, sites where copper levels were low, but concentrations of the copper-antagonist molybdenum were high such as the 'teart' pastures of Somerset.

The drawback of *indicating* pollution is that it only provides a 'snap-shot' picture of the distribution of contaminants at a single moment in time. Variations due to changes in emission rates of a factory, or seasonal differences in deposition due to the weather can not be quantified. This variability can be assessed only by *monitoring* pollution, in other words, by repeating the study on two or more occasions.

Monitoring of metal pollution of the terrestrial environment rarely extends beyond regular analysis of air samples collected in the vicinity of major sources of contamination. Some monitoring of soil con-tamination may also occur but it is only *after* major pollution incidents that most surveys are extended to include biological material which would normally be sold for human consumption (milk, sheep, wheat etc.). Even then, other components of terrestrial ecosystems are rarely considered.

An environmental monitoring scheme for metals which considers only concentrations in air and soil, can not indicate accurately when the potential effects of the pollutants on all components of terrestrial eco-systems should give cause for concern. At present, we do not have sufficient knowledge of the rates of transfer of metals between species at each trophic level, and their effects on growth and reproduction, to enable the extent of ecological disturbance to be predicted. For an example, one need look no further than the aftermath of the Chernobyl disaster.

The heaviest contamination of radioactive fallout from this incident in the U.K. occurred in upland areas in North Wales and Cumbria in late April 1986. Computer models developed for *mineral* soils before the accident predicted that levels of radioactive caesium in sheep would show an initial increase but would decline rapidly in an exponential fashion to background levels within a year. However, since caesium is retained to a much greater extent in *peaty* soils than in lowland mineral soils, at the time of writing (about two years after the incident), there are still sheep which are contaminated with caesium above a level at which they are allowed to be sold for human consumption (although *I* believe they are quite safe to eat!).

The failure to predict the movement of caesium through the biota of the affected areas was due to a lack of basic knowledge of the behaviour of mineral elements in different soil types. Since even less is known about the mobility of metals in plants and animals, it is clear that the distribution and effects of such pollutants on the biota of contaminated terrestrial ecosystems, can not be predicted accurately by monitoring abiotic components alone.

Having concluded that monitoring of the movement of metals through the biota is essential, we need to consider the ways in which this should be done. These practical aspects of biological monitoring are discussed in Section 8.2. Studies in which terrestrial invertebrates have been used as indicators or monitors of metal pollution are reviewed in Section 8.3. In the light of this information, specific recommendations are presented in Section 8.4 on the ways in which pollution of the terrestrial environment should be monitored in the future.

8.2 DEVELOPING A MONITORING SCHEME

8.2.1 Introduction

The fundamental question which needs to be answered before conducting a monitoring programme is whether the analysis of biological materials will provide information on the availability and transfer rates of metals through the biota which could not have been predicted from measurements of levels in air, precipitation and soil. Most publications on this topic describe a terrestrial invertebrate as a 'good monitor' if the concentrations of a metal in the animal are related closely to those in soil or leaf litter. However, if this relationship were consistent in all environments, there would be no need to monitor the contamination of the species since it could be predicted in every case.

Hopkin *et al.* (1986) compared the concentrations of zinc, cadmium, lead and copper in soil, leaf litter and the isopod *Porcellio scaber*, in 89 sites around the Avonmouth smelting works in S.W. England. When the levels in soil, leaf litter and isopods were mapped, the smelting works was implicated clearly as the main source of metal pollution in the region. The maps for cadmium are shown in Fig. 8.1 but a similar concentric arrangement of lines occurred on maps for the other three metals. This raises the question as to why it was necessary to analyse isopods at all if the same information could be obtained from analysis of soil. After all, soil is much easier to collect than isopods!

The answer is provided when the concentrations of metals in the samples from *individual sites* are compared. The huge scatter of points on the plots shown in Fig. 8.2 demonstrates that it is not possible to

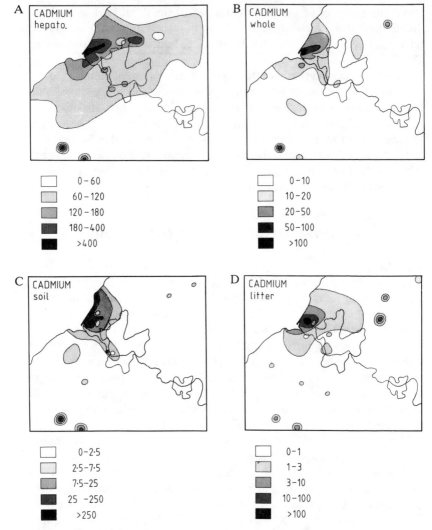

FIG. 8.1 Regional distribution of concentrations of cadmium in the hepatopancreas (A, hepato), whole specimens of the woodlouse *Porcellio scaber* (B, whole), soil (C) and leaf litter (D) in Avon and Somerset, England (see Fig. 5.2). The main area of cadmium pollution was north-west of Bristol at Avonmouth, the site of a primary zinc, lead and cadmium smelting works. Other isolated sites where samples contained higher than background levels of cadmium were associated with disused mine sites (mainly Shipham in the south-west corner) and rubbish tips. (From Hopkin *et al.* 1986.)

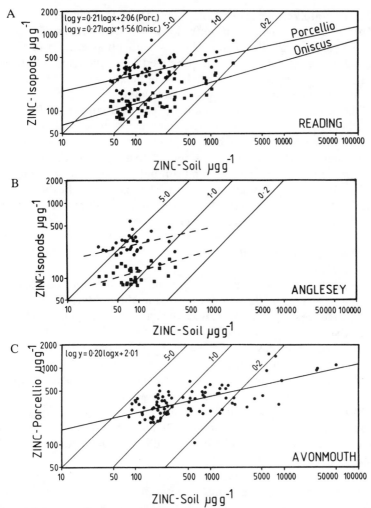

FIG. 8.2 Scatter diagrams relating concentrations of zinc in pooled samples of 12 *Porcellio scaber* (solid circles) or *Oniscus asellus* (solid squares) to concentrations in soil at the same sites in (A) a 30 km × 30 km area centred on Reading, south-east England, (B) the whole island of Anglesey, north Wales, and (C) for *Porcellio* only in the area around Avonmouth shown in Fig. 8.1. Note that despite the large scatter of points, the regression equations for *Porcellio* in A and C are almost identical (regression lines from A superimposed as dotted lines on B). Note also that the concentration of zinc in *Porcellio* is consistently greater than that in *Oniscus*. (A, C, from Hopkin *et al.* in press, b, by permission of the University of Florence; B, previously unpublished).

predict accurately the level of zinc in a pooled sample of twelve specimens of *Porcellio scaber* from the concentration in soil at the same site. Graphs displaying an even greater scatter of points were obtained for cadmium, lead and copper (similar plots were obtained for comparisons of metals in leaf litter and isopods, Fig. 7.8). Thus, although the overall relationships for each metal are fairly constant over large geographic areas (Fig. 8.2), the availability of zinc, cadmium, lead and copper to *Porcellio scaber* in *individual* sites, can be assessed only by direct analysis of the isopods themselves. In other words, we can predict from Fig. 7.8 that cadmium and copper are always assimilated to a much greater extent by *Porcellio scaber* from leaf litter than lead, but we would need to analyse the animals to know by exactly how much. The reasons for the differences in availability of metals to terrestrial invertebrates at different sites have been discussed extensively in Chapter 6.

8.2.2 Practical Considerations

8.2.2.1 Introduction

Having established the need to monitor the concentrations of metals in the biota to determine the extent of contamination of terrestrial ecosystems by a source of pollution, we need to consider the ways in which such a scheme should be conducted. Three factors need to be considered. First, the species of organisms to be sampled, second, the time interval between sampling occasions and third, the distance between sampling sites.

In an ideal world of unlimited funds, manpower and analytical equipment, all the species of plants and animals would be sampled regularly at short time intervals from closely-spaced sites within a large area surrounding the source of pollution. Abundance and diversity of each species at all the sites would be monitored by manual collection, pitfall trapping and extraction from leaf litter and soil (see Section 4.2.2). Needless to say, such a programme would be difficult to establish because of the large cost. A compromise scheme has to be developed which includes a 'representative suite' of species. The time interval and distance between sites would depend on the manpower available but should still allow seasonal variations in the distribution of metals between species to be separated reliably from changes due to an increase or decrease in emissions of metals from the source of pollution.

The following discussion refers exclusively to the use of invertebrates as monitors. The factors which should be considered when deciding on which plants to use in a monitoring programme have been extensively covered by Martin & Coughtrey (1982) and Burton (1986).

8.2.2.2 Invertebrate species

Biological indicating and monitoring is an established method for assessing the extent of metal pollution of marine ecosystems. Practical examples include the 'Mussel Watch' scheme (Davies & Pirie 1980) and monitoring the effects of tin-based anti-fouling paints on marine organisms by its effects on littoral populations of the dog whelk *Nucella lapillus* (Bryan *et al.* 1986a). 'Suites' of invertebrate species are now recommended for monitoring programmes (Phillips 1977, 1980; Bryan *et al.* 1985). However, in the terrestrial environment, apart from suggestions in published papers that particular invertebrate groups might be 'good monitors', the composition of a similar suite of organisms has not been suggested.

Before describing the features which should be possessed by an organism before it can be considered as part of a monitoring scheme, it is important to recognise that considerable differences occur in the concentrations of metals between individuals of the same species within a site. Hopkin *et al.* (1986) settled on 12 as being the minimum number of *Porcellio scaber* on which the mean value for the concentration of a metal in the whole population could be assessed reliably. Individual variation in other species may be greater or smaller than in *Porcellio scaber* requiring the collection of correspondingly more or fewer individuals.

Many terrestrial invertebrates are unsuitable monitors of metal pollution as part of a regular sampling programme. In order to be considered, a species should conform to the following criteria:

(1) The animal should be common in urban and rural sites during at least part of the year, and be easy to collect and identify (preferably in the field).
(2) The animal should contain concentrations of metal(s) which are related to dietary exposure.
(3) The animal should be large enough to allow the concentrations of the relevant metal(s) to be determined accurately in its tissues by a rapid, routine technique.
(4) The animal should be geographically widespread (relative to other species) to allow direct comparison of contamination between widely-spaced areas.

(5) The animal should not migrate over long distances (this excludes all flying insects).
(6) The animal should not be killed by moderate pollution (unless the survey includes population analyses when sensitivity to contamination would be an attribute).
(7) The biology of the animal should be reasonably well understood (i.e. basic feeding habits, fecundity, seasonal abundance etc.) and there should be some information available from laboratory experiments on its response to the metal(s) being monitored.

In practice, species which conform to these criteria are unlikely to be found at all sites on every sampling occasion throughout the year. A common opilionid, for example, would be a useful component of a monitoring scheme but in the U.K., adults of most species are present only during the late summer and autumn. Thus, the final list of monitoring organisms may include twelve species, but perhaps only six of these might be collected from any one site on a specific date.

8.2.2.3 Frequency of sampling

Monthly collections would be sufficient to detect seasonal variations in the concentrations of metals in invertebrates in a contaminated site. However, if sampling was being conducted with this regularity from several sites, the effort involved in analysing the samples would be immense. A compromise might have to be reached so that the majority of sites were visited say every three months, with a few sites being sampled on a monthly basis.

In an aerially contaminated site, a sudden reduction in the input of pollutants may not lead to a rapid decrease in concentrations of metals in the invertebrates. The diet of many species would remain contaminated for some time with metals derived from previous deposition. Thus, the monitoring programme would need to be conducted for several years to allow genuine reductions in levels in the animals to be distinguished from seasonal variations.

8.2.2.4 Area of survey and distance between sampling sites

The area to be monitored should include the source of pollution (if this is known) and should extend beyond the region within which greater concentrations of metals can be detected in the soil and biota relative to 'background levels'.

The distance between sites is subject to the same limitations as frequency of sampling i.e. available manpower. In most studies, the distance travelled by metal pollutants from a point source, and their effects on the biota, have been indicated or monitored in sites along a line (Bengtsson 1986; Hunter *et al.* 1987a, 1987b, 1987c). Such a transect should extend away from a site adjacent to the source of emissions to a site in an uncontaminated area. Ideally, several points along the transect should be monitored, but the number and distance between sites will depend on available manpower and the distance over which the pollutants travel (see Section 5.4.2).

Monitoring sites on a grid pattern is much more informative as it provides information on the regional distribution of metal pollutants. Such an approach using elm leaves (Little & Martin 1972), bags of *Sphagnum* moss (Little & Martin 1974; Gill *et al.* 1975) and soil, leaf litter and isopods (Fig. 8.1) showed clearly that the Avonmouth area was the major source of contamination of the County of Avon with zinc, cadmium, lead and copper. However, the resolution of such a grid will determine whether particular sources of metal pollution are detected. For example, a grid of 5 km was not sufficient to detect a localised source of zinc pollution in Reading, S.E. England, but a resolution of 0.5 km in the town showed that soil and isopods were heavily contaminated with the metal in an area of a few km radius around a galvanising factory (Hopkin *et al.* in press, b).

Monitoring pollution on an area basis also allows a 'surface' to be plotted which describes concentrations of a metal in a particular component of the biota (Fig. 8.3). If the density of the species is known, then the fluxes of the metal through specific components of the contaminated area can be assessed.

8.3 SPECIFIC EXAMPLES

8.3.1 Introduction

All publications which report concentrations or effects of metals on invertebrates collected from the field are relevant to any discussion on biological indicating and monitoring. However, since this literature has already been reviewed in detail in Chapter 7, the only papers referred to in this section are those which have examined specifically the potential of invertebrates as monitoring organisms.

The environmental impact of metal pollution can be assessed by either

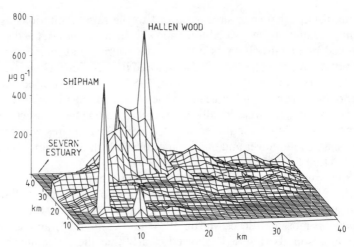

FIG. 8.3 Three-dimensional plot of concentrations of cadmium in the hepatopancreas of *Porcellio scaber* looking in a north-north-easterly direction (compare with the two-dimensional representation of these data in Fig. 8.1A). The highest concentrations of cadmium were in woodlice from Hallen Wood (797 $\mu g\ g^{-1}$) near to the smelting works at Avonmouth, and at Shipham (699 $\mu g\ g^{-1}$), a disused zinc mine which is heavily contaminated with cadmium. Concentrations of cadmium were above background levels over a wide area which extended for 25 km to the east of the smelting works. (From Hopkin *et al.* 1986.)

measuring concentrations in 'indicator organisms' (Section 8.3.2), or by comparing the abundance and diversity of invertebrates in contaminated ecosystems with similar but uncontaminated sites (Section 8.3.3). Invertebrates have been used also to a very limited extent to indicate metalliferous ore deposits (Section 8.3.4).

The use of plants as monitors of metal pollution is not considered here. The topic has been extensively reviewed by Martin & Coughtrey (1982) and more recently by Gailey & Lloyd (1986a, 1986b, 1986c) and Burton (1986) and these publications should be consulted for further information.

8.3.2 Concentrations of Metals in Indicator Organisms
8.3.2.1 Molluscs
Slugs and snails are common in most terrestrial ecosystems and there are several species which fit the criteria outlined in Section 8.2.2.1. Coughtrey & Martin (1976) showed that the concentrations of zinc, cadmium, lead and copper in the snail *Helix aspersa*, were greatest in

sites close to the Avonmouth smelting works (see Section 5.3.2) and that there was a linear relationship between dry weight and metal content of individuals in each population (Coughtrey & Martin 1977a). *Helix pomatia* has been considered also (Meincke & Schaller 1974) but this species is too rare to be of practical use as a monitor in most European countries. Slugs are ideal (Popham & d'Auria 1980), providing that care is taken in identification. Some species can be identified with certainty only by dissection.

8.3.2.2 Earthworms

The use of earthworms as biological monitors of metal pollution has been discussed recently by Bouché (1984) and Ma (1987). They would appear to be excellent candidates as there are few soils which do not contain at least one species of earthworm, the exceptions being heavily polluted and/or acidic environments. In addition, earthworms do not migrate over long distances (Helmke *et al.* 1979). Worms are particularly useful in intensively-farmed agricultural soils where they may be the only macroinvertebrates present.

Apart from care in identification, the other major consideration is the contribution of gut contents to total metal concentrations. If one wishes to separate the contents of the lumen of the digestive tract from assimilated metals in the tissues completely, then dissection and rinsing of the gut is essential. Worms fed on filter paper, or starved, do not necessarily void all of this material.

In the field, most species can be collected by digging to a depth of 30 cm or so with a fork and hand-sorting the soil. The use of formalin is probably most effective but there are two drawbacks to this technique. First, it can take more than 30 minutes for a representative sample of worms to come to the surface. Second, if several sites are to be visited consecutively, the weight of liquid which needs to be carried is prohibitive.

The most common and widely-used species have been *Lumbricus terrestris* (Czarnowska & Jopkiewicz 1978), and the acid-tolerant *Lumbricus rubellus* (Carter *et al.* 1980; Kruse & Barrett 1985). A range of species were compared by Van Hook (1974). Numerous other authors have examined metal concentrations in field populations and this work is reviewed in Section 7.3.3.3.

8.3.2.3 Isopods

The remarkable ability of terrestrial isopods to accumulate metals, and their abundance in most terrestrial ecosystems, has encouraged several researchers to assess their potential as indicators of pollution. Wieser *et al* (1976) discovered an almost perfect correlation between mean concentrations of copper in *Oniscus asellus* and leaf litter in a wide range of sites with a concentration factor of five. A similar concentration factor was found for cadmium in *Oniscus asellus* by Coughtrey *et al*. (1977), who suggested that the slopes of regression lines of metal content against dry weight may be better indicators of contamination than mean levels in the animals.

Hopkin *et al*. (1986) did not find as close a correlation between copper levels in *Porcellio scaber* and leaf litter (Fig. 7.8) as Wieser *et al*. (1976). Furthermore, the concentration factor was much greater when the levels of copper were low in leaf litter (over 100 in a few sites) than when they were high (Fig. 7.8). For zinc, lead and copper, there did not appear to be any substantial advantage in using concentrations in the hepatopancreas to indicate metal pollution instead of whole animals. However, for cadmium, the hepatopancreas showed up as being 'contaminated' over a much wider area than would have been deduced from levels in whole isopods. Dissection may be important therefore where the transfer of metals to predators which eat only soft tissues is being assessed. Small increases in metal levels in the hepatopancreas may be detected also which would not have been apparent in whole animals.

8.3.2.4 Insects

Considering the wealth of information available on the dynamics of metals in insects (Section 7.5), it is surprising that hardly any authors have examined their potential for biological indicating and monitoring of pollution. There are two main reasons for this lack of interest. First, most species which are large enough to analyse are not 'good' accumulators of metals and second, many species are highly seasonal and are common only for short periods of the year. Of course, low assimilation of metal pollutants is of interest in itself and several species of insects should be included in any suite of monitoring organisms, even if they are common for only part of the year.

Insects which have been analysed for their ability to indicate metal pollution include ants (Bengtsson & Rundgren 1984), bees (Bromenshenk *et al*. 1985) and their honey (Jones 1987), and flies (Nuorteva & Häsänen 1972; Koirtyohann *et al*. 1976). Toshkov *et al*. (1974) reported that pollen collected by bees from the vicinity of a copper smelting works contained 788 μg g^{-1} of the metal whereas pollen from control sites had only 50 μg g^{-1} of copper.

More research is required on the seasonal variations of metal concentrations in common insect groups in uncontaminated and contaminated sites. These should include earwigs (Dermaptera), beetles (larvae and adults), dipteran fly larvae, lepidopteran larvae, ants and (in the tropics) termites.

8.3.2.5 Myriapods

Most sites contain several species of centipedes and millipedes, some of which may be extremely common. In the U.K., the centipedes *Lithobius forficatus* and *Lithobius variegatus* are frequently abundant in urban and rural sites respectively and would be good 'representatives' of the carnivorous trophic level (Hopkin & Martin 1983). Lithobiids seem to be particularly common in disused mine sites. Several of the geophilids such as *Haplophilus subterraneus* and *Necrophloeophagus longicornis* are also common although more experience is required in identifying this group.

Millipedes only assimilate a small proportion of the metals which pass through their digestive system (Hopkin *et al*. 1985a). However, inclusion of millipedes in a monitoring scheme would be of interest because they represent the same trophic level as isopods which accumulate metals to a much greater extent, apparently from the same leaf litter diet. In the U.K., millipedes which are widespread and common include *Glomeris marginata*, *Tachypodoiulus niger* and *Cylindroiulus punctatus*.

8.3.2.6 Arachnids

Clausen (1984a, 1984b, 1986) is the only author to have examined directly the use of spiders as biological indicators of metal pollution. Several species were collected from contaminated and uncontaminated sites and the concentrations of lead were compared with levels in lichens. His work highlighted the importance of separating invertebrates into species before analysis rather than pooling taxonomic groups. Substantial differences were found between the concentrations of lead in different species of spiders collected from the same site. The importance of surface contamination was also stressed. Some 25% of the lead 'in' spiders from the most contaminated sites could be washed off. Thus, the possible contribution of such unassimilated metals to whole body concentrations should be considered carefully, particularly in arthropods with a 'hairy' exoskeleton.

Spiders are extremely common in terrestrial ecosystems, but most of the large species are seasonal in abundance in temperate climates. The ground-dwelling lycosids, agelenids and cribellates are present all year round and may be more suitable for comparison of contamination on a monthly basis whereas spiders which are adult during the summer and autumn only, would be more useful for year to year monitoring.

8.3.3 Effects of Metal Pollutants on Abundance and Diversity of Terrestrial Invertebrates

Differences in the abundance and diversity of invertebrate species between uncontaminated and contaminated sites are difficult to attribute to pollution because natural variation is so great. Such effects are only clear adjacent to major sources of metals when whole taxonomic groups may be absent. Even then, because pollutants are rarely emitted singly (e.g. metals + SO_2 + NO_x etc.), it may not be possible to 'pin the blame' for deleterious effects on a specific contaminant. Furthermore, I know of no researchers who have been able to compare the invertebrate ecology of an area before and after a source of metal pollution has been established.

Most studies have consisted of a comparison between sites on a single sampling occasion. Thus, Van Rhee (1967) showed that earthworms were absent from orchard soils which had been contaminated heavily with copper. This led to an accumulation of undecomposed leaf litter. A similar accumulation of litter in woodlands downwind of the Avonmouth smelting works was attributed to a lack of earthworms and millipedes by Hopkin *et al.* (1985a; Table 5.3) and Bengtsson & Rundgren (1982) showed that the abundance and diversity of enchytraeid worms in coniferous forest soils decreased with proximity to a brass mill in Sweden (Table 5.5).

Read (1988) is the only author to have *monitored* the population structure of invertebrates at the species level in woodland sites subject to different degrees of aerial metal contamination. Using pitfall traps emptied on a fortnightly basis, she detected differences between the sites but these were difficult to assign directly to levels of pollution arising from aerial fallout from the Avonmouth smelting works. However, maturation of certain species of carabid beetles appeared to be delayed in the most contaminated woodland due possibly to a reduction in available prey (Read *et al.* 1987).

Regular monitoring of invertebrate populations is clearly desirable for a complete understanding of the effects of metal pollutants on terrestrial ecosystems. However, such an approach is extremely labour-intensive and is unlikely to attract sufficient funding to enable a comprehensive survey to be carried out in the forseeable future.

8.3.4 Indication of Natural Ore Deposits by Terrestrial Invertebrates

Biogeochemical prospecting by measuring concentrations of metals in plants is a well-established method for locating ore deposits (Martin & Coughtrey 1982) but very few authors have attempted this with animals. Termites can be useful indirectly as their tunnelling activity may bring soil from considerable depths to the surface (Prasad & Dunn 1987). In Zimbabwe, for example, sub-surface deposits of zinc have been located after detection of anomalous levels of the metal in termite mounds. Burrows have been observed down to 4.7 m in the walls of disused mine shafts (Watson 1970), and some species may tunnel to a depth of 30 m in search of the water table (D'Orey 1975).

A correlation between concentrations of a metal in a particular species and its natural occurrence in soils and rocks has been much more difficult to substantiate. The suggestion that gold-bearing rocks can be located with reference to levels of the metal in cockchafers (Babička *et al.* 1945) or bees (Razin & Rozhkov 1966) is an attractive idea for industrial sponsorship but is unlikely to be of any practical use. Similarly, attempts to localise the regional source of flying insects based on their 'chemical fingerprints' has met with little success in aphids (Bowden *et al.* 1985a), hornflies (Chamberlain 1977), noctuid moths (Zhulidov *et al.* 1982; Bowden *et al.* 1984; Sherlock *et al.* 1985) and mirid bugs (Cohen *et al.* 1985).

8.4 RECOMMENDATIONS AND CONCLUSIONS

8.4.1 Justification

The most important question we should ask before embarking on biological monitoring of metal pollution is 'will analysis of plants and animals provide us with information on the level of contamination of the biota which could not have been predicted from abiotic sampling?' The factors discussed in this section demonstrate clearly that the answer is 'yes!'

Research to date has shown that the levels of metals in air can be monitored successfully by analysis of materials which have been exposed for short periods at different distances from the source of pollution. Such materials include filters through which air is drawn by suction, and passive collectors of metal-containing particles such as sticky traps (Gailey & Lloyd 1986a, 1986b, 1986c) and moss bags (Little & Martin 1974). These methods 'integrate' pollutant levels over the period of exposure. Concentrations of metals in soils can be monitored by regular collection of samples, a relatively easy task.

The chemical speciation and rates of uptake of different metals by plants, and the routes by which pollutants are assimilated by animals are, however, poorly understood. The extent to which different metals are assimilated from the diet is so variable in different species that only very broad conclusions can be drawn. We could predict, for example, that an increase in the release of cadmium into the environment would be more harmful than release of a similar amount of lead because concentration factors for this metal are consistently less than those for cadmium. However, our knowledge on the assimilation of cadmium and lead in the terrestrial environment is limited to a small number of species of plants and animals and there are many elements (e.g. arsenic, molybdenum and almost all radioactive metals) about which almost nothing is known at all. Colborn (1982), for example, showed that increased levels of molybdenum could be detected in aquatic insects when the concentrations in the water were below the detection limit of his analytical equipment. Thus, contamination of the biota could not have been predicted from abiotic monitoring alone.

With our present state of knowledge, it would not be possible to accurately predict the ecological consequences of, say, a doubling of aerial emissions of metals from a smelting works, application of metal-rich sewage sludge to an agricultural soil, or the long term effects of the dumping of metal-rich mining waste. There is no substitute for measuring the concentrations of metal pollutants directly in carefully chosen representatives of the main trophic levels in terrestrial ecosystems.

8.4.2 Developing a Monitoring Scheme

In this section, suggestions are made on the ways in which a monitoring scheme could be conducted in a temperate region where sites are easily accessible by road. The detailed methodology of such a survey would clearly have to be modified to take account of local conditions. However, broad adoption of these recommendations would result in future surveys being much more comparable, even those carried out in different countries and climates.

The factor which determines the extent of a monitoring scheme is the manpower available for collection and analysis of samples. This puts limits on the amount and type of abiotic and biotic material which can be collected, the intervals of time between sampling and the number of sites which can be visited. As a rough guide, the hypothetical scheme described below in which zinc, cadmium, lead and copper would be monitored in representatives of the biota in 36 sites on a monthly basis, would require two people working full time on the project (assuming analysis was conducted by flame or flameless atomic absorption spectrometry of acid digests). Monitoring of the population dynamics of the biota in this number of sites in the U.K. would require at least one extra person, and probably more in warmer climates where species diversity is much greater.

Suppose the area to be surveyed contained a major source of aerial metal contamination and that the brief of the project was to assess the environmental impact of these emissions on surrounding terrestrial ecosystems. The first consideration would be the number of sites to be monitored. A six by six grid arranged in a square (i.e. 36 sites) would be sufficient to assess the extent of regional contamination.

The distance between sites would depend on the amount of pollution emitted by the works. At Avonmouth (Figs. 8.1, 8.3), sites would have to be at least 5 km apart (i.e. a 25 km by 25 km grid) because the pollution extends for more than 20 km from the smelting works. If the area around the Liverpool copper refinery were to be monitored in sites arranged in the same pattern, the distance between points would need to be only 1 km (or less) since the emissions are carried for a much shorter distance (Fig. 5.16). Similar considerations would apply to disused mining areas where contamination may be very local. The specific location of sites might have to deviate from the grid to take account of access and suitability for collecting invertebrates. In addition, some extra sites adjacent to the pollution source might need to be included so that the most heavily contaminated area could be monitored more closely in the survey. This area is where gross effects of metal contamination (absence of major taxonomic groups, litter accumulation) will be most obvious. Ideally, the sites should be visited on a monthly basis.

At each site, samples of surface soil, leaf litter and foliage from the most common plants should be collected, together with the 12 largest specimens of the most common invertebrates. These are likely to include slugs, isopods, Collembola, earthworms, spiders and at certain times of the year, opilionids, ground beetles and other large insects. Samples should be dried as soon as possible after allowing time for voiding of gut contents by the animals, if this is considered to be necessary. The aim should be to include species which can be found in sufficient numbers at every site from which a 'suite' of organisms can be chosen.

Monitoring levels of metals in vertebrates would be useful but time-consuming as each site would have to be visited more than once for setting of traps. Perhaps a small number of sites could be selected in which more intensive sampling was carried out. Monitoring metal concentrations in the air would be useful also, and this could be done by the use of moss bags suspended a metre or so above ground. Electrically-driven air samplers are expensive, prone to vandalism and require a source of mains electricity if they are to be run for an extended period.

The data collected from such a survey would provide a sequence of 'snapshot pictures' of the distribution of metals throughout the biota in each site. If enough information was available on the population dynamics, feeding behaviour and rates of predation on the different species, it may be possible to calculate fluxes of metals between trophic levels and to highlight the most important pathways of transport in the food web. Only then would it be possible to begin to develop models for predicting the effects of metals on terrestrial ecosystems *before* they were released into the environment.

8.4.3 Conclusions
Biological monitoring of environmental metal contamination is, I believe, the least-developed of the most interesting research areas. It is clear that with our current state of knowledge, the concentration of a metal in abiotic samples is not a good indicator of its 'biological availability'. Put simply, we do not fully understand what happens to metals when they are released into terrestrial ecosystems. We need to know!

In this chapter, I have been deliberately speculative and many of the suggestions are open to criticism. However, I hope that the ideas will provoke environmental scientists into taking a closer look at invert-ebrates when they are designing monitoring schemes.

CHAPTER 9

Metals in Terrestrial Invertebrates at the Cellular Level

9.1 INTRODUCTION

The most important regulatory organs for metals in terrestrial invertebrates are those associated with the digestive system (see Chapter 6). The epithelia of these tissues, which are often only one cell in thickness, provide the only barrier between the food in the lumen of the digestive system and the internal environment of the animal. Therefore, storage mechanisms and/or methods of exclusion have to be extremely efficient because terrestrial invertebrates (unlike aquatic organisms) are not able to excrete elements from the blood into the external medium across the respiratory surfaces if they are taken up to excess. These mechanisms, which include metal-binding proteins and metal-containing granules, are the subject of this chapter.

The earliest attempts at determining the distribution of metals in the tissues of terrestrial invertebrates employed histochemical staining of sections for light microscopy. Unfortunately, although these studies were in most cases executed meticulously, many of the stains used to demonstrate the presence of particular metals have been shown subsequently to bind to a range of elements. Nevertheless, these early observations with the light microscope did provide clues as to which tissues it would be most profitable to study with more sophisticated techniques.

The introduction of the electron microscope has had as dramatic an impact on the study of the subcellular distribution of metals in terrestrial invertebrates as it has on most other branches of biology. Many intracellular accumulations of metals which appeared as featureless dots in the light microscope were discovered to have a striking internal structure (e.g. Figs. 9.6, 9.15, 9.19). Over the last 15 years, X-ray microanalysis has become widely available and it is now possible for biologists to examine simultaneously, the ultrastructure and composition of tissues in the electron microscope. Indeed, the classification system for metal-containing granules proposed in this chapter could not have been formulated without the information provided by this technique.

In this chapter, the factors which control the rates at which metals pass across membranes are discussed. Some examples of the main types of metal-containing proteins and granules in terrestrial invertebrates are presented together with brief descriptions of the techniques which have been employed to study them. A new classification system for metal-containing granules is proposed based on the chemistry of elements accumulated rather than on their two-dimensional appearance in thin section which has led to confusion between the different types in the past. The chapter concludes with a review of the subcellular distribution of metals in the major groups of terrestrial invertebrates.

9.2 PASSAGE OF METALS ACROSS MEMBRANES

The factors which control metal fluxes across membranes have been reviewed recently by Simkiss & Taylor (in press) on whose paper the following discussion is based.

The standard model for the structure of a plasma membrane is a lipid bilayer in which proteins are embedded. Until recently it was thought that the only routes which metals could take through membranes were as ions through channels, or *via* pinocytotic vesicles. In aqueous solutions, ions become surrounded by water molecules. The energy required to get through the membrane is that needed to strip off the water or expand the channel. For example, there is evidence that Mn^{2+} can pass through the calcium channel of the myoepithelial cells of marine polychaete worms (Anderson 1979). Of the transition metals tested, only Mn^{2+} had a sufficiently low enthalpy of hydration to be able to shed its water of hydration and pass through the channel. The other route is by attachment of metals to coated pits in the membrane followed by pinocytosis. Thus, iron is transported from the surface of each mammalian liver cell at 10^5 transferrin binding sites which are recycled back to the surface every 4 min (Morgan & Baker 1986).

Recently, Simkiss (1983a) has shown that class B metals such as copper, zinc, cadmium and mercury form complexes in saline solutions that are very lipid soluble. Thus, as well as the two routes of metal uptake into cells described above, a third pathway, that of direct passage through the membrane, may exist also.

The passage of major metals into particular cells is controlled by the properties of the membrane exposed to the solution in which the ions are dissolved (Green *et al*. 1980). Thus, sodium-transporting membranes contain a high concentration of sodium channels while iron-accumulating cells possess a large number of transferrin receptor molecules at their surface. Intracellularly, these metals are integrated into precise metabolic pathways and storage proteins (Simkiss & Mason 1984; Ettinger *et al*. 1986; Pattison & Cousins 1986). Control of passage of many of the class B metal ions, particularly those which do not have any biological function (e.g. mercury, cadmium), may not be possible if the elements pass through the membrane in an unrestricted fashion. This could have been the stimulus for the evolution of detoxification systems for surplus metals which are stored until such time as they can be excreted.

9.3 METAL-CONTAINING PROTEINS

9.3.1 Analytical Techniques
The simplest technique for determining the subcellular distribution of metals in the organs of terrestrial invertebrates is to homogenise the tissue, separate the cellular debris from the cytosol by centrifugation and determine the concentrations of metals in the fractions by an analytical method such as atomic absorption spectrometry (Section 4.4.2). Although somewhat crude, this simple approach can provide useful information. For example, when the hepatopancreas of the garden snail *Helix aspersa* was treated in this way, the largest proportion of total zinc was detected in the insoluble pellet whereas most of the cadmium was associated with the soluble fraction of the homogenate (Cooke *et al*. 1979). The distribution of metals between the different types of cell debris is relatively easy to determine by X-ray microanalysis in an electron microscope (Section 9.4.2.3). The distribution of metals in the soluble fraction is more difficult to elucidate and involves the separation of proteins of different molecular weights.

The most widely used technique for the separation of proteins of different molecular weights is gel-permeation chromatography. The supernatant from a centrifuged sample of homogenised tissue is mixed with a buffer and applied to the top of a column of gel, typically Sephadex G-75 (Pharmacia Chemicals). As the solute passes down the gel, its movement depends on the bulk flow of the mobile phase and the Brownian motion of the solute molecules which cause their diffusion both into and out of the stationary phase. The separation in gel-permeation chromatography depends on the different abilities of the various sample molecules to enter pores which contain the stationary phase. Large molecules which never enter the stationary phase move through the gel fastest. Smaller molecules, which can enter the gel pores, move more slowly through the column since they spend a proportion of their time in the stationary phase. Molecules are eluted therefore in order of decreasing molecular size.

The absorbance of ultraviolet light of specific wavelengths is recorded as the proteins leave the base of the column and are collected in a sequence of fractions by an automatic sampler. Prior calibration of the column with reference proteins of known molecular weights, enables the molecular weights of proteins of interest in particular fractions to be calculated. The concentrations of metals or activity of isotopes can be determined in the samples by a range of chemical and physical techniques such as flame and flameless atomic absorption spectrometry, neutron activation analysis, liquid scintillation counting and gamma spectrometry. The two sets of results are compared and the molecular weights of proteins which bind particular metals can be determined (Fig. 9.1).

A comprehensive review of all the metal-containing proteins which are likely to occur in terrestrial invertebrates is beyond the scope of this book. There are, for example, more than 150 enzymes which require zinc as an essential part of their structure. Brief descriptions of some metal-containing proteins which are of particular importance to terrestrial invertebrates are given below to indicate the wide range of processes in which they are involved. Numerous reviews have been published on the topic. The texts edited by Harrison (1985) provide a good starting point.

Metal-containing proteins can be split into two groups, those which exist specifically to sequester metals (e.g. metallothionein, ferritin) and those which require metal ions as an essential part of their structure (e.g. haemocyanin, haemoglobin, carbonic anhydrase).

9.3.2 Metallothionein
Metallothioneins are nonenzymatic low molecular weight ($c.$ 10^4

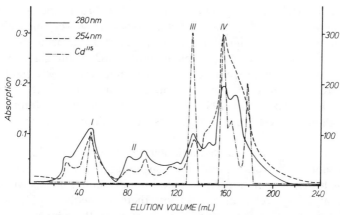

FIG 9.1 Typical elution profile of the cytoplasmic fraction of the stonefly *Eusthenia spectabilis* providing evidence for the existence of a cadmium-binding protein. ^{115}Cd incorporated *in vivo* is associated almost exclusively with a low molecular weight fraction (IV) absorbing strongly at 254 nm. The ^{115}Cd peak (III) probably results from aggregation of low molecular weight proteins causing them to elute with an apparently high molecular weight. Fraction IV has a high E254/E280 ratio (1.39) indicative of metallothioneins, whilst all other fractions absorbed more strongly at 280 nm. Identical extracts from control animals showed no evidence of a similar low molecular weight fraction. The molecular weight of this cadmium-binding protein as determined by gel-permeation chromatography was about 10^4 Daltons. (From Everard & Swain 1983, by permission of Pergamon Journals Ltd.)

Daltons) proteins of high sulphur and metal content (Hunziker & Kägi 1985). Metallothioneins have been isolated from a wide range of organisms since their discovery in the kidney cortex of the horse (Margoshes & Vallee 1957). When highly purified, naturally-occurring vertebrate metallothionein contains 0.1% copper, 2.2% zinc, 5.9% cadmium, 0.2% iron, 16.3% nitrogen and 9.3% sulphur (Kägi & Vallee 1961). The large amount of sulphur is part of the amino acid cysteine which comprises some 30% of the weight of the protein. Metallothionein-metal complexes are formed by the interaction of three sulphydryl groups of cysteine with one atom of the class B or borderline metals mercury, cadmium, zinc, silver or copper (Cherian & Goyer 1978). Production of the protein can be induced by all these metals. For example, Maroni *et al.* (1986) demonstrated that radioactive cysteine was incorporated into *Drosophila* metallothionein following cadmium administration.

Metallothionein-like proteins have been detected in a wide range of marine invertebrates (e.g. Olafson *et al.* 1979; Frankenne *et al.* 1980; Ridlington *et al.* 1981; Armitage *et al.* 1982). However, the only terrestrial invertebrates in which these proteins have been found are molluscs (Cooke *et al.* 1979; Ireland 1981, 1982; Dallinger & Wieser 1984b), earthworms (Suzuki *et al.* 1980; Yamamura *et al.* 1981), stoneflies (Everard & Swain 1983; Fig. 9.1) and larvae of dipteran flies (Aoki & Suzuki 1984; Aoki *et al.* 1984; Maroni & Watson 1985; Maroni *et al.* 1986). Some species may not possess metallothioneins (Suzuki *et al.* 1984).

In invertebrates, metallothioneins have been found only intracellularly. Several authors have suggested that their primary function is to detoxify non-essential metals and essential metals which are assimilated by cells in excess of their immediate physiological requirements. This conjecture is supported by the observations that metal-free metallothioneins could not be detected in the hepatopancreas of the snail *Helix aspersa* (Cooke *et al.* 1979) and that the protein is synthesized only in response to a metal 'insult' rather than being stored for potential influxes of cadmium, copper, zinc, mercury or silver. The protein acts also as a homeostatic reservoir for essential metals. Zinc released from metallothionein, for example, can be incorporated into carbonic anhydrase (Hartmann & Weser 1984). Categorical evidence for a detoxificative function for invertebrate metallothioneins is still lacking. Indeed, it has been suggested that complexes of non-essential metals such as cadmium and mercury might be metabolised more slowly than essential metal complexes so that the accumulation of non-essential metals gives rise to an apparent role in detoxification (Simkiss & Mason 1983).

Further information about metallothioneins can be found in the reviews by Cherian & Goyer (1978), Webb (1979), Elinder & Nordberg (1985), Enger *et al.* (1986), Vašák (1986) and Kägi (1987).

9.3.3 Ferritin
Iron is extremely insoluble in water at physiological pH's and must be bound in the soluble protein transferrin before transport between tissues. Iron is stored intracellularly in ferritin (Harrison 1977; Crichton 1982). Ferritin consists of a shell of apoferritin protein subunits with an aggregate molecular weight of 450,000 Daltons. Within this shell, up to 4,500 atoms of iron can be stored as ferric oxyhydroxide (Simkiss & Mason 1983).

Intracellular deposits of ferritin have been observed in a number of aquatic invertebrates (see Steeves 1968; Jones *et al.* 1969; Moore & Rainbow 1984). In terrestrial invertebrates, deposits of ferritin have been unequivocally observed only in the cells of the midgut of homopteran insects (Bielenin & Weglarska 1965; Gouranton 1967; Gouranton & Folliot 1968; Cheung & Marshall 1973; Kimura *et al.* 1975) and lepidopteran larvae (Locke & Leung 1984). Since iron is an essential element for all animals, deposits of ferritin will doubtless be detected in a much wider range of terrestrial invertebrates.

Ferritin is often used as an ultrastructural marker (Smith *et al.* 1969).

9.3.4 Haemocyanin
Haemocyanins are blue, copper-containing respiratory proteins found dissolved in the haemolymph of molluscs (Sminia & Vlugt van Daalen 1977) and some groups of arthropods including isopods (Sevilla & Lagarrigue 1979; Terwilliger 1982), spiders (Foelix 1982) and centipedes (Rajulu 1969). In terrestrial molluscs, these proteins are synthesized in the pore cells of the connective tissues (Sminia & Vlugt van Daalen 1977).

The oxygen-binding site in haemocyanin is a pair of copper atoms bound directly to amino acid side chains (Volbeda & Hol 1986). In molluscs, there is one atom of copper per 25,000 Daltons of protein whereas in arthropods, the ratio is slightly higher at 1: 35000 (Bonaventura & Bonaventura 1980). In molluscs, the native haemocyanins found in the haemolymph occur generally as giant cylindrical molecules with molecular weights as large as 9,000,000 Daltons. Large haemocyanin molecules occur also in arthropods but the maximum molecular weight of these is only (!) 3,500,000 Daltons (Bannister 1977; Mangum 1980).

The integrity of haemocyanin is maintained by calcium and magnesium ions in the blood (Burton 1972b; Sevilla & Lagarrigue 1979). Most of the zinc in the blood is also associated with haemocyanin (Zatta 1984). It is vital that other divalent cations which might interfere with the stabilising activities of these metals are not 'allowed' to circulate in the body fluids. The rapid removal of injected metals from the blood into the hepatopancreas has been demonstrated in the snail *Helix aspersa* (Simkiss 1981) although in the wild, excessive amounts of metals are prevented from entering the blood by the binding activities of the epithelial cells of the digestive organs.

9.3.5 Haemoglobin
Haemoglobin consists of an iron porphyrin compound 'haem' associated with a protein 'globin'. Haem is composed of four pyrrole rings joined

with methane groups to form a super ring with an atom of ferrous iron in the centre attached to the pyrrole nitrogens. The haem component of the molecule is a constant feature of all haemoglobins but the globin portion varies in different species.

Among terrestrial invertebrates, haemoglobin occurs in the blood of earthworms and a few species of molluscs (Terwilliger & Terwilliger 1985) and insects (e.g. chironomid midges). The main difference between haemoglobin and haemocyanin (apart of course from the different metal ions which form part of their structure) is that the oxygen-binding capacity of haemoglobin is about three times greater than that of haemocyanin. This feature is of advantage to earthworms which spend most of their time underground where oxygen levels may on occasion be quite low, particularly after heavy rain.

In earthworms, haem is synthesized in the chloragogenous tissue which surrounds the intestine (MacRae & Bogorad 1958; Delkescamp 1964a, 1964b; Ireland & Fischer 1979). Histochemical tests suggested originally that the chloragosomes (Fig. 9.9) were rich in iron (Matsumoto 1960). However, more recent studies by X-ray microanalysis have failed to detect significant amounts of this metal (Fig. 9.11A) so the specific site of haem synthesis within the chloragogenous tissue has still to be found.

9.3.6 Carbonic Anhydrase
Carbonic anhydrase is a catalyst for the formation of calcium carbonate (Maren 1967) and is of particular importance in millipedes, isopods and snails which possess a calcareous exoskeleton or shell. This enzyme must contain zinc if it is to function correctly. The substitution of a 'foreign' metal ion into its structure results in a drastic reduction in the ability of carbonic anhydrase to hydrate carbon dioxide (Table 2.2). Such results emphasise the importance of the strict regulation of the movement of similar and contrasting metal ions within animals. It is clear that a mechanism must exist to maintain concentrations of zinc at the optimum for carbonic anhydrase activity while preventing similar metal ions from reaching levels which would allow active competition for zinc-binding sites in the enzyme.

9.4 METAL-CONTAINING GRANULES

9.4.1 Introduction
Studies on the digestive and excretory organs of terrestrial invertebrates

by light and electron microscopy, histochemistry and X-ray micro-analysis, have shown that these tissues may contain accumulations of material in which a wide range of metals can be detected. Several attempts have been made to classify these accumulations or 'granules', the most recent being those of Brown (1982), Taylor & Simkiss (1984) and Hopkin (1986). In the first part of this section the methods which have been used to study these granules are described, in particular, the technique of X-ray microanalysis which has been responsible for most recent significant discoveries. In the second part, a new system is proposed which classifies metal-containing granules into four types based on their structure and composition.

9.4.2 Analytical Techniques
9.4.2.1 Histochemistry
Several attempts have been made to demonstrate the presence of metals in the tissues of terrestrial invertebrates by histochemistry. This is legitimate if the stains used are metal-specific. For example, Hryniewiecka-Szyfter (1972, 1973) in his studies on the distribution of metals in the hepatopancreas of the woodlouse *Porcellio scaber*, was able to show using prussian blue to stain for iron and rubeanic acid for copper, that these two metals were contained in granules in two different cell types in the gland. Iron was shown to occur exclusively in the B cells whereas copper was found only in the S cells, a distinction confirmed subsequently by X-ray microanalysis (Hopkin & Martin 1982b).

Unfortunately, most stains are not as specific as prussian blue and rubeanic acid. For example, dithizone has been used to demonstrate that the earthworm *Eisenia foetida* from a contaminated site accumulates lead in the body wall (Wielgus-Serafinska & Kawka 1976). Dithizone, however, may be used as a histochemical stain for zinc (Pearse 1972) which is present naturally throughout the tissues of uncontaminated worms (Ireland & Richards 1977). The silver-sulphide method of Timm (1958) has also been used extensively for demonstrating the presence of 'heavy metals' in tissues (Fig. 9.17).

Now that more sophisticated analytical techniques are available, histochemistry should be reserved for preliminary studies on the gross distribution of metals within organs or whole animals. Such methods may enable relatively small areas of metal accumulation to be identified which might otherwise be overlooked (Sumi *et al.* 1984; Marigomez *et al.* 1986a, 1986b). Modern image analysers make this procedure much easier (Lytton *et al.* 1985). Such areas can then be examined in more detail by

X-ray microanalysis in an electron microscope.

Recent developments in histochemistry include 'autometallography' (Danscher 1984) and the use of immuno-labelled colloidal gold to localise metallothionein proteins (Clarkson *et al.* 1984).

9.4.2.2 Autoradiography

A powerful technique for determining the subcellular distribution of individual metals is autoradiography. A suitable radioactive isotope of the metal of interest is administered either in the food or by injection. Animals are killed at intervals after feeding and tissues are removed, fixed and sectioned. The sections are placed in contact with a photographic emulsion, often for several weeks, and the radiation given off as a result of the decay of the isotope precipitates grains of silver in the emulsion. The film is developed and the sites where the radioactive isotope occurs show up as black dots which can be localised within the cells by comparison with cells in the light microscope. Schoetti & Seiler (1970), for example, were able to show by autoradiography that ^{65}Zn ingested by the slug *Arion rufus* was stored in the basophil cells of the hepatopancreas.

There are a number of advantages and disadvantages associated with the use of autoradiographic techniques for determining the subcellular distribution of metals in individual tissues. The main disadvantage is that several metals of environmental and physiological interest (e.g. ^{64}Cu) have radioisotopes of a very short half life. The method can also be difficult to use in a semi-quantitative or quantitative fashion and the potential for multi-element studies is limited. Nevertheless, autoradiography can be extremely useful when undertaken in conjunction with studies on the uptake, dynamics and retention of radioactive metals at the whole animal and organ levels (see Section 4.6).

9.4.2.3 Electron microscopy and X-ray microanalysis

Electron microscopy has revealed that metal-containing granules in terrestrial invertebrates, which may appear as featureless dots in the light microscope, are often of striking appearance in thin section. Histochemical tests have been carried out on such material to attempt to demonstrate the presence of particular metals but the stains used in most cases are as non-specific as those used in light microscopy. In recent years X-ray microanalysis has transformed the study of the subcellular distribution of metals in animals. Indeed it is unlikely that this area of research would have developed at all without the widespread commercial

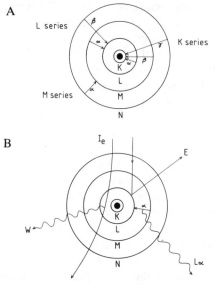

FIG. 9.2 Schematic diagrams to show production of X-rays by atoms in the path
of an electron beam. (A) Orbital transitions which produce X-rays usually
detected by energy-dispersive X-ray microanalysis. (B) Energy from the electron
beam is transferred to an electron which is excited and ejected from the atom
(E). An electron in the M shell fills the space in the L shell. During the transition,
the energy lost (α = the potential difference between the two shells) is emitted
as an X-ray photon (Lα). Non-specific 'background' or 'white' radiation (W)
is produced during non-specific interactions between incident electrons (I_e) and
the atom. (After several authors.)

availability of this technique since the mid-1970s.

Modern X-ray microanalysers are highly sophisticated instruments
with the capability to analyse thin sections with a spacial resolution of
20 to 30 nm and to detect as little as one attogram (10^{-18} g) of an element
(Hall 1979). However, the methods of specimen preparation are still
some years behind the development of the analytical hardware, the main
problem being the retention of elements in their *in vivo* positions while
preserving enough contrast for ultrastructural features to be resolved.

To perform X-ray microanalysis, a narrow beam of electrons is focussed on to the specimen in an electron microscope. The atoms of the
specimen are excited to such a degree that electrons are ejected from
their orbital shells (Fig. 9.2). The spaces created are filled by electrons
falling in from outer orbits towards the nucleus. During this process,

these electrons lose energy which is emitted as X-ray photons. Different elements give off X-rays of different energies so by 'collecting' X-ray photons emitted over a period of time (typically 100 seconds) and comparing the energies with those in standard tables, the presence of particular elements in the specimen can be elucidated.

The two most widely-used techniques for the detection of X-rays are wavelength-dispersive and energy-dispersive X-ray microanalysis. In wavelength-dispersive X-ray microanalysis, the wavelengths of the X-rays are calculated from the angle through which they are refracted as they pass through a crystal. A detector is positioned at specific points in the spectrum of X-rays and the number of X-ray photons of a specific energy are counted. The resolution of this technique is very good but it has the disadvantage of being able to examine only a small part of the X-ray spectrum representing one element at a time.

In energy-dispersive X-ray microanalysis, the X-ray photons enter a silicon lithium-drifted crystal which acts as a semiconductor producing electronic pulses, the voltages of which are proportional to the energy of the X-rays which produced them. The pulses are sorted by an analyser into memory channels which each have a range, typically, of 50 electron volts (eV). X-rays are collected for about 100 seconds and the accumulated 'counts' in each memory channel displayed as a histogram with X-ray energy divided into channels on the horizontal scale, and number of X-ray counts on the vertical scale. The final X-ray spectrum consists of peaks superimposed on a background of non-characteristic radiation produced from interactions between electrons in the vicinity of the nuclei of atoms within the specimen. The elements from which the peaks are derived can be determined from reference tables of X-ray energies. Most of the elements of interest to biologists emit X-rays in the range one to 25 kilo electron volts (keV). Some typical X-ray spectra are shown in Figs. 9.6B, 9.11 and 9.14B, C.

The main disadvantage of energy-dispersive X-ray microanalysis when compared with the wavelength-dispersive method is its poorer resolution. Peaks from some elements which are close together can not be resolved due to the statistical spread of X-rays about the mean. For example, two of the X-ray peaks for lead at 2342 eV ($M\alpha$) and 2442 eV ($M\beta$) interfere with the only peaks for sulphur at 2307 eV ($K\alpha_1$), 2322 eV ($K\alpha_4$) and 2465 eV ($K\beta$) so the presence or absence of sulphur from specimens containing lead is difficult to confirm. However, this disadvantage is outweighed by the ability of energy-dispersive X-ray microanalysis to display simultaneously the whole X-ray spectrum enabling a rapid determination of the complete elemental composition of the specimen to be made in a very short time.

The technique can also be used in conjunction with a scanning transmission detector (STEM) to show the distribution of a single element over a large area of the specimen. A 'window' is delimited to include an X-ray peak of the element of interest. When an X-ray with an energy within this range enters the detector, a dot appears on the screen in the exact position from which it was emitted. After a short period of time, the individual dots form an 'elemental map' which can be compared with an electron micrograph of the same area (Fig. 9.8). Recent developments in computer software enable different elements to be displayed in different colours and for an almost limitless array of analytical programmes to be applied to the data.

Almost all the studies on the subcellular distribution of metals in terrestrial invertebrates have been performed with the energy-dispersive method because of its ease of use and greater commercial availability. Consequently, the discussion which follows is concerned mostly with this type of X-ray microanalysis. Further details can be found in the reviews by Chandler (1977, 1978), Russ (1978), Duncumb (1979), Hall (1979), Hayat (1980), Moreton (1981), Hall & Gupta (1984) and Revel *et al.* (1984). For beginners to the technique, the excellent introductory book by Morgan (1985) is recommended.

The ideal thin section for X-ray microanalysis would retain all the elements in their *in vivo* positions, be perfectly preserved and exhibit excellent contrast in the electron microscope. However at present, these are mutually-exclusive ideals. The typical 'conventional' procedure for the preparation of sections for transmission electron microscopy involves fixation in gluteraldehyde, staining in osmium tetroxide, dehydration in alcohols, embedding in resin, sectioning on to water and staining with aqueous solutions of uranium and lead salts. It has been shown in several studies that these procedures cause massive disruption of intracellular ionic gradients and result in the loss of all diffusible elements from the tissues (reviewed by Morgan 1980). Consequently, almost all studies by X-ray microanalysis on the subcellular distribution of metals in terrestrial invertebrates have analysed metal-containing granules from which, it is assumed, there is no extraction of elements during preparation.

There is evidence, however, that material can be extracted from these inclusions during sectioning. More than 50% of the calcium and phosphorus in the calcium phosphate granules in the hepatopancreas of the shore crab *Carcinus maenus* diffuses into the water of the knife boat of the ultramicrotome within a few minutes of sectioning (Hopkin 1980). Furthermore, there can be differential extraction of elements so that the ratios in the granules change depending on the time between sectioning and removal from the knife bath. The removal of elements from the calcium phosphate granules is due probably to resin not penetrating their structure. Under normal circumstances, the granules are highly insoluble but when sectioned, the internal layers are exposed to water and material can leach out. Such extraction during sectioning has led some authors to conclude (probably erroneously) that these calcium phosphate granules are highly soluble *in vivo* (Meyran *et al.* 1986).

In contrast, I have found that there is no loss of elements from the 'copper' granules in the hepatopancreas of isopods when these are left to float in the knife bath for as long as 30 min. These observations emphasise that papers in which quantification of X-ray microanalysis has been attempted without due regard to possible diffusion and redistribution effects should be treated with caution (Morgan 1979).

'Conventionally' prepared material is sectioned and analysed usually without staining with osmium, uranium or lead salts because the X-ray peaks derived from these elements may interfere with those from elements of interest in the tissues. Such unstained material lacks contrast, although this can be enhanced electronically by the use of a scanning transmission unit (STEM). Most analyses carried out on tissues prepared in this way can provide information only on the qualitative composition of metal-containing granules because of the extraction and redistribution effects described above (Morgan 1979). It is now generally accepted that the only legitimate technique for the analysis of diffusible elements in biological tissues is to freeze and section at very low temperature and analyse tissue on a cold stage in the electron microscope, preferably without freeze-drying (Gupta & Hall 1981; Barnard 1982; Hall & Gupta 1982). Such sections are extremely difficult to prepare and very few laboratories have been able to master the techniques involved.

Even frozen-hydrated sections are subject to artifacts if sufficient care is not taken during preparation. For example, early studies with frozen sections suggested that mitochondria were calcium-rich organelles. However, the presence of calcium has been shown subsequently to be an artifact resulting from insufficiently cold conditions within the sectioning chamber of the cryomicrotome (Barnard 1982). The level of calcium in mitochondria is now known to be quite low (Somlyo *et al.* 1985). To prevent redistribution of elements, specimens must be frozen extremely rapidly, even to the extent of firing them with a gun into liquid nitrogen! (Chang *et al*. 1980).

A technique which is likely to be used increasingly for the X-ray microanalysis of a range of tissues is that of low-temperature ashing of specimens using an active oxygen plasma generated by radiofrequency at low temperature (Mason 1983). Moderately thin sections are mounted on silicon monoxide films for transmission electron microscopy, or solid supports for scanning electron microscopy. The organic material is removed as oxides of carbon leaving the inorganic residues (Mason & Nott 1980). This material forms a 'skeleton' of ash in which most cellular organelles can be recognised. X-ray microanalysis is facilitated because the concentration of inorganic material relative to organics is raised bringing some elements within the detection limits of the X-ray microanalyser. This technique is particularly useful for visualising metal-containing granules.

Quantification of X-ray microanalysis of bulk specimens in the scanning electron microscope requires the use of complex computer programs which are beyond the scope of this book. However, if material is frozen and fractured so that a large area of the internal structure of the tissue is exposed (Figs. 9.8, 9.12), scanning electron microscopy in conjunction with qualitative X-ray microanalysis is useful for highlighting areas which it may be profitable to section and analyse in greater detail in the transmission electron microscope.

For thin sections, quantification is more straightforward and is achieved by calculating the ratio of the X-ray counts from an element to the background of white radiation on which the peaks are superimposed. This method relies on the fact that the amount of white radiation produced is linearly proportional to the mass of material through which the electrons are passing. A full description of this technique known as the 'Hall Method' (after its originator Dr T Hall) is given by Tapp & Hockaday (1977). Since the Hall method has been employed by most authors who have attempted quantitative X-ray microanalysis on terrestrial invertebrates, it is described briefly below.

The total white radiation in an X-ray spectrum gives a measure of the total excited mass in the section and takes into account the specimen thickness and shape of the excited volume. In practice, a portion of the white radiation (W) is recorded (Fig. 9.3). The ratio of the X-ray counts for a particular element (I_x) to W is proportional to the ratio of elemental mass to total mass or in other words, to the elemental concentration.

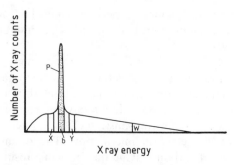

Fig. 9.3 An idealised energy-dispersive X-ray spectrum showing a single element peak superimposed on the background radiation:

No. of counts for element (P) = peak (P + b, shaded) − background (b)

Background (b) is estimated from counts measured at *x* and *y*. The number of counts in a portion of the white radiation (*W*) are recorded also. When two or more peaks overlap or are close together, *x* and *y* are positioned as close to the peaks as possible.

$$C_x = k \frac{I_x}{W}$$

where C_x is the relative concentration or 'mass fraction' of the element. The measurement of the absolute concentration requires the calculation of *k* which is determined from standards. These can take a number of forms. The microscope can be calibrated using sections into which a number of elements have been incorporated, or a reference element (typically cobalt) is dissolved in the resin and the number of X-ray counts from this standard can be compared with elements in the tissue. Knowing the relative sensitivity of the analytical equipment to elements other than cobalt enables the concentrations of all elements to be determined in the section.

In studies with frozen-hydrated material, a physiological solution into which the tissues have been placed before freezing can be used as the internal standard. The relative peak heights for the different elements can be compared with their absolute concentrations determined by a technique such as atomic absorption spectrometry.

The use of X-ray microanalysis for the localisation of histochemical stains holds exciting possibilities since methods can be developed which do not necessarily have to have a visible reaction product (for reviews see Bowen & Ryder 1978; Sumner 1983). X-ray microanalysis is useful also for confirming the presence of ultrastructural markers such as colloidal gold and thorium oxide which can be difficult to visualise in sections, particularly if the tracers are mixed with gut contents containing other electron-dense material (Hopkin & Nott 1980).

Microdroplets can be analysed by spraying tissue digests on to grids coated with a transparent supporting film (Morgan *et al.* 1975; Nott & Mavin 1986). Although the detection limit of energy-dispersive X-ray microanalysis for most elements is quite high (about 0.1% of the excited volume), the method is useful for determining the relative concentrations of major ions since a spectrum is obtained of all the elements present.

9.4.3 Classification of Metal-containing Granules in Terrestrial invertebrates

9.4.3.1 Introduction

There is no generally-agreed term to describe accumulations of metal-containing material in invertebrates although the authors of the most recent reviews on the subject have elected to call them 'granules' (Simkiss 1976a, 1976b; Icely & Nott 1980; Mason & Nott 1981; Brown 1982; Taylor & Simkiss 1984; Hopkin 1986). For the sake of consistency, this is the term used in this book. Most dictionaries define a granule as a 'small grain' which suggests a discrete hard spherical body. Many metal-containing granules possess these properties which makes them difficult to section for transmission electron microscopy and X-ray microanalysis. However, it must be remembered that many of the 'granules' described in this section are no more than membrane-bound accumulations of homogeneous material.

In this section, it is proposed that metal-containing granules in terrestrial invertebrates should be classified into four types. *Type A* granules consist of concentric layers of calcium and magnesium phosphates which may contain other class A and borderline metals such as manganese and zinc. *Type B* granules are of a much more heterogeneous appearance in thin section but always contain large amounts of sulphur in association with class B and borderline metals including cadmium, copper, mercury and zinc. *Type C* granules are composed almost exclusively of iron and probably represent intracellular breakdown products of unwanted ferritin. *Type D* granules are generally much larger than the first three types and are composed of concentric layers of calcium carbonate. Types A, B and C are formed in membrane-bound vesicles within cells but type D granules are produced extracellularly. Metals other than calcium have not been detected in type D granules, nevertheless, they have been

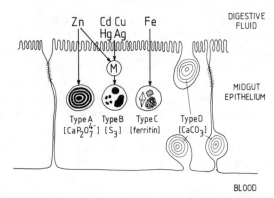

FIG. 9.4 Schematic diagram of the four types of metal-containing granules found in the midgut of invertebrates. Types A, B and C are intracellular and are contained within a membrane. Type D granules are extracellular. Metals may be transported to granules by metallothionein proteins (M). (From Hopkin 1986, by permission of the Nederlandse Entomologische Vereniging.)

included in this scheme to avoid confusion with type A granules. The classification is summarised in Fig. 9.4. The strength of this scheme is the division of granules into different types based on the chemistry of the metals accumulated rather than on the two-dimensional appearance in thin section which has led to confusion in the past.

In formulating this classification, results of histochemical tests for metals have generally been omitted because of their non-specificity, as have X-ray microanalytical studies in which the presence of metals in granules has been demonstrated by very small peaks in X-ray spectra. These can easily arise from contamination of the specimen during preparation, or from materials used in the construction of the analytical instrument.

9.4.3.2 Type A granules

Type A granules are composed of concentric layers of amorphous calcium and magnesium ortho-and pyrophosphates with a small (<10%) organic component (Howard *et al.* 1981; Taylor *et al.* 1986). Class A metals such as potassium and manganese and the borderline metal zinc may be present also (Mason & Simkiss 1982) but class B metals such as cadmium, copper and mercury have not been detected.

Type A granules are usually spherical, less than 5 µm in diameter and are intracellular within a membrane-bound vacuole. They are extremely hard and insoluble and are difficult to section for transmission electron microscopy. Material torn from the granules during ultramicrotomy (e.g. Fig. 9.6) accumulates on the edge of the knife and leaves unsightly scratch marks across the sections.

All the type A granules are of the same size and internal structure in each columnar cell of the midgut of the silkworm *Bombyx mori* (Turbeck 1974) and basophil cell of the hepatopancreas of the garden snail *Helix aspersa* (Mason & Simkiss 1982). This suggests that within a single cell, all the type A granules grow synchronously by the deposition of concentric layers of calcium and magnesium phosphates. Indeed, examination of published micrographs reveals that invariably, the two-dimensional appearance of type A granules within individual cells can be derived from a single three-dimensional model (Fig. 9.5). If such considerations are not taken into account, it is possible to misinterpret sections of type A granules. For example, Green (1979) concluded that there were empty vacuoles in the midgut cells of the New Zealand glow worm when these almost certainly contained granules which had not been sectioned.

Examples of type A granules include:-

1. Concentrically-structured granules in the basophil cells of the hepatopancreas of the snail *Helix aspersa* (Mason & Simkiss 1982; Fig. 9.6).
2. 'Mineral concretions' in the intestinal cells of Collembola (Humbert 1974a, 1977, 1978, 1979; Van Straalen *et al.* 1987).
3. 'Spherocrystals' in the ileal cells of the gut of the cockroach (Jeantet *et al.* 1980; Fig. 9.15).
4. 'Spherites' in the cells of the Malpighian tubules of Diptera (Green 1979).
5. 'Excretory concretions' in the intestinal cells of nematodes (Jenkins *et al.* 1977).
6. Concentrically-structured granules in the hepatopancreas of the spider *Dysdera crocata* (Fig. 9.19 and front cover).

The concentrically structured chloragosomes in the chloragogeneous tissue which surrounds the gut of earthworms accumulate class A and borderline metals (Morgan & Morris 1982; Figs. 9.9, 9.11A) but can not be considered as typical type A granules since they contain sulphur and a large percentage (*c.* 80%) of organic matter (see Section 9.5.2).

Suggested functions of type A granules include:-

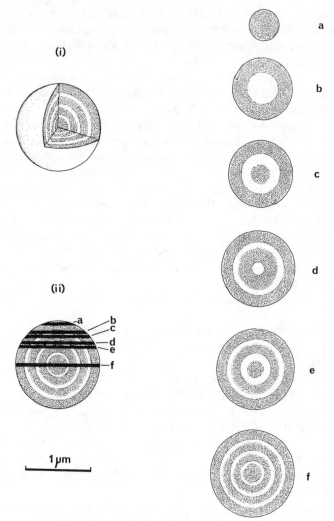

FIG. 9.5 Schematic diagram to show how sections of a concentrically structured type A granule (a–e) can be derived from planes (ii) through a model (i) based on section f. The section thickness is 60 nm. (From Hopkin & Nott 1979, by permission of Cambridge University Press.)

A

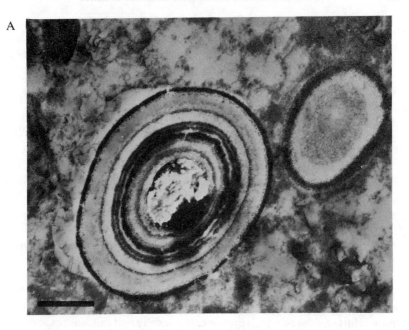

B

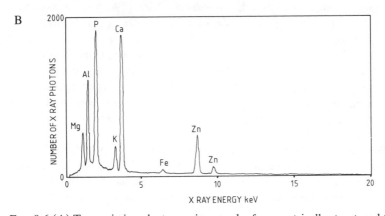

FIG. 9.6 (A) Transmission electron micrograph of concentrically structured type A granules in a basophil cell of the hepatopancreas of the garden snail *Helix aspersa* from a site near to the Avonmouth smelting works (see Table 7.1). Scale bar = 0.5 μm. (B) Energy-dispersive X-ray spectrum derived from an electron-dense ring of an unstained granule similar to that shown in (A). The main constituents of the granule are calcium, magnesium, phosphorus, zinc, potassium and a trace of iron. The peak for aluminium is derived from the specimen support grid. (A and B, previously unpublished.)

1. Storage-excretion of class A and borderline metals in molluscs (Simkiss 1981), Collembola (Humbert 1974a, 1977, 1978, 1979), cockroaches (Jeantet *et al.* 1980), Homoptera (Gouranton 1968), silkworm larvae (Waku & Sumimoto 1974) and nematodes (Jenkins *et al.* 1977).

2. Calcium reserves in the eggs of millipedes for embryonic growth (Crane & Cowden 1968), in molluscs for shell repair (Abolinš-Krogis 1965) and reproduction (Fournié & Chétail 1982a, 1982b), for construction of the pupal case in Homoptera (Cheung & Marshall 1973; Marshall & Cheung 1973b) and Diptera (Grodowitz & Broce 1983; Krueger *et al.* 1987) and to supply calcium to the ovaries of *Calliphora* (Taylor 1984).

3. Phosphate reserves in molluscs (Campbell & Boyan 1976), tapeworms (von Brand *et al.* 1965) and nematodes (Jenkins *et al.* 1977).

4. 'Buffering' calcium concentrations in cells of amphipods (Meyran *et al.* 1986).

Type A granules are clearly able to accumulate potentially toxic class A and borderline metals such as manganese, zinc and lead. However, there is little evidence which proves categorically that this was responsible for their evolution although Thomas & Ritz (1986) have managed recently to induce the formation of type A granules in barnacles by exposing them to zinc. It is possible that type A granules were evolved originally for some other physiological purpose(s) such as calcium storage (Simkiss 1976a, 1976b — see Raeburn (1987) for a discussion of intracellular calcium regulation) and were, because of their charged oxygen atoms which act as powerful ligands for class A metals, able to detoxify as insoluble deposits, class A or borderline metals which entered cells in excess of requirements.

The ability to form type A granules may have 'preadapted' some groups of terrestrial invertebrates such as snails, to survive in sites which are heavily contaminated with zinc (Cooke *et al.* 1979), although this metal may be bound also in metallothionein proteins and type B granules.

Type A granules are probably initiated in small vacuoles which form in response to a rise in intracellular levels of class A and borderline metals (Taylor *et al.* 1986).

9.4.3.3 Type B granules

Type B granules have a much more heterogeneous structure than type

A granules. Their appearance in thin section ranges from membrane-bound accumulations of electron-dense material which resemble lysosomes, to dense granules with a diameter of more than 5 μm which may be torn from the tissue during sectioning. However, the composition of type B granules is more consistent than their appearance and they always contain sulphur in association with class B and borderline metals such as cadmium, copper, mercury, zinc, lead and iron. In cells in which there are no type A granules, type B granules may also contain a small amount of calcium (Hopkin & Martin 1982b). Type B granules are insoluble in water and alcohol and are composed mainly of organic material. On heating to 500 °C, the lighter organic matrix disappears leaving a residue of metals (Fig. 9.18).

Examples of type B granules include:-

1. 'Cadmosomes' or 'debris vesicles' in the chloragogeneous tissue surrounding the gut of earthworms (Morgan & Morris 1982).
2. 'Copper granules' in the S cells of the hepatopancreas of isopods (Hopkin & Martin 1982b; Fig. 9.14A, B).
3. 'Lysosomes' in the midgut cells of the cockroach (Jeantet *et al.* 1980; Fig. 9.16A).
4. 'Copper and sulphur-containing granules' in the fat body cells of homopteran insects (Marshall 1983).
5. 'Copper' granules in the cupriophilic cells of the midgut of *Drosophila* (Tapp & Hockaday 1977; Maroni *et al.* 1986).
6. 'Concretions' in the midgut cells of the housefly (Sohal *et al.* 1977; Figs. 9.17, 9.18).
7. 'Copper' granules in the hepatopancreas of the spider *Dysdera crocata* (Fig. 9.19).

Suggested functions of type B granules include:-

1. Storage-excretion of class B and borderline metals in earthworms (Morgan & Morris 1982), isopods (Hopkin & Martin 1982b, 1984a), cockroaches (Jeantet *et al.* 1980), Homoptera (Marshall 1983), *Drosophila* (Tapp & Hockaday 1977) and the housefly (Sohal *et al.* 1977).
2. Copper reserves in Homoptera (Marshall 1983).

The evidence for a detoxificative function for type B granules is stronger than for type A granules. Their heterogeneous appearance in thin section, ability to accumulate a wide range of elements and the presence of acid phosphatase (Sohal & Lamb 1977; Tapp & Hockaday

1977) suggests that they have their origin in lysosomes. Furthermore, the presence of large amounts of sulphur in association with cadmium, copper, mercury and zinc has led several authors to suggest that type B granules are composed largely of residues of metallothionein proteins (Jeantet *et al*. 1980; Morgan & Morris 1982; Hopkin & Martin 1984a). Metallothioneins are synthesized in response to the presence of cadmium, copper, mercury or zinc within cells and have a half life of only a few days (Simkiss & Mason 1983). Non-essential metals such as cadmium and mercury, and essential metals such as copper and zinc which are assimilated in excess of requirements, can be detoxified by incorporation into type B granules for permanent storage in an insoluble form (e.g. isopods, Hopkin & Martin 1982b; houseflies, Sohal *et al*. 1977; cockroaches, Jeantet *et al*. 1980), or excreted by the normal processes of cellular degeneration and replacement (e.g. clothes moth larvae, Waterhouse 1952). Type B granules may occur together with type A granules in the same cell in cockroaches (Jeantet *et al*. 1980; Fig. 9.15) and spiders (Fig. 9.19).

Gibbs *et al*. (1981) put forward the interesting suggestion that marine polychaetes retain type B granules in their tentacles deliberately as an anti-predation mechanism. Hopkin & Martin (1985b) made a similar suggestion to explain the tendency of the spider *Dysdera crocata* to store large amounts of zinc and copper.

9.4.3.4 Type C granules
Type C granules may be polyhedral inclusions with an obvious crystalline structure (Cheung & Marshall 1973) or loosely-bound deposits of flocculent material (Fig. 9.14A). Type C granules, therefore, can bear a superficial resemblance to type B granules although they can be easily distinguished by X-ray microanalysis as they are so rich in iron. At high magnification in the electron microscope, type C granules may possess a crystalline structure similar to that of ferritin (Cheung & Marshall 1973), the protein which stores iron within the body. However, most type C granules probably represent intracellular breakdown products of ferritin known as haemosiderin (Taylor & Simkiss 1984) which has no crystalline structure.

The difference between type B and C granules may be less distinct in animals which have been exposed to extremely high levels of metals. For example, the type C granules in the B cells of the hepatopancreas of isopods from sites heavily contaminated with metals may contain zinc and lead as well as iron (Hopkin & Martin 1982b). Conversely, the granules which are rich in iron in the midgut cells of the housefly (Sohal & Lamb 1977) always contain some copper and large amounts of sulphur and are probably type B granules produced in response to the high level of iron (0.1%) included in the sucrose diet of the insects.

Examples of type C granules include:-

1. 'Iron granules' in the B cells of the hepatopancreas of isopods (Hopkin & Martin 1982b; Fig. 9.14A, C).
2. Deposits of ferritin in the midgut cells of homopteran insects (Bielenin & Weglarska 1965; Gouranton 1967; Cheung & Marshall 1973; Kimura *et al.* 1975) and lepidopteran larvae (Locke & Leung 1984).

Type C granules probably function primarily as intracellular stores of iron whereas type B granules provide a method for the detoxification of unwanted elements, which may include iron if this is present at very high levels in the diet. The iron in type C granules is probably stored initially as ferritin although this may not be apparent if the crystalline structure of the protein is altered by reagents used in the preparation of sections for electron microscopy. There may be a distinction between type C granules containing waste iron which is not remobilised and granules containing ferritin from which iron can be released.

9.4.3.5 Type D granules

Type D granules are different from the other three types in several ways. They are extracellular, composed of calcium carbonate and may be very large, exceeding 1 mm in diameter in the calciferous glands of earthworms (Morgan 1981). When small (less than about 20 μm in diameter), type D granules are typically spherical and are composed of concentric layers of material. This has led to confusion in some groups of terrestrial invertebrates with type A granules which have a similar two-dimensional appearance in thin section (Simkiss & Mason 1983).

The calcium carbonate may be in an amorphous or crystalline form. In the calciferous glands of earthworms, for example, the type D granules are initially spherical (Fig. 9.10A), but as they increase in size, they amalgamate to form massive cuboidal crystals (Fig. 9.10B).

Type D granules are relatively undamaged when sectioned for electron microscopy although they are quite soluble and a considerable amount of material is probably extracted during preparation. Metals other than calcium have not been detected in significant amounts.

Examples of type D granules include:-

1. 'Calcium granules' in the connective tissue of snails (Simkiss 1976b).
2. 'Calcium granules' in the arterial walls of slugs (Fretter 1952).
3. 'Mineralised secretory granules' in the calciferous glands of earthworms (Morgan 1981; Fig. 9.10).
4. 'Excretory concretions' in the excretory ducts of trematode parasites (Erasmus 1967; Martin & Bils 1964).

Most authors agree that type D granules are involved in the regulation of the pH of extracellular fluids by the conversion of carbon dioxide through the catalytic activity of carbonic anhydrase.

9.5 SPECIFIC EXAMPLES

9.5.1 Molluscs

The hepatopancreas is the most important organ of accumulation of metals in terrestrial molluscs (Section 7.2). It is composed of three main cell types, the digestive, excretory and basophil cell. The basophil cell contains numerous concentrically-structured type A granules composed of calcium and magnesium phosphates (Fig. 9.6; Abolinš-Krogis 1965, 1970a, 1970b, 1976; Walker 1970; Burton 1972; Martoja & Martoja 1973; Campbell & Boyan 1976; Fournié & Chétail 1982b; Janssen 1985). This has led to the basophil cells being called 'calcium cells' by some of these authors. The intestine is thought not to be important in nutrient assimilation and may function mainly as a water-reabsorbing organ (Forester 1977).

The type A granules in the basophil cells of the hepatopancreas of the garden snail *Helix aspersa* are composed of an equimolar mixture of calcium and magnesium ions present as ortho-phosphates (PO_4^{3-}) and pyrophosphates ($P_2O_7^{4-}$)(Howard *et al.* 1981; Taylor *et al.* 1983). Pyrophosphate ions are formed in large quantities in most anabolic cells but are hydrolysed rapidly by pyrophosphatase, an enzyme which occurs in every cell so far examined (Mason & Simkiss 1982). Pyrophosphatases are eminently suitable for a metal detoxification system for class A metals because they are produced easily *in vivo* and form very insoluble salts with class A and some borderline metal ions at physiological pH's. Indeed, the metal binding constants of Mg^{2+} and Ca^{2+} are 6300- and 3700-fold greater for pyrophosphate than for orthophosphate (Howard *et al.* 1981).

The most detailed experiments on the dynamics of radioactive metals in terrestrial molluscs have been those carried out by Simkiss, Taylor and co-workers on *Helix aspersa*. Their findings are particularly important because they demonstrate the fundamental differences in the ways in which class A and class B metals are metabolised.

In the main series of experiments (Simkiss 1981; Simkiss & Taylor 1981), two groups of snails were taken. The first group were injected *via* a cannulated optic tentacle with equimolar solutions of pairs of metal chlorides labelled with radioactive isotopes. The metals used were $^{54}Mn^{2+}$, $^{59}Fe^{2+}$, $^{60}Co^{2+}$, $^{65}Zn^{2+}$, $^{85}Sr^{2+}$, $^{109}Cd^{2+}$, and $^{203}Hg^{2+}$. After six hours, a sample of blood and the whole hepatopancreas was removed from each animal. A sample of type A granules was prepared by homogenisation and repeated resuspension of one half of the piece of hepatopancreas. The level of radioactivity was then measured in the sample of granules, in the other half of the hepatopancreas and the blood sample. The results were expressed as ratios of the relative uptake of the two isotopes in the tissue or tissue fraction (Table 9.1).

Samples of blood were obtained from the second group of snails and the granules removed from the hepatopancreas in the same manner. The blood samples were spiked with pairs of radioisotopes, at the same concentration as the first group of snails, equilibrated with gas composed of 5% carbon dioxide and 95% oxygen, and shaken for six hours with a sample of the isolated granules. The granules were then washed and counted in a separate blood sample.

The discrimination ratios between the two isotopes for each measurement were expressed as the observed ratio (O.R.) of each sample to its precursor. Thus, for two isotopes X and Y, the O.R. for a tissue (t) exposed to blood (b) was:-

$$O.R. = \frac{X_t/Y_t}{X_b/Y_b}$$

The O.R.'s for each set of measurements in these experiments are shown in Table 9.1. The metal shown in the first column is expressed as a molar concentration factor over the other metals. Thus in Table 9.1A, manganese is concentrated 3.4 times more than iron in the hepatopancreas, 63.5 times more than zinc and 352 times more than strontium. A comparison of these ratios enabled a hierarchy to be produced for the ions which indicated the approximate preference of the three 'systems' for handling each ion (Table 9.1).

TABLE 9.1

Experiments with radioisotopes on the garden snail *Helix aspersa* (from Simkiss 1981, by permission of the Company of Biologists Ltd)

A.   Observed molar ratios for the accumulation of metal ion pairs in hepatopancreas tissue treated *in vivo*.

(Note that the larger the number and the further the metal is to the right, the weaker the cellular accumulation. An 'iso metal' diagonal divides the table and theoretically all values to the left of this should be below 1.0 and all those to the right should be over 1.0. The values to the left of this 'iso metal' line are the reciprocals of the values for the same metal pairs to the right of this diagonal.)

Hierarchy	Mn	>	Fe	>	Cd	>	Hg	>	Zn	>	Co	>	Sr
Mn	—		3.4		35.7		35.5		63.5		92[a]		352
Fe	0.29		—		1.1		2.8		3.3		19.2		36.2
Cd	0.03		0.87		—		10.9		3.2		11.3		39.6
Hg	0.03		0.35		0.09		—		0.44		0.64		4.9
Zn	0.02		0.31		0.31		2.3		—		4.4		2.7
Co	0.01[a]		0.05		0.08		1.6		0.22		—		8.0
Sr	0.003		0.03		0.03		0.10		0.37		0.12		—

[a] Single samples.

B. Observed molar ratios for the accumulation of metal ion pairs in granules formed *in vivo*.

Hierarchy	Mn >	Fe >	Zn >	Co >	Cd >	Hg >	Sr
Mn	—	6.3	56	n.d.	35.7	265	4 075
Fe	0.16	—	3.5	16.6	3.3	3.8	28.6
Zn	0.02	0.28	—	9.0[a]	17.8	12.8	2.9
Co	n.d.	0.06	0.11[a]	—	1.6	2.9	20.6
Cd	0.03	0.30	0.06	0.63	—	5.3	2.4
Hg	0.004	0.26	0.08	0.35	0.19	—	1.5
Sr	0.0002	0.03	0.34	0.05	0.42	0.66	—

[a] Single samples. n.d., no data available.

C. Observed molar ratios for the accumulation of metal ion pairs in granules exposed to labelled blood *in vitro*.

Hierarchy	Mn >	Fe >	Sr >	Hg >	Cd >	Zn >	Co
Mn	—	0.46	1.44	5.6	11.7	20.3	13.4
Fe	2.2	—	5.5	4.7	20.9	6.1	28.1
Sr	0.69	0.18	—	3.9	10.8	3.2	12.3
Hg	0.18	0.21	0.26	—	0.82	1.5	4.0
Cd	0.09	0.05	0.09	1.22	—	0.74	1.6
Zn	0.05	0.16	0.32	0.70	1.4	—	1.5[a]
Co	0.07	0.04	0.08	0.25	0.62	0.67[a]	—

[a] Single samples.

TABLE 9.2

Percentage of metals that are ultrafilterable from blood samples compared with their relative uptake into hepatopancreas tissue of *Helix aspersa* (from Simkiss 1981, by permission of the Company of Biologists Ltd)

	Mn	*Co*	*Zn*	*Cd*
Ultrafilterable	32%	27%	9%	8%
Uptake relative to Mn	100%	1.1%	1.6%	2.8%

The differences between the hierarchies in Tables 9,1A, 9.1B and 9.1C could be a reflection of the extent to which metals are bound to blood proteins and other ligands when they are injected into snails. However, in other experiments, this has been shown not to be the case and there is clearly no connection between the extent of binding in the blood and the uptake of metals relative to manganese in the hepatopancreas (Table 9.2). For example, four times as much manganese is bound in the blood (32%) as cadmium (8%)(Table 9.2) whereas 35.7 times more manganese is taken up into hepatopancreas tissue than cadmium (Table 9.1A). Clearly, differences in the extent of binding of the metals by the blood are not the explanation for the discrimination at the cellular level.

The major differences between the results in Tables 9.1A, 9.1B and 9.1C should be due to the fact that *in vitro* (Table 9.1C), the granules are simply absorbing the metal on to their surfaces whereas *in vivo* (Table 9.1B), the ratios of the ions will be changed according to the activities of the cytoplasm of the basophil cells in the hepatopancreas.

The hierarchy in Table 9.1A is thus a measure of the total accumulating systems in the cell. Part of these effects are due clearly to the incorporation of metals on to granules but granules can be isolated separately to give the hierarchy shown in Table 9.1B. The differences between the positions of metals in these two hierarchies must therefore be due to some additional cytoplasmic accumulating system. Examination of the figures in the tables shows that total tissues (Table 9.1A) preferentially take up cadmium and mercury which have 'moved' from behind cobalt to in front of zinc when compared with Table 9.1B. Simkiss (1981) interpreted this as evidence for a separate protein binding system that can accumulate cadmium, mercury and zinc in some cells of the hepatopancreas. The cadmium and mercury accumulation is probably accounted for in the whole tissue experiments by metallothionein proteins.

The cellular processes involved in the formation of metal granules can be analysed further by comparing the *in vivo* series (Table 9.1B) with the corresponding data obtained from the *in vitro* series (Table 9.1C). Differences between the two series can be interpreted as being due to selective cytoplasmic activities. It is clear, comparing the two sets of data, that the cells of the hepatopancreas exert a very strong discrimination against strontium entering from the blood.

The cells of the hepatopancreas preferentially take up zinc and cobalt ions and deposit them in granules and this should be considered in addition to the strong preference which all the series show for manganese and iron.

So, in terrestrial gastropod molluscs, there appear to be two systems for removing metals from the blood by storage in the hepatopancreas. First, the metallothionein proteins (Dallinger & Wieser 1984b; Berger *et al.* 1986) which, with their abundant thiol groups, bind class B type metals such as mercury, cadmium and copper and the type A granules which, with their pyrophosphate ions with charged oxygen atoms, act as powerful ligands for class A metals such as calcium, potassium, magnesium and manganese. Zinc is borderline and is able to be bound by either system.

Mason & Simkiss (1982) examined the ultrastructure of the hepatopancreas of *Helix aspersa* after intravascular injections of manganese. A large dose of manganese induced the loss from the basophil cells of all the granules, which were voided in the faeces. The subsequent replacement of granules was studied after a further injection of manganese. The granules were reformed as concentric layers of calcium magnesium pyrophosphates but the injected manganese accumulated on the inside of the membrane-bound vesicle in which each granule was forming. The mineralisation process apparently requires the formation of a nucleus of calcium and magnesium pyrophosphate before the manganese can be mineralised within the vacuole. 48 hours after the second injection, 90% of the manganese in the body was in the hepatopancreas and about 30% of this metal was bound firmly in the granules (Table 9.3). These were again voided rapidly *via* the faeces. The loss of granules must have been due to the injection of manganese since under normal conditions, the granules are lost extremely slowly from the hepatopancreas. Indeed, the half life of zinc incorporated into the type

TABLE 9.3

Subcellular distribution of ^{54}Mn in the hepatopancreas of *H. aspersa* at various times after an intravascular injection (from Mason & Simkiss 1982, by permission of Academic Press, Inc.)

Time after injection (hours)	Percentage of total activity			Percentage of whole body ^{54}Mn in granules	Percentage of ^{54}Mn lost from tissues during preparation for electron microscopy
	Granules	Membranes and cell debris	Cytosol and microsomes		
0.5	28	47	25	2.5	81.8
2	16	51	33	2.4	49
8	16	51	33	7.0	25
24	31	50	19	27.0	25
48	33	50	17	33.0	22
64	0	63	37	0	49
72	0	78	22	0	37

A granules in the hepatopancreas of the slug *Arion rufus*, is of the order of 50 days (Schoetti & Seiler 1970). Lead may be present also in the type A granules in *Arion ater* (Ireland 1984b).

Mason & Simkiss (1982) have suggested that the inner surface of the membrane surrounding the type A granules synthesizes pyrophosphates and may possess a number of enzymes involved in controlling the accumulation of anions at the site of granule formation. This concept is in keeping with the observation that intracellular granules form most rapidly in snails that are well-fed and therefore producing pyrophosphate by cellular anabolism (Simkiss *et al.* 1982). When *Helix aspersa* is fed on a diet to which manganese has been added, the manganese accumulates on the outside of granules already present rather than displacing calcium (Greaves *et al.* 1984; Taylor *et al.* 1986, 1988; Fig. 9.7).

Simkiss (1981) has pointed out that there might be discrimination for metal ions by the digestive epithelia. Indeed, it is interesting to compare the results of his studies with those of other workers who have administered metals to molluscs in the food. One of the major differences is in the behaviour of strontium. Simkiss (1983b), in experiments with $^{54}Mn^{2+}$ and $^{85}Sr^{2+}$, has confirmed the earlier observation that manganese is taken up rapidly from the blood into the type A granules in the basophil cells of the hepatopancreas of *Helix aspersa* but that in contrast, strontium accumulates in the sites of shell formation. Since strontium can be considered as a 'biochemical analogue' of calcium, it is clear that the calcium which forms type A granules must enter the basophil cells of the hepatopancreas from the lumen, particularly since isolated granules absorb strontium almost as efficiently as manganese (Table 9.1C). This conjecture is supported by the work of Fretter (1953) who showed that the type A granules in the basophil cells of the hepatopancreas of *Arion hortensis* accumulated large amounts of ^{90}Sr when the slugs were fed on lettuce labelled with this tracer. In contrast, zinc and cadmium are stored by the hepatopancreas whether they are present in the food (Schoetti & Seiler 1970; Cooke *et al.* 1979; Ireland 1981, 1982; Dallinger & Wieser 1984a) or the blood (Simkiss 1977, 1981). The detoxification system for cadmium, however, is not able to prevent tissue damage when levels of the metal in the diet are very high (Russell *et al.* 1981).

These experiments raise a number of interesting questions. First, would a massive influx of zinc into the basophil cells of the hepatopancreas be sequestered initially by metallothionein proteins before being bound in type A granules, especially in the light of the fact that synthesis of these proteins in direct response to elevated levels of zinc

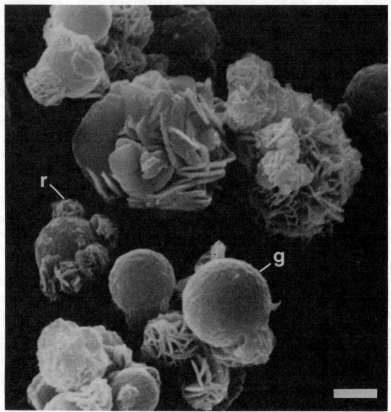

FIG. 9.7 Scanning electron micrograph of type A calcium pyrophosphate granules from a homogenate of the hepatopancreas of the snail *Helix aspersa*. The snails were fed on a diet rich in manganese. These granules normally have a smooth surface (g) but in this preparation, many are covered with 'rosettes' (r) of manganese-rich material. Scale bar = 0.5 µm. (Previously unpublished micrograph of a specimen supplied by Dr Marina Taylor.)

has not been demonstrated in molluscs (Simkiss & Mason 1983)? Second, can metallothioneins be synthesised in response to an influx of cadmium and if so, is the metal metabolised by storage-excretion in type B granules? Thirdly, how does the hepatopancreas respond to high levels of copper in the diet since this metal tends to be distributed much more evenly among the tissues than other metals (Coughtrey & Martin 1976a; Dallinger & Wieser 1984a)?

Despite the considerable number of studies on the efficiency of metal-based molluscicides at controlling snail and slug pests (Godan 1983), very few attempts have been made to assess the ultrastructural basis for their effects. The only work to date of which I am aware has been that of Ryder & Bowen (1977a, 1977b). These authors demonstrated by X-ray microanalytical mapping that copper sulphate is absorbed para-cellularly by the foot of slugs (Fig. 9.8) and that this molluscicide may also pass into the eggs.

9.5.2 Earthworms

Studies on the subcellular distribution of metals in earthworms have been made almost exclusively on the chloragogenous tissue, an assemblage of cells surrounding the gut. The chloragogenous cells or 'chloragocytes' perform a number of functions which include the storage of metabolic wastes (Fischer 1977). This material is discharged into the coelomic fluid by the rupture of the chloragocytes and can be stored in large 'waste nodules' (see Fig. 7.4) or excreted *via* the nephridia in each segment (Edwards & Lofty 1977). The chloragocytes contain two types of metal-containing granule. The first type, the 'chloragosome', is a concentrically-structured inclusion (Fig. 9.9). The second type, the 'debris vesicle', is a membrane-bound deposit of heterogeneous material.

Pooled samples of chloragosomes isolated from *Lumbricus terrestris* contain 3% phosphate, 2.3% calcium, 1 to 3% zinc and 0.2 to 0.4% magnesium on a dry weight basis (Prentø 1979). A freeze-dried prep-aration of a single chloragosome from *Allolobophora longa* has been shown, by quantitative X-ray microanalysis, to contain 0.9% zinc, 2.7% calcium, 3.5% phosphorus, 2.0% sulphur, 0.04% iron and 0.25% pot-assium (Morgan & Winters 1982). The presence of some or all of these elements in the chloragosomes of a range of species has been confirmed broadly by other workers (Matsumoto 1960; Ireland & Richards 1977; Wróblewski *et al.* 1979; Fischer & Trombitás 1980; Morgan 1981, 1982; Morgan & Morris 1982; Richards *et al.* 1986; Fig. 9.11A). Lead has been detected also in the chloragosomes of earthworms from the spoil tips of disused mine sites (Ireland & Richards 1977; Ireland 1978, 1983; Morgan & Morris 1982; Morgan 1982) and roadside verges (Gullvag 1978). The elemental composition of the chloragosomes, together with their concentric structure, suggests that they are type A granules. However, they can not be regarded as typical representatives due to the presence of sulphur and large amounts of organic matter which may constitute more than 80% of their dry weight.

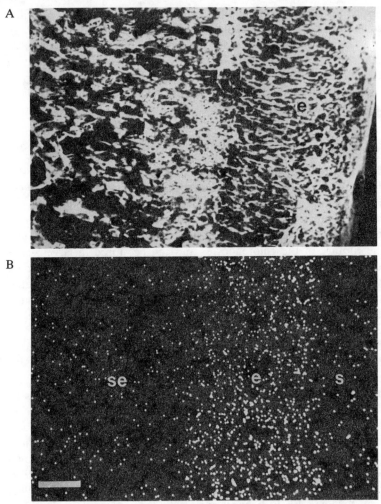

FIG. 9.8 (A) Scanning electron micrograph of part of a freeze-fractured surface of the foot of the slug *Agriolimax reticulatus* after immersion of the live animal for 30 min in a solution containing 1 mg ml^{-1} copper as the sulphate. (B) X-ray elemental map for copper of the specimen shown in A. The copper is retained in the epithelium (e) where the density of X-ray counts is much greater than in the sub-epithelial tissue (se) or the specimen stub (s). Scale bar = 100 μm. (From Ryder & Bowen 1977a, by permission of Academic Press, Inc.)

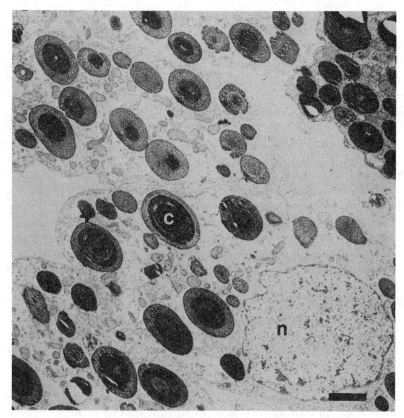

Fig. 9.9 Transmission electron micrograph of chloragocytes in the chloragogenous tissue of the earthworm *Lumbricus terrestris*. The chloragosomes (c) are composed of concentric layers of material which contain calcium, phosphorus, zinc and lead (see Fig. 9.11A). n, nucleus. Scale bar = 2 μm. (From Morgan 1981, by permission of Springer–Verlag.)

The debris vesicles in the chloragosomes of *Dendrobaena rubida* and *Lumbricus rubellus* collected from the spoil heaps of a disused mine, contained cadmium in association with sulphur in the ratio (Cd:S) of about 1:3 (Morgan & Morris 1982). This suggests that debris vesicles contain residues of metallothionein proteins and should be classified as type B granules. Indeed, metallothionein-like proteins have been detected in *Eisenia foetida* (Suzuki *et al.* 1980; Yamamura *et al.* 1981) and *Lumbricus terrestris* (Ireland 1983). A detailed study on the presence of metallothioneins in earthworms has been conducted recently by Morgan (1987). Morgan & Morris (1982) suggested that the debris vesicle should be termed the 'cadmosome' because of its affinity for cadmium, although it is probable that other class B metals such as mercury would be stored in the debris vesicles if present in sufficient quantities in the diet. Lead may also enter the debris vesicles if present in high enough concentrations in the food (Richards & Ireland 1978). Thus, the debris vesicles must be the end product of a lysosome-based excretory system which sequesters unwanted material, including metallothionein residues, for subsequent excretion by rupture of the chloragocytes into the coelomic fluid.

At the anterior end of the gut of most species of earthworms are a number of pouches, the calciferous glands (Fig. 7.3). These contain numerous spherical type D granules composed entirely of calcium carbonate (Prentø 1979; Figs. 9.10A, 9.11B). The type D granules may amalgamate and form massive cuboidal crystals with a diameter of 1 mm or more (Morgan 1981; Fig. 9.10B). The calciferous glands are thought to regulate the pH of the food by release of the type D granules into the lumen of the gut. This suggestion is supported by the observation that worms which are able to survive in acid soils, have much more active calciferous glands than those which are acid-intolerant (Morgan 1982).

A proportion of this calcium released into the lumen of the gut may be recycled by transport across the epithelial cells of the intestine into the chloragosomes, released into the coelomic fluid by the rupture of the chloragocytes and be absorbed across the basement membrane of the cells of the calciferous glands (Prentø 1979). This process has been demonstrated experimentally using strontium which, when injected into the coelomic cavity of *Lumbricus terrestris,* can be detected subsequently in the type D granules in the calciferous glands but not in the chloragosomes (Morgan 1981).

Attempts have been made to prevent losses of elements from chloragosomes during preparation for electron microscopy by smearing samples of fresh chloragogenous tissue directly on to coated specimen grids and allowing them to dry in air (Fischer & Trombitás 1980; Morgan 1982; Morgan & Morris 1982). In the transmission electron microscope, chloragosomes and debris vesicles can be recognised within the smear and X-ray microanalysis has confirmed the presence of elements which have been detected in 'conventionally prepared' chloragogenous tissue. Using this technique, Morgan (1982) produced quantitative X-ray microanalytical data which suggested that the Ca:P ratios were sig-

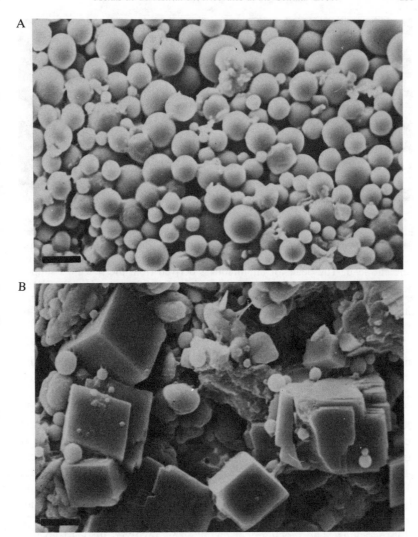

FIG. 9.10 Scanning electron micrographs of extra-cellular type D granules in the calciferous glands of the earthworm *Lumbricus terrestris*. The granules initially form spherites of calcium carbonate (A) which eventually produce cuboidal crystals (B). Only calcium and traces of potassium can be detected by X-ray microanalysis (see Fig. 9.11B). Scale bars (A) 5 μm, (B) 10 μm. (From Morgan 1981, by permission of Springer–Verlag.)

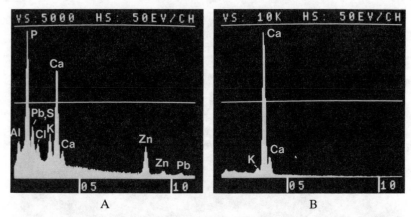

FIG. 9.11 (A) X-ray spectrum of a single chloragosome from the earthworm *Lumbricus rubellus*. These granules are composed mostly of organic material (*c.* 80%). The main inorganic constituents are calcium, phosphorus, potassium, zinc and lead. The peak for aluminium is derived from the specimen support grid. (From Morgan & Morris 1982, by permission of Springer–Verlag.) (B) X-ray spectrum of a type D calcium carbonate granule from the calciferous glands of *Lumbricus terrestris* (see Fig. 9.10). The only metals detected are calcium with a trace of potassium. (From Morgan 1981, by permission of Springer–Verlag.)

nificantly higher in the chloragosomes of species which possessed active calciferous glands than species in which these organs were relatively inactive.

Very little work has been carried out into the biochemical basis of the deleterious effects of metals on earthworms. There is evidence that high levels of lead in the tissues results in a reduction in the activity of amino-laevulinic acid dehydratase, an enzyme involved in haem synthesis (Ireland & Fischer 1979) and leads to depletion of stores of glycogen in the chloragogenous tissues (Ireland & Richards 1977; Richards & Ireland 1978). However, nothing is known of the ways in which metals are transported by the epithelial cells of the gut from the lumen into the chloragocytes. The biochemical mechanisms involved in this transport would repay further study.

9.5.3 Crustacea
9.5.3.1 Isopods
The most important organ of storage of metals in terrestrial isopods is the hepatopancreas. Indeed, in isopods from contaminated sites, the concentrations of zinc, cadmium, lead and copper in this organ are the

highest so far recorded in the soft tissues of any terrestrial animal (Hopkin & Martin 1982a, 1984a). This remarkable ability has led several authors to examine the ultrastructure of the hepatopancreas to try and elucidate how such large amounts of metals are stored.

The hepatopancreas of terrestrial isopods consists of four blind-ending tubules (six in some littoral species such as *Ligia oceanica*) which open into the anterior end of the gut. The epithelium is composed of two cell types, the B and S cells (Clifford & Witkus 1971; Vernon *et al.* 1974) which are derived from mitotically dividing cells at the distal ends of the tubules (Donadey & Besse 1972; Bettica *et al.* 1984). The B cells contain extensive deposits of glycogen and lipid (Patanè 1934; Szyfter 1966; Alikhan 1972b; Storch 1982) and are thought to be involved in the secretion of digestive enzymes and the storage of food material (Storch 1982, 1984). The main function of the S cells is the storage of metals (Prosi *et al.* 1983).

In terrestrial isopods from uncontaminated sites, two main types of intracellular granules have been observed in the hepatopancreas. Type B granules, composed of homogeneous material containing copper and sulphur (and a little calcium), occur in the S cells (Wieser 1968; Wieser & Klima 1969; Alikhan 1972a; Hopkin & Martin 1982b; Marcaillou *et al.* 1986; Dallinger & Prosi, 1988; Prosi & Dallinger, 1988). Type C granules, composed of more loosely-bound deposits of flocculent material in which only iron can be detected, are stored in the B cells (Hryniewiecka-Szyfter 1972, 1973). Type A and D granules have not been detected in terrestrial isopods.

In isopods from sites which are contaminated with metals from mining or smelting operations, the 'copper' granules may contain zinc, cadmium and lead and the 'iron' granules may accumulate zinc and lead (Figs. 9.12, 9.13, 9.14). The copper granules of *Oniscus asellus* from serpentine areas on the Lizard peninsula in Cornwall, England, which I have examined by X-ray microanalysis, may contain cobalt and nickel also.

The remarkable ability of the S cells of terrestrial isopods to store large amounts of metals can be demonstrated vividly by rearing juvenile specimens of *Oniscus asellus* from the same brood on uncontaminated leaf litter, or leaves collected close to the Avonmouth smelting works (Section 5.3.2). After only six weeks, the difference in the extent to which granules have accumulated in the S cells of the hepatopancreas of isopods subjected to the two types of food is striking (Fig. 9.13).

The copper granules of isopods are insoluble in water, alcohol and acetic acid (Patanè 1934) and in animals from uncontaminated sites,

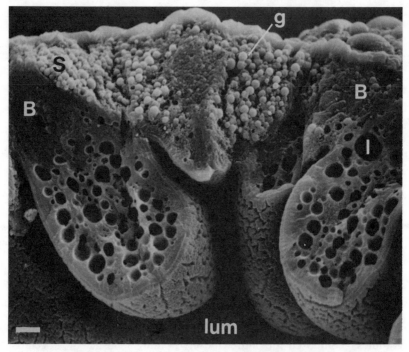

FIG. 9.12 Scanning electron micrograph of a transversely fractured tubule from the hepatopancreas of the woodlouse *Oniscus asellus* from a site contaminated by zinc, cadmium, lead and copper. The type B 'copper' granules (g) are prominent within the S cells (S). The spaces (l) in the B cells (B) are left by lipid droplets dissolved out during preparation. lum, lumen. Scale bar = 5 μm. (From Hopkin & Martin 1982b, by permission of Longman Group UK Ltd.)

contain about 80% of the total copper in the hepatopancreas (Wieser & Klima 1969). The results of extensive chemical tests on type B granules in amphipods and barnacles led Icely & Nott (1980) and Walker (1977) to conclude that the copper was associated with an organic compound (probably residues of metallothionein proteins) and was not combined with sulphur as the sulphide. The iron granules are probably stores of breakdown products of ferritin.

Concentrations of lead and copper in the hepatopancreas of individual specimens of *Oniscus asellus* from mine sites may exceed 2.5% and 3.4% of the dry weight respectively with no apparent ill effects (Hopkin & Martin 1982a). However, when the concentrations of zinc and cadmium in the hepatopancreas of this species exceed about 1.5% and 0.5% of

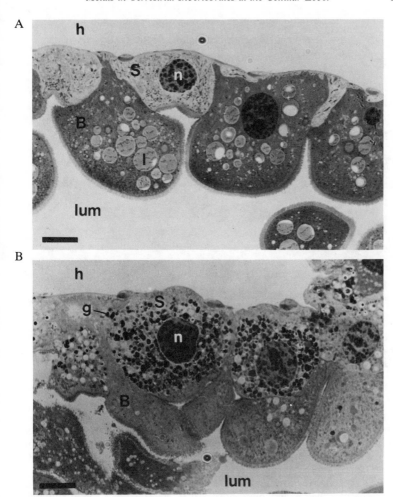

FIG. 9.13 Light micrographs of B cells (B) and S cells (S) in the hepatopancreas of two specimens of the woodlouse *Oniscus asellus*, 6 weeks after release from the brood pouch of the same female from an uncontaminated site. The cytoplasm of the S cells of the juvenile reared on leaf litter contaminated with zinc, cadmium, lead and copper by a smelting works (9.13B) contains far more type B 'copper' granules (g) than the S cells of the juvenile reared on uncontaminated food (9.13A). h, haemocoel; l, lipid droplet; lum, lumen of hepatopancreas tubule; n, nucleus. Scale bars = 10 μm. (From Hopkin & Martin 1984a, by permission of the Zoological Society of London.)

A

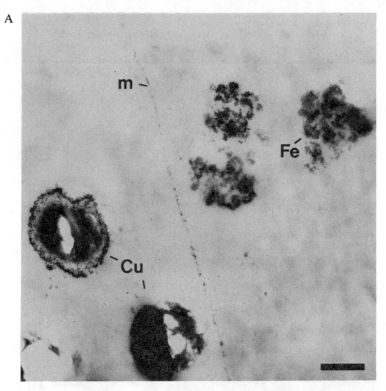

FIG. 9.14 (A) Transmission electron micrograph of the border between an S and B cell in the hepatopancreas of a moribund specimen of the woodlouse *Oniscus asellus* from a site contaminated with zinc, cadmium and lead (unstained). The intracellular membranes are lined with electron-dense material (m) which contains zinc and lead. Scale bar = 0.5 μm. The type B 'copper' granules (Cu) in the S cell contain zinc, cadmium, lead, copper, sulphur and a trace of calcium (9.14B). The type C 'iron' granules (Fe) in the B cell contain iron, zinc, lead and traces of calcium and phosphorus (9.14C). The peak for aluminium is derived from the specimen support grid. (A from Hopkin & Martin 1982b, by permission of Longman Group UK Ltd; B and C previously unpublished.)

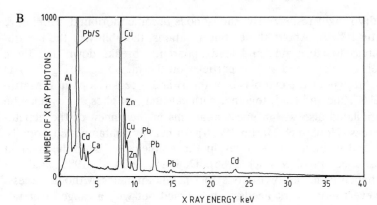

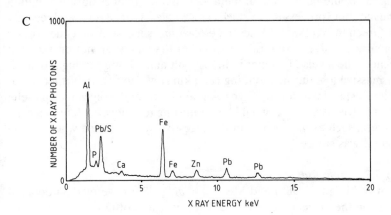

FIG. 9.14—*contd*

the dry weight respectively, the isopods become moribund (Hopkin & Martin 1982b). Above these concentrations, the ability of the hepatopancreas to store zinc and lead apparently breaks down and these metals are deposited as fine particles on the membranes (Fig. 9.14A) and throughout the cytoplasm of the B and S cells (Hopkin & Martin 1982b). Zinc and lead, together with calcium and phosphorus, may be accumulated also within microorganisms in the lumen of the hepatopancreas (Hopkin & Martin 1982b). In heavily contaminated isopods, lead and cadmium may occur in the epithelial cells of the hindgut (Prosi 1983; Prosi & Back 1986). Dallinger & Prosi (1986) analysed the hepatopancreas of isopods for metallothioneins without success. However, more work needs to be carried out with a range of metals under different conditions before the absence of these proteins can be confirmed.

Perhaps the most interesting aspect of metal physiology to study in isopods would be the fate of metal-containing granules during moulting (the moulting process in Crustacea as a whole has been reviewed recently by Greenaway (1985)). Wieser (1965b) has suggested that the copper granules dissolve during this process but the metal can not be excreted because the levels of copper in intermoult animals are too high to allow for massive loss during moulting (Hopkin *et al.* 1985b). A further factor of interest is the absence of type A granules from the digestive epithelia of terrestrial isopods. It would be interesting to follow the fate of a class A metal such as manganese in the digestive system to see where in the body it was stored.

9.5.3.2 Amphipods

Graf & Meyran (1985) and Meyran *et al.* (1986) have made a detailed study on the ultrastructure of the digestive caeca of *Orchestia cavimana* during the moult cycle. The animals have the ability to store calcium in concentrically structured granules in *extracellular* channels in the caecal cells. These granules are dissolved after moulting and the calcium released is used for mineralising the new exoskeleton. The granules contain calcium and phosphorus so would appear to be type A in form. However, their extracellular location puts them in the type D category. Perhaps this is one of the few cases (the original function?) where type A granules are involved directly in temporary storage of calcium. The situation in *Orchestia* is an intriguing one and it would be of great interest to compare calcium regulation at the ultrastructural level during moult in a series of marine, freshwater and fully terrestrial species.

9.5.4 Insects

9.5.4.1 Introduction

The distribution of metals in the cells of insects has been studied for a longer period than any other group of terrestrial invertebrates. Early in this century, Steudel (1913) fed iron saccharate and iron lactate to a range of insects and demonstrated, in a series of beautiful coloured illustrations of histochemically stained sections, that the cells of the midgut of the ant lion *Myrmeleon*, the carabid beetle *Carabus*, the bumble bee *Bombus* and the hornet *Vespa*, accumulated this element in granular deposits but that the metal could not be detected in the midgut of the cockroach *Periplaneta*, the cockchafer *Melolantha* or the mole cricket *Gryllotalpa*. This work spawned many other histochemical studies of a similar nature on the distribution of iron in insects which were summarised by Lotmar (1938).

In the 1940s and 1950s, the most detailed observations made to date by light microscopy on metals in insects were carried out by Waterhouse and colleagues on *Lucilia*, the blowfly parasite of sheep (Waterhouse 1940, 1945a, 1950; Waterhouse & Stay 1955), the fruit fly *Drosophila* (Poulson 1950a, 1950b; Poulson & Bowen 1952; Poulson *et al.* 1952) and the larvae of the clothes moth *Tineola* (Waterhouse 1952).

With the widespread introduction of electron microscopes into laboratories during the 1960s, researchers were able to examine the fine structure of the digestive and excretory tissues which had been shown, by light microscopy and histochemistry, to accumulate metals. During the late 1960s and early 1970s, French researchers were active in applying the newly developed technique of X-ray microprobe analysis to examine simultaneously the ultrastructure and composition of the metal-containing inclusions which were being discovered within cells. While these studies lacked sophistication due to the infancy of the technique, they laid a foundation for more recent detailed work.

Since the advent of commercial energy-dispersive X-ray microanalysers in the mid 1970s, the number of publications on metals in insects in which this technique has been used has increased greatly and a clearer picture has emerged of the ways in which these elements are stored. More recently, attention has turned to metal-binding proteins, particularly metallothioneins (Everard & Swain 1983; Aoki & Suzuki 1984; Aoki *et al.* 1984), glycoproteins (Clubb *et al.* 1975) and ommochromes, a group of pigments derived from the amino acid tryptophan (Martoja & Truchet 1983). A discussion of the methods of storage of metals in digestive and excretory epithelia of insects is included in the review by Martoja & Ballan-Dufrançais (1984).

9.5.4.2 Thysanura (bristletails)

A small number of studies were made by French workers during the early 1970s on *Petrobius maritimus*, the common supra-littoral thysanuran. The midgut cells contain numerous type A granules (Fain-Maurel *et al.* 1973; Cassier *et al.* 1974). These, and similar granules in the fat body, Malpighian tubules and epidermis, are composed of calcium, magnesium, potassium, iron, zinc and sodium (Martoja 1972, 1974).

9.5.4.3 Collembola

The midgut cells in representatives of the family Tomoceridae contain numerous type A granules in which calcium, magnesium and sodium have been detected (Humbert 1974a, 1977, 1978, 1979). The granules are formed in the rough endoplasmic reticulum, grow by accretion of concentric layers and are capable of binding a wide range of class A and borderline elements from the diet including aluminium, silicon, iron, uranium, lead, strontium and manganese. These granules may perform a detoxificative function since the whole epithelium of the gut, together with the metal-containing granules, is lost during moulting which is frequent in Collembola (Joosse & Verhoef 1983; Van Straalen *et al.* 1987). The resistance of *Orchesella cincta* to high concentrations of lead in the diet may be due to the ease with which it can excrete the metal bound in granules when it moults (Joosse & Buker 1979).

9.5.4.4 Orthoptera

The Malpighian tubules of crickets contain intracellular type A calcium phosphate 'spherocrystals' of up to 2 µm in diameter (Berkaloff 1958; Bell & Anstee 1977) in which potassium, magnesium and manganese have been detected also (Ballan-Dufrançais & Martoja 1971; Lhonoré 1971). More recent work on locusts has shown that copper, zinc, silver

---

FIG. 9.15 Transmission electron micrograph of 'spherocrystals' (s, type A granules) and 'lysosomes' (ly, type B granules) in the ileal cells of the cockroach *Blatella germanica* fed on a diet enriched with mercury. c, cuticular lining of the ilium; m, mitochondrion. Scale bar = 1 µm. (From Jeantet *et al.* 1980, by permission of Éditions Scientifiques Elsevier.)

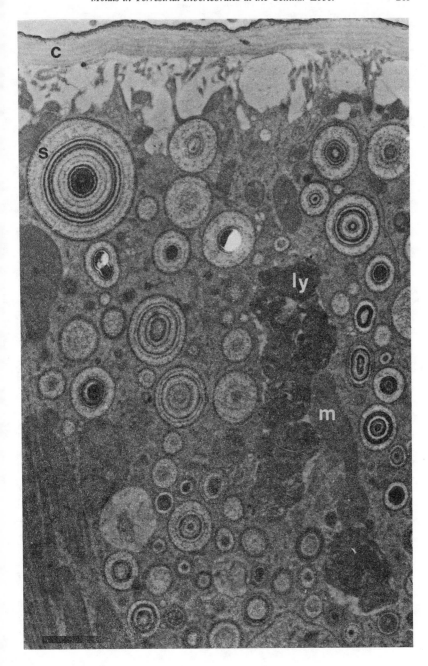

and cadmium injected in solution into the blood of *Locusta migratoria* are bound to ommochromes in the nephrocytes around the heart (Martoja & Truchet 1983). The authors suggested that ommochromes perform a function attributed to metallothioneins in other species of animals. However, the way in which locusts bind metals which enter the blood from the lumen of the gut needs to be investigated before the absence of metallothioneins can be confirmed.

9.5.4.5 Dictyoptera

Knowledge of the ultrastructural distribution of metals in cockroaches has increased significantly in recent years as a result of research by French workers. Preliminary studies by Ballan-Dufrançais (1972, 1974) and Jeantet *et al.* (1977) have shown that various organs of cockroaches, in particular the ileum, Malpighian tubules and fat body, contain intracellular inclusions in which a range of metals including zinc, cadmium and iron can be detected. Three more recent publications by these authors have concentrated on the dynamics of mercury accumulation by the cells of the ileum (Ballan-Dufrançais *et al.* 1979, 1980; Jeantet *et al.* 1980).

In these studies, specimens of *Blatella germanica* were fed on honey which contained 20 μg g^{-1} of methyl mercury chloride (CH_3HgCl), or mercury chloride ($HgCl_2$) for periods of from six hours to a week before fixation for electron microscopy. Metal-containing inclusions which formed in the cells of the gut were examined subsequently by quantitative X-ray microanalysis.

The cockroaches showed no obvious ill effects of the treatment during the experiments. Two types of inclusion were observed in the ileal cells of control animals, type A granules (Fig. 9.15) in which calcium, magnesium, iron, potassium, zinc and phosphorus were detected, and 'lysosomes' with the properties of type B granules which occasionally contained electron-dense inclusions of copper, zinc and sulphur. In

---

FIG. 9.16 Transmission electron micrographs of metal-containing granules in the ileal cells of the cockroach *Blatella germanica* fed on a diet enriched with mercury. (From Jeantet *et al.* 1980, by permission of Éditions Scientifiques Elsevier.) (A) The type B 'lysosomes' contain dense inclusions (Hg) composed of mercury, zinc, copper and sulphur. Scale bar = 0.5 μm. (B) 'Mineral concretions' ('type B core', 'type A periphery') from a cockroach exposed to mercuric chloride for 7 days. Outer concentric layers contain calcium and phosphorus whereas the electron-dense cores contain mercury, zinc, copper and sulphur. Scale bar = 0.5 μm.

A

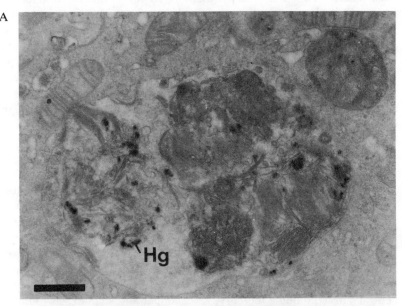

B

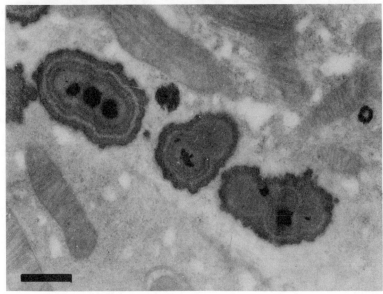

cockroaches fed on mercury-contaminated food, the electron dense inclusions in the 'lysosomes' were much more numerous and contained mercury (Fig. 9.16A). The type A granules never contained this metal although a zinc-rich layer was occasionally observed on the periphery which may have formed from zinc displaced by mercury from the type B granules. Semi-quantitative X-ray microanalysis showed that there was never more than one atom of mercury and/or zinc to every three atoms of sulphur while the atomic ratio of copper to sulphur was as high as two. Molecular storage of the metals can therefore be represented by:-

$$(Hg \leftrightarrow Zn \leftrightarrow Cu) \, S_3 + Cu_n$$

The mercury, zinc and some copper were probably bound to sulphydryl groups (S_3) in low molecular weight metallothioneins. Copper not in this form may have been associated with another protein of higher molecular weight. Consequently, n can vary independently of the metallothionein storing process.

In the ileal cells of cockroaches exposed to mercury for more than four days, a third type of inclusion was formed, probably from residues of type B granules. These 'mineral concretions' were composed of a central core containing mercury, zinc, copper and sulphur, surrounded by irregular concentric (sealing?) layers of material in which calcium and phosphorus were detected (Fig. 9.16B). These 'hybrid' granules composed of a 'type B core' and a 'type A periphery', were not extruded and may be stored permanently in the ileal cells. It is probable that if silver and cadmium were included in the diet, they would follow the same detoxification pathway as mercury and be stored in type B granules also.

Such an efficient system for the detoxification of potentially toxic metals may enable cockroaches to eat a much more heterogeneous diet than most other insects and may have been one of the factors which 'pre-adapted' them for the exploitation of human habitats.

9.5.4.6 Isoptera

The cells of the digestive epithelia and pericardium of termites possess concentrically-structured granules in which a wide range of elements have been detected (Martoja & Martoja 1973; Martoja *et al.* 1975; Alibert & Martoja 1976). These include class A and borderline metals such as calcium, magnesium, potassium, manganese, iron, zinc and sodium.

Several species of termites possess a very rich symbiotic gut microflora and fauna and it would be interesting to examine whether the micro-organisms accumulate metals in the same way as those in the hepato-pancreas of the isopod *Oniscus asellus* (Hopkin & Martin 1982b).

9.5.4.7 Hemiptera

The Order Hemiptera is split into two sub-orders, the Heteroptera and the Homoptera. The only study on Heteroptera in which granules have been observed was that of Wigglesworth & Saltpeter (1962) on the blood-sucking bug *Rhodnius prolixus* (the 'white rat' of insect physiologists!). These authors showed that cells in the upper part of the Malpighian tubules contained concentrically-structured granules. The appearance of these inclusions in thin section suggests that they are type A granules although this will only be confirmed following analysis of their composition by X-ray microanalysis. Far more attention has been paid to the Homoptera which includes the cicadas (Cicadidae), plant-or frog-hoppers (Cercopidae), aphids (Aphidae) and scale insects (Coccoidea).

One of the pioneering studies on intracellular type A granules was made by Gouranton (1967) on the larvae and adults of the plant-hoppers *Philaenus spumaris* and *Aphrophora alni* (originally observed in *P. spumaris* early in this century by Licent (1912)). Gouranton showed that the granules in the midgut of cells of the larvae and adults of the two species had a diameter of up to 2 μm, were concentrically-structured and contained calcium, magnesium, iron, carbonates and phosphates, but did not contain uric acid or guanine. The granules were initiated as small spheres in the rough endoplasmic reticulum, increased in size by the addition of material to their outer surface and were expelled eventually into the lumen and excreted. This process was observed in both larval and adult stages.

More recent studies on the chemical composition of the type A granules in cicadas and plant-hoppers have confirmed that they are composed of calcium and magnesium phosphates. In the larvae, some of the granules may be passed eventually out of the body, dissolved, and the calcium incorporated into calcium carbonate in the dwelling tube (Cheung & Marshall 1973; Marshall & Cheung 1973a).

The fat body of plant-hoppers contains typical type B granules composed of copper and sulphur in an atomic ratio of 3:2 (Marshall 1983). Copper, it was suggested, is stored in these animals either as an end product of copper detoxification, or as a store to aid synthesis of copper-containing enzymes by the mycetial symbionts.

Type C granules were first detected in Homoptera in the midgut cells of the scale insect *Quadraspidiotus* (Bielenin & Weglarska 1965). More recent studies incorporating transmission electron microscopy and X-ray microanalysis have confirmed the presence of iron and demonstrated that the type C granules are composed of ferritin (Cheung & Marshall 1973; Kimura *et al.* 1975).

Considering their economic importance as pests, it is surprising that so little work has been carried out on the ultrastructural distribution of metals in aphids. Forbes (1964) described the fine structure of the gut of *Myzus persicae* and included a micrograph of a type A granule but made no comment as to its function. At about the same time, a comprehensive study by light microscopy was performed by Ehrhardt (1965a, 1965b) on 25 species. The midgut cells of most species contained numerous type A granules of up to 5 μm in diameter composed of calcium and magnesium phosphates and carbonates. When class A or borderline metals such as barium, strontium, iron, cobalt and manganese were added to the diet, they were incorporated rapidly into the type A granules in the midgut epithelium.

9.5.4.8 Lepidoptera

The most comprehensive study by light microscopy on the uptake and cellular distribution of metals in any terrestrial invertebrate was made by Waterhouse (1952) on the larvae of the clothes moth *Tineola biselliella*. The study was initiated to assess the potential of a range of elements as insecticides when incorporated into wool. However, none of the 30 metallic and five non-metallic elements tested had any effect on the larvae.

In the lumen of the gut, the disulphide bonds of the cysteine present in the wool were reduced due to the extreme alkalinity of the digestive secretions (pH 10). The resulting sulphydryl groups formed insoluble complexes with the metals capable of forming sulphides such as zinc, iron, cadmium, thallium, cobalt, nickel, tin, lead, antimony, bismuth, arsenic, copper, tellurium, osmium, mercury, silver, palladium, platinum and gold. These passed harmlessly along the digestive tract and were voided in the faeces. Metals were bound also to amino acids and polypeptides liberated by digestion of the food, or present in the digestive secretions forming highly-dispersed colloidal solutions.

The midgut epithelium of *Tineola* is composed of two cell types, the 'goblet' cells and the 'columnar' cells. The metal colloids were taken up and stored as sulphur-rich type B granules in a large vacuole in each goblet cell. Class A elements incapable of forming insoluble sulphides such as barium, calcium and magnesium, were deposited as phosphates in type A granules in the columnar cells. During moulting, the entire midgut epithelium, together with any accumulated metals, is cast off and regenerated. Clothes moth larvae are therefore able to detoxify a wide range of metals and non-metals because of the unusual micro-environment of their gut.

More recent studies by electron microscopy on the silkworm *Bombyx mori* have confirmed the observations of Waterhouse on *Tineola* and have shown that type A granules composed of calcium phosphate are present in the columnar cells but not in the goblet cells of the larval midgut (Waku & Sumimoto 1971; Turbeck 1974). The Malpighian tubules in the larval and pupal stages of this species also contain type A granules but these must be lost during metamorphosis as they can not be detected in adult silkworms. Formation of type A granules has been induced in the Malpighian tubules of *Galleria mellonella* (Le Garff & Barbier 1982).

When cadmium and zinc are injected orally as chlorides into the larvae of *Bombyx mori,* the metals cause substantial cell damage and are dispersed throughout the organs of the body (Akai 1975; Tanaka *et al.* 1982). However, in these experiments the metals were administered in such hugh amounts (30 µl of a 2% solution of cadmium chloride in a single dose) that the results are of little relevance in understanding how the caterpillars would respond to metals in even the most heavily polluted sites. Some tissues are quite sensitive to relatively low levels of cadmium. The gonads of larvae of *Scotia segetum,* for example, suffered cellular damage when their diet was supplemented with 50 µg g^{-1} of cadmium chloride (Zelenayova & Weismann 1985).

Locke & Leung (1984) conducted a comprehensive study by electron microscopy on the induction and distribution of ferritin in larvae of *Calpodes ethlius* (Hesperiidae). In contrast to vertebrate cells where ferritin is confined to the cytosol and lysosomes, Locke & Leung were able to demonstrate that in *Calpodes,* ferritin occurred in the vacuolar system of the smooth endoplasmic reticulum (SER) and not in the cytosol. Ferritin was found naturally in the rough endoplasmic reticulum and SER of cells at the hind end of the midgut, in pericardial cells and in the 'yellow region' of the Malpighian tubules. Additional ferritin could be induced by loading the gut or haemolymph with iron. Overloading with iron caused ferritin to be secreted into the lumen of the gut. Locke & Leung proposed that the SER in these cells functions in iron homeostasis by holding ferritin for loading and unloading as it moves to and from the reticulum at the cell surface where it can be maximally exposed to extracellular fluid flow. Transferrin may play a part in this process also (see Section 9.3.3).

9.5.4.9. Diptera

The dynamics of metal accumulation has been studied more intensively in Diptera than in any other insect order. Indeed, some 20 research papers relevant to this topic have been published since the pioneering work of Waterhouse (1940) on the uptake of iron by the larvae of *Lucilia cuprina,* a blowfly parasite of sheep. Studies have concentrated on the midgut and Malpighian tubules, the organs most involved in uptake storage and excretion of metals.

The housefly *Musca domestica* has been studied in the greatest detail (Sohal 1974; Sohal *et al.* 1976, 1977; Sohal & Lamb 1977, 1979). The key experiments in this work involved feeding adult flies on sucrose diets supplemented with salts of zinc, calcium, copper and iron. Subsequent analyses of tissues showed that calcium and zinc were stored mainly in the Malpighian tubules whereas copper and iron were found primarily in the midgut. The metals were stored in both these organs, intracellularly, in type B granules rich in sulphur and phosphorus. These granules increased in number with age of the flies until they occupied a considerable proportion of the volume of the cytoplasm in a 40-day-old fly (Fig. 9.17). Sohal *et al.* (1976) classified the intracellular granules in the Malpighian tubules into three categories but these probably represent stages in the development of a single type. The organic nature of the type B granules can be confirmed by heating them to 500 °C when only a residue of metals remains (Fig. 9.18).

In addition to the type B granules, the lumen of the Malpighian tubules contained concentrically-structured type A granules composed of calcium and phosphorus (Krueger *et al.* 1987). These granules were probably formed within the cells of the Malpighian tubules and expelled into the lumen. Calcium-containing granules which develop extra-

---

FIG. 9.17 Transmission electron micrograph of type B granules in the midgut epithelium of the housefly *Musca domestica* stained by the Timm's sulphide–silver method for the localisation of 'heavy' metals recommended by Pearse (1972). The grains of silver (Ag) are associated only with the granules. lum, lumen. Scale bar = 0.5 μm. (From Sohal *et al.* 1977, by permission of Longman Group UK Ltd.)

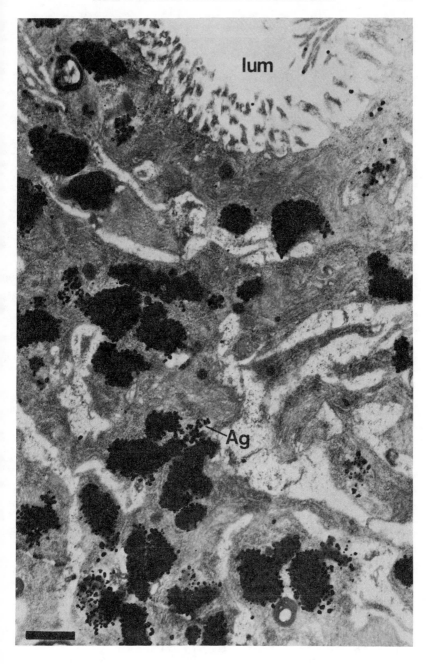

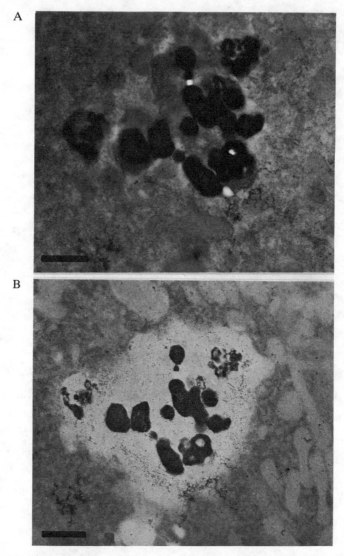

A

B

FIG. 9.18 Transmission electron micrographs of unstained type B granules in the midgut epithelium of the housefly *Musca domestica* before (A) and after (B) incineration at 500°C. Note that the dense material in which most of the metals are located is not lost. Scale bars = 0.2 μm. (From Sohal *et al.* 1977, by permission of Longman Group UK Ltd.)

cellularly are normally composed of calcium carbonate (type D) rather than calcium phosphate (however see Section 9.5.3.2 for a possible exception to this rule in amphipods).

In the Calliphoridae, an extensive series of observations have been made by light microscopy on the uptake of iron and copper by the larvae of *Lucilia cuprina*, a blowfly parasite which develops in open sores on sheep (Waterhouse 1940, 1945a, 1950; Waterhouse & Stay 1955). Copper was stored by the cells throughout the length of the midgut. Iron, however, was accumulated only in the mid-midgut where conditions in the lumen were very acidic (pH 3 to 4) and where iron was present in the ferrous state (Fig. 6.3). An efficient system for controlling iron is essential in an animal which feeds largely on the blood of vertebrates with its high iron content. Waterhouse (1940) suggested that insecticides soluble at pH 3 to 4 could be applied to infected sores. These would be harmless to sheep but would be absorbed in the acidic portion of the midgut of *Lucilia*. It is not known whether this suggestion was followed up commercially. Waterhouse & Wright (1960) showed that the copper and iron in the midgut could be stored in two different cell types. However, their studies did not include X-ray microanalysis of the inclusions observed.

When calcium and magnesium are added to the diet of *Lucilia*, the metals form type A granules in the lumen of the Malpighian tubules. Type A granules composed of calcium and magnesium phosphates occur also in the epithelial cells of the Malpighian tubules of the mosquito *Culex* (Houk 1977) and the New Zealand glow worm *Arachnocampa* (Green 1979). These observations lend weight to the theory that the type A granules in the lumen of the Malpighian tubules of *Musca domestica* (Sohal *et al.* 1976) were formed in the cells before expulsion into the lumen. Similar granules in the Malpighian tubules of the closely related *Calliphora* contain the class A and borderline metals calcium, phosphorus, magnesium, potassium and zinc (Martoja & Seureau 1972) which may be excreted, or may store calcium for subsequent use by the ovaries (Taylor 1984a).

The midgut of the flesh fly *Sarcophaga*, another member of the Calliphoridae, contains two types of intracellular granule (Nopanitaya & Misch 1974). The composition of these inclusions has not been determined but their ultrastructural appearance suggests that they are type A and type B granules. The midgut of *Sarcophaga* is also important in the storage of cadmium. Indeed, 90% of the metal accumulated by the larvae from a cadmium-enriched diet is retained in this organ (Aoki

& Suzuki 1984; Aoki *et al.* 1984). The metal is bound in metallothionein proteins, the residues of which are probably stored in type B granules in the cells of the midgut (Maroni & Watson 1985; Maroni *et al.* 1986).

Two cell types have been recognised in the midgut of the fruit fly *Drosophila*, the copper-containing 'cupriophilic' cells and the non copper-containing 'interstitial' cells (Poulson 1950a; Poulson & Bowen 1952; Poulson *et al.* 1952; Filshie *et al.* 1971). In cupriophilic cells, copper is stored with sulphur in type B granules, sometimes together with zinc (Tapp 1975; Tapp & Hockaday 1977). Large amounts of iron may be stored in these granules also if the diet is enriched with this element (Poulson & Bowen 1952). Deposits of copper are detected rarely in the other cell type, the interstitial cell (Filshie *et al.* 1971). Type A granules occur in the lumen of the Malpighian tubules of *Drosophila* but have not been observed in the cells of the midgut (Wolburg *et al.* 1973). Damage to chromosomes has been observed in *Drosophila* exposed to cadmium chloride in the diet (Kogan *et al.* 1978).

The genetic basis of tolerance to cadmium and copper in *Drosophila* is currently being studied in great detail by a number of workers (Wallace 1982; Lastowski-Perry *et al.* 1985; Maroni *et al.* 1986a, 1986b, 1987; Otto *et al.* 1986, 1987; Mokdad *et al.* 1987). In wild populations, tolerance may have evolved through selective pressures resulting from the application of copper-containing fungicides to fruit. The mechanism appears to consist of duplication of the gene which codes for metallothionein so that more of the detoxifying protein is synthesised by the larvae in response to a 'metal insult'.

The midgut cells of most adult Diptera do not undergo mitosis so the metals which are accumulated remain in the epithelium until the flies die. Such a system of storage-excretion (similar to that in isopods and cockroaches) may utilise less energy than would be required to continuously break down and replace the cells.

9.5.4.10 Hymenoptera

The midgut epithelium of the ant *Formica polyctena* contains con-centrically-structured type A granules in which calcium, potassium, magnesium, manganese, iron and zinc have been detected by electron microprobe and secondary ionic emission microanalysis (Jeantet *et al.* 1974). Sixteen other metals have been detected in trace amounts (Martoja *et al.* 1975; Jeantet *et al.* 1977) but their apparent presence in granules may have been artifactual due to specimen contamination or the nature of the analytical methods employed. These observations need to be confirmed, especially since large numbers of type A granules are voided periodically into the lumen of the midgut by apocrine secretion which would provide a route for the excretion of unwanted metals.

Intracellular concentrically-structured granules have been observed also in the Malpighian tubules of the bee *Melipona quadrifasciata* but their elemental composition has not been determined (Mello & Bozzo 1969).

Few authors have examined the effect of metal-based pesticides on the ultrastructure of invertebrates. In one of the studies on insects, Berndt (1974) showed that the reduction in fecundity of colonies of the Pharoah's Ant *(Monomorium pharaonis)* treated with sodium arsenate, was due to breakdown of the cells of the germanium in the reproductive tissues of queens. Detailed knowledge of the specific sites of action of toxic metals may enable more effective insecticides to be developed.

It has been known for many years that iron is stored in the cells of the midgut of bees when the metal is added to their diet (Lotmar 1938). However, it has been suggested recently that the metal could function as an aid to navigation (Kuterbach *et al.* 1982; Kuterbach & Walcott 1986a, 1986b). The body wall of each abdominal segment and the fat body of *Apis mellifera* contain cells which are rich in granules containing iron (type C granules?) which may enable the bees to detect changes in position relative to the earth's magnetic field.

9.5.4.11 Coleoptera
There are more species in the Coleoptera than in any other insect order. It is surprising, therefore, that hardly any work on the uptake and cellular distribution of metals in beetles has been carried out. Preliminary observations on the fat body, midgut and Malpighian tubules of five species have demonstrated the presence of intracellular granules in which calcium, potassium and iron have been detected (Bayon & Martoja 1973, 1974). The composition suggests classification as type A granules although more detailed studies by electron microscopy are required before this can be confirmed.

9.5.5 Myriapods
Concentrically-structured type A granules composed predominantly of calcium and phosphorus have been observed in the midgut of julid millipedes (Hubert 1974, 1975, 1978a, 1978b, 1979; Martoja *et al.* 1977). Traces of manganese, copper, magnesium and zinc have been detected also. These granules are lost periodically when the entire midgut epi-

thelium is replaced during moulting (Hubert 1979). The 'hepatic tissue' which forms a layer around the midgut, may contain granules rich in iron (Hubert 1979). Other tissues, including the fat body and Malpighian tubules contain dense inclusions but their composition has not been studied in detail.

The eggs of millipedes contain concentrically-structured type A granules composed of calcium and phosphorus (Petit 1970) which may represent sites of storage supplying the mineral needs of the developing embryo (Crane & Cowden 1968).

The only published observations on the distribution of metals in centipedes have demonstrated the presence of concentrically-structured 'lime bodies' in the Malpighian tubules of *Lithobius forficatus* (Fuller 1966) and *Scolopendra subspinipes* (Wang & Wu 1948). However, I have recently detected type A granules containing calcium, phosphorus, zinc and lead in the cells of the midgut of *Lithobius variegatus*, which are voided into the lumen of the digestive tract and expelled in the faeces.

The suggestion by Needham (1960) that the connective tissue pigment lithobioviolin was copper-based, was later withdrawn (Bannister & Needham 1971).

9.5.6 Arachnids
Concentrically-structured granules have been observed in the digestive epithelia of mites (Wright & Newell 1964), scorpions (Martoja & Martoja 1973) and the opilionid *Phalangium opilio* (Becker & Peters 1985a, 1985b) but no ultrastructural studies directly concerned with metals in these groups have been performed.

In spiders, the hepatopancreas of *Dysdera crocata* contains numerous type A and type B granules which may occur in the same cell (Fig. 9.19). In spiders I have collected from urban localities, the type A granules contain calcium, phosphorus, zinc and lead and the type B granules, copper and sulphur. The presence of lead in the type A granules in the hepatopancreas of *Coelotes terrestris* has been demonstrated recently by Ludwig & Alberti (1988). In *Dysdera crocata*, large numbers of these metal-containing granules are lost in the faeces after every meal. The ultrastructure of the digestive tract of spiders has been reviewed recently by Collatz (1986).

9.5.7 Parasites
9.5.7.1 Cestodes (tapeworms)
One of the most detailed studies on the formation and composition of

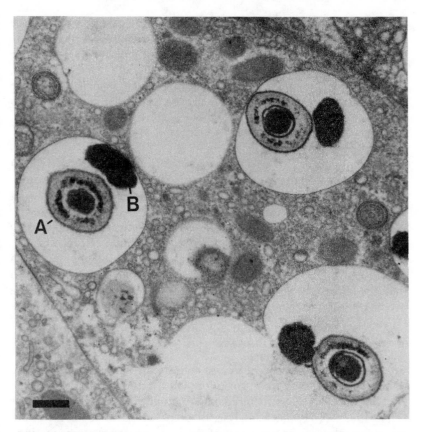

FIG 9.19 Transmission electron micrograph of type A (A) and type B (B) granules in the hepatopancreas of the spider *Dysdera crocata*. The type A granules contain calcium, phosphorus, zinc and lead, and the type B granules copper and sulphur. The cell is breaking down at the end of a feeding cycle allowing the metal-containing granules and other waste material to be voided in the faeces. Scale bar = 1 μm. (Previously unpublished.)

type A granules in any animal was made in a series of papers by Von Brand and colleagues on tapeworms (Von Brand *et al.* 1960, 1965; Scott *et al.* 1962; Nieland & Von Brand 1969). The granules, which occupy a considerable volume of the proglottids, are composed mainly of amorphous calcium and magnesium phosphates although a wide range of other elements may occur in trace amounts also (Kegley *et al.* 1969).

The granules in *Mesocestoides corti* in the gut of mice, incorporate phosphorus or strontium when these elements are added as the chloride to the drinking water of their hosts (Von Brand *et al.* 1965). It would be interesting to repeat these experiments adding zinc, silver, mercury, lead, cadmium, copper etc. to the diet of the host animals to compare the ultrastructural response of tapeworms to borderline and class B metals.

9.5.7.2 Trematodes (flukes)

The excretory ducts of trematodes contain large numbers of type D granules composed of calcium carbonate (Martin & Bils 1964; Erasmus 1967). These extracellular granules may be involved in the fixation of carbon dioxide and the buffering of host acids.

Drugs containing antimony have proved most effective in suppressing the fecundity of the adult stage of *Schistosoma mansoni* which lives in the hepatic portal vein of man and causes the debilitating and often fatal disease bilharzia. X-ray microanalysis of treated parasites has shown that the metal accumulates in the vitelline gland and prevents the production of eggs which are the cause of most of the tissue damage (Erasmus 1974).

9.5.7.3 Nematodes

The intestinal cells of *Trichuris suis* collected from the digestive caeca of pigs, and *T. muris* from the colon and caecum of laboratory mice, contain concentrically-structured type A granules (Jenkins 1973; Jenkins *et al.* 1977). X-ray microanalysis of cryosections of these inclusions has demonstrated the presence of calcium, phosphorus, magnesium, potassium and traces of iron. The granules may function as reservoirs of phosphates, or for the storage and subsequent excretion of waste products since the Trichuroidea lack a conventional tubular or glandular excretory system.

CHAPTER 10

Concluding Remarks

It is usual in review books to conclude by saying that more research is required on the topic in question. While this is clearly the case for the areas covered in this text, what is needed is new approaches with experiments designed to answer specific questions. Otherwise we shall end up in ten years time with a quantitative (i.e. twice as much literature) rather than a qualitative increase in our knowledge of the ecophysiology of metals in terrestrial invertebrates. In this final chapter, some suggestions are made of research areas which are most in need of further study.

On a *global level*, the overriding problem of metals as pollutants is their persistence (Nriagu 1988). Once released into the environment, they are not broken down into harmless components as is the case with, for example, sewage effluent. If pollutant metals were dispersed evenly throughout the biosphere then toxicity would not be a problem. Such dispersal has clearly occurred with lead, which has been detected in polar snow and ice (Section 3.1), but just because an increase in concentrations of lead in plant and animal tissues has taken place during the last century does not necessarily give cause for concern. Metal pollutants cause problems only when they become concentrated locally such as in the soil (Chapter 5), or in the case of mercury in Minimata Bay in Japan, in a specific food item (Section 2.4) (for an amusing and thought-provoking discussion of this topic, see Mackay 1988).

At the *ecosystem* level, the soil/litter microcosm is the main 'sink' of metal pollutants. A considerable amount of information is available on the distribution of metals such as zinc, cadmium, lead and copper in the terrestrial biota but studies are required on other elements which are currently released such as arsenic and mercury, and most radioisotopes. Interactions with other pollutants such as 'acid rain' are clearly important in affecting the movement of metals in soils (Martin & Coughtrey 1987) and should be considered also when assessing the toxic effects of metals in sites which are subject to pollution from a range of industrial concerns. Biological monitoring of the dispersal and effects of pollutants (Chapter 8) should be conducted around all industrial sites which release metals into the environment. This will enable the distribution of metals throughout the biota to be estimated but fluxes can be calculated only from information on the ecology of the plants and animals concerned. Use of new analytical techniques which give a 'spectrum' of all elements present (such as ICPES — Section 4.4) is to be encouraged as these may provide an early warning of the build up of 'novel' pollutant metals which might be missed with a more conventional method such as atomic absorption spectrometry.

Novel research areas might include field and laboratory measurements of fluxes of metals through the soil and litter profiles using microcosms (Anderson & Ineson 1982; Anderson *et al.* 1983; Bengtsson *et al.* 1988), and the possibility that plants withhold essential trace metals from pests to limit their growth. The latter suggestion is not as far-fetched as it might sound since plants have been shown to withhold major nutrients from their leaves in response to defoliating insects (Lawton 1986). Indeed, the early experiments of Creighton (1938) showed that the growth rates of several species of insects feeding on zinc-and copper-deficient plants were drastically reduced. Such effects could be particularly important in sparsely-vegetated ecosystems such as deserts and polar regions.

At the *species* level, it is still not known whether any terrestrial invertebrate (with the exception of *Drosophila* — Section 9.5.4.9) has evolved genetic tolerance to metals in the diet, despite the fact that this phenomenon has been demonstrated clearly in plants (Section 5.1.2). Studies such as those by Hopkin & Martin (1984b) on centipedes, have suggested that some invertebrates in contaminated sites possess tolerance, but breeding experiments in the laboratory need to be conducted before this can be proved to be inheritable rather than phenotypic. It should be stressed that tolerance is most likely to have developed in populations which have been subjected to very heavy contamination for many years.

Perhaps the most intriguing question which needs to be answered is why closely-related species fed on the same diet accumulate the same metals to different concentrations in their tissues, despite there being no apparent differences in the structure and function of their digestive systems? (e.g. in isopods — Hames & Hopkin, in press). Further studies are needed at the *cellular* level to elucidate whether discrimination is taking place at the cell membrane exposed to the digestive fluids and whether rates of turnover of metals differ between species. In herbivores and detrivores, such differences could be due, in part, to the composition of the intestinal microflora. The influence of microorganisms on metal uptake in terrestrial invertebrates is in need of much more research.

In the same way as *fluxes* of metals through ecosystems are difficult to calculate from static pictures of concentrations in different components, the dynamics of metal movements at the cellular level are difficult to interpret from preserved cells. We know broadly the sites of intracellular storage of particular metals (storage proteins and the type A, B and C granule system described in Chapter 9) but the specific transport routes and rates of metal transfer along them have yet to be elucidated. Studies are needed with radioisotopes followed by autoradio-graphy, and immuno-staining techniques such as those used to detect metallothioneins by Clarkson *et al.* (1984). Once such routes have been identified, it may be possible to develop a new generation of insecticides which interfere with such pathways.

The theme which has been stressed throughout this book is that regulation of metal uptake in terrestrial invertebrates is as essential to their survival as water balance, efficient digestion of food and excretion of nitrogenous wastes. There are still invertebrate Phyla and metals about which next to nothing is known and exciting discoveries are waiting to be made by 'metalworkers' who approach their research in the right way. By the 'right way', I mean planning experiments designed to answer specific questions, the results of which provide information which helps us to understand how the animals live in the field rather than under a narrow set of (often unrealistic) laboratory parameters.

Molar-Weight Conversion Table for Some Metals Mentioned in the Text

Metal	a.w.	Z	g cm^{-3}	1 μmol =	1 μg =
Ca	40.08	20	1.54	40.08 μg	24.95 nmol
Cr	52.00	24	7.20	52.00 μg	19.23 nmol
Mn	54.94	25	7.44	54.94 μg	18.20 nmol
Fe	55.85	26	7.88	55.85 μg	17.90 nmol
Co	58.93	27	8.90	58.93 μg	16.97 nmol
Ni	58.71	28	8.90	58.71 μg	17.03 nmol
Cu	63.55	29	8.93	63.55 μg	15.74 nmol
Zn	65.37	30	7.14	65.37 μg	15.30 nmol
Cd	112.40	48	8.65	112.4 μg	8.897 nmol
Hg	200.59	80	13.59	200.6 μg	4.985 nmol
Pb	207.19	82	11.34	207.2 μg	4.826 nmol

a.w., atomic weight; Z, atomic number; g cm^{-3}, density.

Suppliers of Certified Biological Reference Materials

National Physical Laboratory
Teddington
Middlesex TW11 0LW
UK

Marine Analytical Standards Program
National Research Council
Atlantic Research Laboratory
1411 Oxford Street
Halifax
Nova Scotia
Canada B3H 3Z1

Office of Standard Reference Materials
Room B311
Chemistry Building
National Institute of Standards and Technology
Gaithersburg
Maryland 20899
USA

References

Abbasi, S.A. & Soni, R. (1983). Stress-induced enhancement of reproduction in earthworm *Octochaetus pattoni* exposed to chromium (VI) and mercury (II) — implications in environmental management. *Int. J. Environ. Stud.* **22**: 43-48.

Abo-Elghar, M.R. & Radwan, H. (1971). Toxicological, chemosterilant and histopathological effects of triphenyl-tin-hydroxide on *Spodoptera littoralis* Boisd. *Acta Phytopathol. Acad. Sci. Hung.* **6**: 261-280.

Abolinš-Krogis, A. (1965). Electron microscope observations on calcium cells in the hepatopancreas of the snail *Helix pomatia*, L. *Ark. Zool.* **18**: 85-92.

Abolinš-Krogis, A. (1970a). Electron microscope studies of the intracellular origin and formation of calcifying granules and calcium spherites in the hepatopancreas of the snail *Helix pomatia*, L. *Z. Zellforsch. Mikrosk. Anat.* **108**: 501-515.

Abolinš-Krogis, A. (1970b). Alterations in the fine structure of cytoplasmic organelles in the hepatopancreatic cells of shell-regenerating snail, *Helix pomatia* L. *Z. Zellforsch. Mikrosk. Anat.* **108**: 516-529.

Abolinš-Krogis, A. (1976). Ultrastructural study of the shell-repair membrane in the snail *Helix pomatia* L. *Cell Tiss. Res.* **172**: 455-476.

Adams, J. (1984). The habitat and feeding ecology of woodland harvestmen (Opiliones) in England. *Oikos* **42**: 361-370.

Adriano, D.C. (1986). *Trace Elements in the Terrestrial Environment*. Berlin, Springer-Verlag.

Ahmad, M. & Khan, N.H. (1980). Chemical repellants for *Dysdercus koenigii* (Fabr.). *Ind. J. Entomol.* **42**: 820-821.

Akai, H. (1975). Characteristic ultrastructural changes in the midgut cells of *Bombyx* larvae following administration of cadmium chloride. *Appl. Entomol. Zool.* **10**: 67-76.

Akao, A. (1939). Beiträge zum Wachstumsphänomen des Seidenspinners. Die verschiedenen aufbauenden und katalytischen Elemente und deren biologische Bedeutung während des Wachstums. *J. Biochem.* **30**: 303-349.

Akey, D.H. & Beck, S.D. (1971). Continuous rearing of the pea aphid *Acyrthosiphon pisum*, on a holidic diet. *Ann. Entomol. Soc. Am.* **64**: 353-356.

Akey, D.H. & Beck, S.D. (1972). Nutrition of the pea aphid, *Acyrthosiphon pisum*: requirements for trace metals, sulphur and cholesterol. *J. Insect Physiol.* **18**: 1901-1914.

Alary, J., Bourbon, P., Esclassan, J., Lepert, J.C., Vandaele, J. & Klein, F. (1983). Zinc, lead and molybdenum contamination in the vicinity of an electric steelworks and environmental response to pollution abatement by bag filter. *Water Air Soil Pollut.* **20**: 137-145.

Alibert, J. & Martoja, R. (1976). Bioaccumulation minérale et purique chez les termites. *Insectes Sociaux* **23**: 49-74.

Alikhan, M.A. (1972a). Haemolymph and hepatopancreas copper in *Porcellio laevis* Latreille (Porcellionidae, Peracarida). *Comp. Biochem. Physiol.* **42A**: 823-832.

Alikhan, M.A. (1972b). Changes in the hepatopancreas metabolic reserves of *Porcellio laevis* Latreille during starvation and the moult cycle. *Am. Midl. Nat.* **87**: 503-513.

Allaway, W.H. (1975). Soil and plant aspects of the cycling of chromium, molybdenum and selenium. In: *International Conference. Heavy Metals in the Environment.* Vol. 1, pp. 35-47. Toronto, 1975. Edinburgh, CEP Consultants.

Allen, G., Nickless, G., Wibberley, B. & Pickard, J. (1974). Heavy metal particle characterization. *Nature* **252**: 571-572.

Andersen, C. (1979). Cadmium, lead and calcium content, number and biomass, in earthworms (Lumbricidae), from sewage sludge treated soil. *Pedobiologia* **19**: 309-319.

Andersen, C. (1980). Lead and cadmium content in earthworms (Lumbricidae) from sewage sludge amended arable soil. In: *Proceedings of the 7th International Conference of Soil Biology.* pp. 148-156. Syracuse, N.Y., July 1979. U.S. Environmental Protection Agency Report No. EPA-560/13-80-038. Office of Pesticide and Toxic Substances, E.P.A., Washington, D.C.

Andersen, C. & Laursen, J. (1982). Distribution of heavy metals in *Lumbricus terrestris*, *Aporrectodea longa* and *A. rosea* measured by atomic absorption and X-ray fluorescence spectrometry. *Pedobiologia* **24**: 347-356.

Anderson, A.W. & Taylor, T.H. (1926). The slug pest. *Bull. Univ. Leeds Dep. Agric.* **143**: 1-14.

Anderson, J.M. & Bignell, D.E. (1982). Assimilation of 14C-labelled leaf fibre by the millipede *Glomeris marginata* (Diplopoda: Glomeridae). *Pedobiologia* **23**: 120-125.

Anderson, J.M. & Healey, I.N. (1972). Seasonal and inter-specific variations in major components of the gut contents of some woodland Collembola. *J. Anim. Ecol.* **41**: 359-368.

Anderson, J.M. & Ineson, P. (1982). A soil microcosm system and its application to measurements of respiration and nutrient leaching. *Soil Biol. Biochem.* **14**: 415-416.

Anderson, J.M. & Ineson, P. (1983). Interactions between soil arthropods and microbial populations in carbon, nitrogen and mineral nutrient fluxes from decomposing leaf litter. In: *Nitrogen as an Ecological Factor* (J. Lee & S. McNeill, Eds.) pp. 413-432. Oxford, Blackwell.

Anderson, J.M. & Ineson, P. (1984). Interactions between microorganisms and soil invertebrates in nutrient flux pathways of forest ecosystems. In: *Invertebrate-Microbial Interactions* (J.M. Anderson, A.D.M. Rayner & D. Walton, Eds.) pp. 59-88. British Mycological Society Symposium No. 6, Cambridge University Press.

Anderson, J.M., Ineson, P. & Huish, S.A. (1983). Nitrogen and cation mobilization by soil fauna feeding on leaf litter and soil organic matter from deciduous woodlands. *Soil Biol. Biochem.* **15**: 463-467.

Anderson, M. (1979). Mn^{2+} ions pass through Ca^{2+} channels in myoepithelial cells. *J. Exp. Biol.* **82**: 227-238.

Anderson, R.A. (1981). Nutritional role of chromium. *Sci. Total Environ.* **17**:

13-29.

Andrews, S.M. & Cooke, J.A. (1984). Cadmium within a contaminated grass-land ecosystem established on a metalliferous mine waste. In: *Metals in Animals* (D. Osborn, Ed.) pp. 11-15. Institute of Terrestrial Ecology Symposium No. 12. Abbots Ripton, Institute of Terrestrial Ecology.

Anke, M., Groppel, B. & Kronemann, H. (1984). Significance of newer essential trace elements (like Si, Ni, As, Li, V...) for the nutrition of man and animal. In: *Trace Element Analytical Chemistry in Medicine and Biology*. Vol. 3 (P. Brätter & P. Schramel, Eds.) pp. 421-464. Berlin, Walter De Gruyter.

Aoki, Y. & Suzuki, K.T. (1984). Excretion of cadmium and change in the relative ratio of iso-cadmium-binding proteins during metamorphosis of fleshfly (*Sarcophaga peregrina*) larvae. *Comp. Biochem. Physiol.* **78C**: 315-317.

Aoki, Y., Suzuki, K.T. & Kubota, K. (1984). Accumulation of cadmium and induction of its binding protein in the digestive tract of fleshfly (*Sarcophaga peregrina*). *Comp. Biochem. Physiol.* **77C**: 279-282.

Applebaum, S.W., Jankovic, M. & Birk, Y. (1961). Studies on the midgut amylase activity of *Tenebrio molitor* L. larvae. *J. Insect Physiol.* **7**: 100-108.

Armitage, I.M., Otvos, J.D., Briggs, R.W. & Boulanger, Y. (1982). Structure elucidation of the metal binding sites in metallothionein by ^{113}Cd NMR. *Fed. Proc.* **41**: 2974-2980.

Aronssohn, F. (1911). Sur la composition minérale de l'abeille. *C.R. Acad. Sci. Paris*, **152**: 1183-1184.

Ash, C.P.J. & Lee, D.L. (1980). Lead, cadmium, copper and iron in earthworms from roadside sites. *Environ. Pollut.* **22A**: 59-67.

Auclair, J.L. & Srivastava, P.N. (1972). Some mineral requirements of the pea aphid *Acyrthosiphon pisum* (Homoptera: Aphididae). *Can. Entomol.* **104**: 927-936.

Ausmus, B.S., Dodson, G.J. & Jackson, D.R. (1978). Behaviour of heavy metals in forest microcosms. III. Effects on litter-soil carbon metabolism. *Water Air Soil Pollut.* **10**: 19-26.

Avery, R.A., White, A.S., Martin, M.H. & Hopkin, S.P. (1983). Concentrations of heavy metals in common lizards (*Lacerta vivipara*) and their food and environment. *Amphibia-Reptilia* **4**: 205-213.

Babich, H., Bewley, R.J.F. & Stotzky, G. (1983). Application of the ecological dose concept to the impact of heavy metals on some microbe-mediated ecologic processes in soil. *Arch. Environ. Contam. Toxicol.* **12**: 421-426.

Babička, J., Komárek, J.M. & Némec, B. (1945). Gold in animal bodies. *Acad. Tcheque. Sci. Bull. Int., Cl. Sci. Math. Nat. Med.* **45**: 131-137.

Bakir, F., Damluji, S.F., Amin-Zaki, L., Murtadha, M., Khaladi, A., Al-Rawi, N.Y., Tikriti, S., Dhahir, H.I., Clarkson, T.W., Smith, J.C. & Doherty, R.A. (1973). Methyl mercury poisoning in Iraq. An inter-university report. *Science* **181**: 230-241.

Balicka, N., Wegrzyn, T. & Czekanowska, E. (1977). Microorganisms as indices of environmental pollution by smelting industry. *Acta Microbiol. Pol.* **26**: 301-308.

Ballan-Dufrançais, C. (1972). Ultrastructure d'iléon de *Blattella germanica* L. (Dictyoptère). Localisation, genèse et composition des concrétions minérales intracytoplasmiques. *Z. Zellforsch. Mikrosk. Anat.* **133**: 163-179.

Ballan-Dufrançais, C. (1974). Accumulations minérales et puriques chez trois espèces d'insectes Dictyoptères. *Cellule* **70**: 315-330.

Ballan-Dufrançais, C. & Martoja, R. (1971). Analyse chimique d'inclusions minérales par spectrographie des rayons X et par cytochimie. Application à quelques organes d'insectes Orthoptères. *J. Microsc. (Paris)* **11**: 219-248.

Ballan-Dufrançais, C., Jeantet, A.Y. & Quintana, C. (1979). Toxicologie du méthyl-mercure chez un insecte: mise en évidence du métal dans les lysosomes par microsonde électronique. *C.R. Acad. Sci. Paris* **288D**: 847-849.

Ballan-Dufrançais, C., Ruste, J. & Jeantet, A.Y. (1980). Quantitative electron probe microanalysis on insect exposed to mercury. I. Methods. An approach on the molecular form of the stored mercury. Possible occurrence of metallothionein-like proteins. *Biol. Cellulaire* **39**: 317-324.

Bannister, J.V. (1977). Ed. *Structure and Function of Haemocyanin*. Berlin, Springer-Verlag.

Bannister, W.H. & Needham, A.E. (1971). Connective tissue pigment of the centipede *Lithobius forficatus* L. *Naturwissenschaften* **58**: 58-59.

Barbour, A.K. (1977). Environmental control techniques for the production, use and disposal of cadmium. In: *Proceedings of the First International Cadmium Conference*. San Francisco, 1977. pp. 100-104. Dorset, Drogher Press.

Barnard, T. (1982). Thin frozen-dried cryosections and biological X-ray microanalysis. *J. Microsc.* **126**: 317-332.

Bartlett, B.B. (1964a). Toxicity of some pesticides to eggs, larvae and adults of the green lacewing *Chrysopa carnea*. *J. Econ. Entomol.* **57**: 366-369.

Bartlett, B.B. (1964b). The toxicity of some pesticide residues to adult *Amblyseius hibisci* with a compilation of the effects of pesticides on phytoseiid mites. *J. Econ. Entomol.* **57**: 559-563.

Bayon, C. & Martoja, R. (1973). Données histophysiologiques sur les accumulations de métaux et de déchets puriques des Coléoptères. I. Phytophage terrestre, *Leptinotarsa decemlineata* Say. *Arch. Zool. Exp. Gén.* **114**: 173-185.

Bayon, C. & Martoja, R. (1974). Données histophysiologiques sur les accumulations de métaux et de déchets puriques des Coléoptères. II. Espèces à régime carné, *Ablattaria laevigata*, *Silpha obscura* (Silphidae), *Agabus bipustulatus* (Dytiscidae), *Adalia bipunctata* (Coccinellidae). *Arch. Zool. Exp. Gén.* **115**: 621-639.

Beck, L. & Brestowsky, E. (1980). Selection and consumption of different litter foliage by *Oniscus asellus*. *Pedobiologia* **20**: 428-441.

Becker, A. & Peters, W. (1985a). Fine structure of the midgut gland of *Phalangium opilio* (Chelicerata, Phalangida). *Zoomorphol.* **105**: 317-325.

Becker, A. & Peters, W. (1985b). The ultrastructure of the midgut and the formation of peritrophic membranes in a harvestman, *Phalangium opilio* (Chelicerata, Phalangida). *Zoomorphol.* **105**: 326-332.

Becker, G. (1978). Giftwirkung einiger organischer Zinnverbindungen auf Eilarven von *Hylotrupes bajulus* (L.). *Mater. Org.* **13**: 123-128.

Beeby, A. (1978). Interaction of lead and calcium uptake by the woodlouse *Porcellio scaber* (Isopoda: Porcellionidae). *Oecologia* **32**: 255-262.

Beeby, A. (1985). The role of *Helix aspersa* as a major herbivore in the transfer of lead through a polluted ecosystem. *J. Appl. Ecol.* **22**: 267-275.

Beeby, A. & Eaves, S.L. (1983). Short-term changes in Ca, Pb, Zn and Cd concentrations of the garden snail *Helix aspersa* Müller from a Central London car park. *Environ. Pollut.* **30A**: 233-244.

Bell, D.M. & Anstee, J.H. (1977). A study of the Malpighian tubules of *Locusta migratoria* by scanning and transmission electron microscopy. *Micron* **8**: 123-134.

Bender, J.A. (1975). Trace metal levels in beach dipterans and amphipods. *Bull. Environ. Contam. Toxicol.* **14**: 187-192.

Bengtsson, G. (1986). The optimal use of life strategies in transitional zones or the optimal use of transition zones to describe life strategies. *Proceedings of the Third European Congress of Entomology* (H.H.W. Velthuis, Ed.) pp. 193-207. Amsterdam, August 1986. Amsterdam, Nederlandse Entomologische Vereniging.

Bengtsson, G. & Gunnarsson, T. (1984). A micromethod for the determination of metal ions in biological tissues by furnace atomic absorption spectrophotometry. *Microchem. J.* **29**: 282-287.

Bengtsson, G. & Rundgren, S. (1982). Population density and species number of enchytraeids in coniferous forest soils polluted by a brass mill. *Pedobiologia* **24**: 211-218.

Bengtsson, G. & Rundgren, S. (1984). Ground-living invertebrates in metal-polluted forest soils. *Ambio* **13**: 29-33.

Bengtsson, G., Berden, M. & Rundgren, S. (1988). Influence of soil animals and metals on decomposition processes: a microcosm experiment. *J. Environ. Qual.* **17**: 113-119.

Bengtsson, G., Gunnarsson, T. & Rundgren, S. (1983a). Growth changes caused by metal uptake in a population of *Onychiurus armatus* (Collembola) feeding on metal polluted fungi. *Oikos* **40**: 216-225.

Bengtsson, G., Gunnarsson, T. & Rundgren, S. (1985a). Influence of metals on reproduction, mortality and population growth in *Onychiurus armatus* (Collembola). *J. Appl. Ecol.* **22**: 967-978.

Bengtsson, G., Gunnarsson, T. & Rundgren, S. (1986). Effects of metal pollution on the earthworm *Dendrobaena rubida* (Sav.) in acidified soils. *Water Air Soil Pollut.* **28**: 361-383.

Bengtsson, G., Nordström, S. & Rundgren, S. (1983b). Population density and tissue metal concentration of lumbricids in forest soils near a brass mill. *Environ. Pollut.* **30A**: 87-108.

Bengtsson, G., Ohlsson, L. & Rundgren, S. (1985b). Influence of fungi on growth and survival of *Onychiurus armatus* (Collembola) in a metal polluted soil. *Oecologia* **68**: 63-68.

Berger, B., Dallinger, R. & Wieser, W. (1986). Cadmium balance and metal-lothionein in *Arianta arbustorum* L. *Proceedings of the Third European Congress of Entomology* (H.H.W. Velthuis, Ed.) p. 325. Amsterdam, August 1986. Amsterdam, Nederlandse Entomologische Vereniging.

Berglund, S., Davis, R.D. & L'Hermite, P. (1984). *Utilisation of Sewage Sludge on Land: Rates of Application and Long-Term Effects of Metals*. New York, D. Reidel.

Berkaloff, A. (1958). Les grains de sécrétion des tubes de Malpighi de *Gryllus domesticus* (Orthoptère). *C.R. Acad. Sci. Paris* **246**: 2807-2809.

Berndt, K.P. (1974). Die Wirkungsweise von Natriumarsenat auf Laborkolonien der Pharaoameise *Monomorium pharaonis*. *Biol. Zentralbl.* **93**: 425-448.

Bernhard, M., Brinckman, F.E. & Sadler, P.J. (1986). Eds. *The Importance of Chemical 'Speciation' in Environmental Processes*. Berlin, Springer-Verlag.

Berrow, M.L. & Webber, J. (1972). Trace elements in sewage sludges. *J. Sci. Food Agric.* **23**: 93-100.

Bertini, I., Luchinat, C., Maret, W. & Zeppezauer, M. (1986). Eds. *Zinc Enzymes*. Boston, Birkhauser.

Bettica, A., Shay, M.T., Vernon, G. & Witkus, R. (1984). An ultrastructural study of cell differentiation and associated acid phosphatase activity in the hepatopancreas of *Porcellio scaber*. *Symp. Zool. Soc. Lond.* **53**: 199-215.

Bewley, R.J.F. (1980). Effects of heavy metal pollution on Oak leaf microorganisms. *Appl. Environ. Microbiol.* **40**: 1053-1059.

Bewley, R.J.F. (1981). Effects of heavy metal pollution on the microflora of pine needles. *Holarct. Ecol.* **4**: 215-220.

Beyer, W.N. (1981). Metals and terrestrial earthworms (Annelida: Oligochaeta). In: *Workshop on the Role of Earthworms in the Stabilization of Organic Residues* (M. Appelhof, Ed.) pp. 137-150. Vol. 1, Proceedings. Kalamazoo, April 1980. Michigan, Beech Leaf Press.

Beyer, W.N. & Anderson, A. (1985). Toxicity to woodlice of zinc and lead oxides added to soil litter. *Ambio* **14**: 173-174.

Beyer, W.N. & Moore, J. (1980). Lead residues in eastern tent caterpillars (*Malacosoma americanum*) and their host plant (*Prunus serotina*) close to a major highway. *Environ. Entomol.* **9**: 10-12.

Beyer, W.N., Chaney, R.L. & Mulhern, B.M. (1982). Heavy metal concentrations in earthworms from soil amended with sewage sludge. *J. Environ. Qual.* **11**: 381-385.

Beyer, W.N., Cromartie, E. & Moment, G.B. (1985a). Accumulation of methylmercury in the earthworm *Eisenia foetida* and its effect on regeneration. *Bull. Environ. Contam. Toxicol.* **35**: 157-162.

Beyer, W.N., Hensler, G. & Moore, J. (1987). Relation of pH and other soil variables to concentrations of Pb, Cu, Zn, Cd and Se in earthworms. *Pedobiologia* **30**: 167-172.

Beyer, W.N., Miller, G.W. & Cromartie, E.J. (1984). Contamination of the O_2 soil horizon by zinc smelting and its effect on woodlouse survival. *J. Environ. Qual.* **13**: 247-251.

Beyer, W.N., Pattee, O.H., Sileo, L., Hoffman, D.J. & Mulhern, B.M. (1985b). Metal contamination in wildlife living near two zinc smelters. *Environ. Pollut.* **38A**: 63-86.

Bhattacharyya, A. & Medda, A.K. (1981a). Effect of cyanocobalamin and cobalt chloride on lipid content of silk gland of *Bombyx mori* L., race Nistari. *Sci. Cult. (India)* **47**: 140-141.

Bhattacharyya, A. & Medda, A.K. (1981b). Effect of cyanocobalamin and cobalt chloride on glycogen content of silk gland of *Bombyx mori* L., race Nistari. *Sci. Cult. (India)* **47**: 268-270.

Bhattacharyya, A. & Medda, A.K. (1983). Histochemical studies on the effects of cyanocobalamin and cobalt chloride on the alkaline and acid phosphatase activity in silk gland of silkworms (*Bombyx mori* L.), race Nistari. *Zool. Jahrb.* **110**: 403-410.

Bielenin, I. & Weglarska, B. (1965). The occurrence of iron in the alimentary canal of *Quadraspidiotus ostreaeformis* Curt. (Homoptera, Coccidea, Aspidiotini). *Acta Biol. Cracow., Ser. Zool.* **8**: 9-16.

Bignell, D.E. (1984). The arthropod gut as an environment for microorganisms. In: *Invertebrate-Microbial Interactions* (J.M. Anderson, A.D.M. Rayner & D. Walton, Eds.) pp. 205-227. British Mycological Society Symposium No. 6, Cambridge University Press.

Bignell, D.E., Oskarsson, H. & Anderson, J.M. (1980). Specialization of the hindgut wall for the attachment of symbiotic micro-organisms in a termite *Procubitermes aburiensis* (Isoptera, Termitidae, Termitinae). *Zoomorphol.* **96**: 103-112.

Bischoff, B. (1982). Effects of cadmium on microorganisms. *Ecotoxicol. Environ. Safety* **6**: 157-165.

Bisessar, S. (1982). Effect of heavy metals on microorganisms in soils near a secondary lead smelter. *Water Air Soil Pollut.* **17**: 305-308.

Blackman, G.G. & Bakker, J.A.F. (1975). Resistance of the sheep blowfly *Lucilia cuprina* to insecticides in the Republic of South Africa. *J. S. Afr. Vet. Ass.* **46**: 337-339.

Blust, R., Van der Linden, A. & Decleir, W. (1985). Microwave-aided dissolution of a biological matrix in AAS sample cups prior to graphite furnace analysis. *Atomic Spectrosc.* **6**: 163-165.

Bocock, K.L. (1963). The digestion of food by *Glomeris*. In: *Soil Organisms* (J. Doekson & J. Van der Drift, Eds.) pp. 85-91. Amsterdam, Elsevier-North Holland.

Bodine, J.H. & Fitzgerald, L.R. (1948). The copper content of an egg and its distribution during the development of an embryo (Orthoptera). *J. Exp. Biol.* **109**: 187-195.

Bolter, E. & Butz, T.R. (1975). Heavy metal mobilization by natural organic acids. In: *International Conference. Heavy Metals in the Environment.* Vol. 2, pp. 353-362. Toronto, 1975. Edinburgh, CEP Consultants.

Bolton, P.J. & Phillipson, J. (1976). Burrowing, feeding, egestion and energy budgets of *Allolobophora rosea* (Savigny)(Lumbricidae). *Oecologia* **23**: 225-245.

Bonaventura, J. & Bonaventura, C. (1980). Hemocyanins: relationships in their structure, function and assembly. *Am. Zool.* **20**: 7-17.

Bond, H., Lighthart, B., Shimabuku, R. & Russel, L. (1976). Some effects of cadmium on coniferous soil and litter microcosms. *Soil Sci.* **121**: 278-287.

Bouché, M.B. (1984). Ecotoxicologie des lombriciens. II. Surveillance de la contamination des milieux. *Acta Oecologia, Oecologia Appl.* **5**: 291-301.

Boutron, C.F. (1986). Atmospheric toxic metals and metalloids in the snow and ice layers deposited in Greenland and Antarctica from prehistoric times to present. *Adv. Environ. Sci. Technol.* **17**: 467-505.

Bowden, J., Digby, P.G.N. & Sherlock, P.L. (1984). Studies of elemental composition as a biological marker in insects. I. The influence of soil type and host-plant on elemental composition of *Noctua pronuba* (L.) (Lepidoptera: Noctuidae). *Bull. Entomol. Res.* **74**: 207-226.

Bowden, J., Sherlock, P.L., Digby, P.G.N., Fox, J.S. & Rhodes, J.A. (1985a).

Studies of elemental composition as a biological marker in insects. II. The elemental composition of apterae of *Rhopalosiphum padi* (L.) and *Metopolophium dirhodum* (Walker) (Hemiptera: Aphididae) from different soils and host plants. *Bull. Entomol. Res.* **75**: 107-120.

Bowden, J., Sherlock, P.L. & Digby, P.G.N (1985b). Studies of elemental composition as a biological marker in insects. III. Comparison of apterous and alate cereal aphids, especially *Rhopalosiphum padi* (L.) (Hemiptera: Aphididae) from oats and wheat, and from oats infected with or free from barley yellow dwarf virus. *Bull. Entomol. Res.* **75**: 477-488.

Bowen, H.J.M. (1975). Residence times of heavy metals in the environment. In: *International Conference. Heavy Metals in the Environment.* Vol. 1, pp. 1-19. Toronto, 1975. Edinburgh, CEP Consultants.

Bowen, H.J.M. (1979). *Environmental Chemistry of the Elements*. New York, Academic Press.

Bowen, I.D. & Ryder, T.A. (1978). The application of X-ray microanalysis to histochemistry. In: *Electron Probe Microanalysis in Biology* (D.A. Erasmus, Ed.) pp. 183-211. London, Chapman & Hall.

Brätter, P. & Schramel, P. (1983). *Trace Element Analytical Chemistry in Medicine and Biology*, Vol 2. Berlin, Walter de Gruyter.

Brätter, P. & Schramel, P. (1984). *Trace Element Analytical Chemistry in Medicine and Biology*, Vol. 3. Berlin, Walter de Gruyter.

Breymeyer, A. & Odum, E.P. (1969). Transfer and bioelimination of tracer ^{65}Zn during predation by spiders on labelled flies. In: *Proceedings of the Second National Symposium on Radioecology* (D.J. Nelson & F.C. Evans, Eds.) pp. 715-720. CONF-670503. Michigan, Ann Arbor Science Publishers.

Bristowe, W.S. (1958). *The World of Spiders*. London, Collins.

Bromenshenk, J.J., Carlson, S.R., Simpson, J.C. & Thomas, J.M. (1985). Pollution monitoring of Puget Sound with honey bees. *Science* **227**: 632-634.

Bronner, F. & Coburn, J.W. (1981). Eds. *Disorders of Mineral Metabolism.* Vol. 1. *Trace Minerals*. New York, Academic Press.

Brookes, P.C., McGrath, S.P. & Heijnen, C. (1986). Metal residues in soils previously treated with sewage sludge and their effects on growth and nitrogen fixation by blue-green algae. *Soil Biol. Biochem.* **18**: 345-353.

Brooks, M.A. (1960). Some dietary factors that affect ovarial transmission of symbiotes. *Proc. Helminth. Soc. Wash.* **27**: 212-220.

Brooks, R.R., Morrison, R.S., Reeves, R.D., Dudley, T.R. & Akman, Y. (1979). Hyperaccumulation of nickel by *Alyssum* Linnaeus (Cruciferae). *Proc. Roy. Soc. Lond.* **203B**: 387-403.

Brown, B.E. (1982). The form and function of metal-containing 'granules' in invertebrate tissues. *Biol. Rev.* **57**: 621-667.

Brown, K.W., Thomas, J.C. & Slowey, J.F. (1983). The movement of metals applied to soils in sewage effluent. *Water Air Soil Pollut.* **19**: 43-54.

Bruce, W.G. (1942). Zinc oxide: a new larvicide for use in the medication of cattle for the control of hornflies. *J. Kansas Entomol. Soc.* **15**: 105-107.

Bryan, G.W. (1984). Pollution due to heavy metals and their compounds. *Marine Ecology*. Vol. 5 (O. Kinne, Ed.) pp. 1289-1431. Chichester, John Wiley & Sons.

Bryan, G.W., Gibbs, P.E., Hummerstone, L.G. & Burt, G.R. (1986a). The decline of the gastropod *Nucella lapillus* around South-West England: evidence for the effect of tributyltin from antifouling paints. *J. Mar. Biol. Ass. U.K.* **66**: 611-640.

Bryan, G.W., Hummerstone, L.G. & Ward, E. (1986b). Zinc regulation in the lobster *Homarus gammarus*: importance of different pathways of absorption and excretion. *J. Mar. Biol. Ass. U.K.* **66**: 175-199.

Bryan, G.W., Langston, W.J., Hummerstone, L.G. & Burt, G.R. (1985). *A Guide to the Assessment of Heavy-Metal Contamination in Estuaries using Biological Indicators*. Occasional Publications of the Marine Biological Association of the United Kingdom, No. 4.

Buchauer, M.J. (1973). Contamination of soil and vegetation near a zinc smelter by zinc, cadmium, copper and lead. *Environ. Sci. Technol.* **7**: 131-135.

Bull, K.R., Roberts, R.D., Inskip, M.J. & Goodman, G.T. (1977). Mercury concentrations in soil, grass, earthworms and small mammals near an industrial emission source. *Environ. Pollut.* **12**: 135-140.

Burkitt, A., Lester, P. & Nickless, G. (1972). Distribution of heavy metals in the vicinity of an industrial complex. *Nature* **238**: 327-328.

Burnett, A.W., Mason, W.H. & Rhodes, S.T. (1969). Reingestion of feces and excretion rates of Zn^{65} in *Popilius disjunctus* versus *Cryptocercus punctalatus*. *Ecology* **50**: 1094-1096.

Burton, M.A.S. (1986). *Biological Monitoring of Environmental Contaminants (Plants)*. Monitoring & Assessment Research Centre, King's College, London.

Burton, R.F. (1972). The binding of alkaline earth ions by the haemocyanin of *Helix pomatia*. *Comp. Biochem. Physiol.* **41A**: 555-565.

Byrd, D.S., Gilmore, J.T. & Lea, R.H. (1983). Effect of decreased use of lead in gasoline on the soil of a highway. *Environ. Sci. Technol.* **17**: 121-122.

Campbell, J.W. & Boyan, B.D. (1976). On the acid-base balance of gastropod molluscs. In: *The Mechanisms of Mineralization in the Invertebrates and Plants* (N. Watanabe & K. Wilbur, Eds.) pp. 109-133. University of South Carolina Press.

Cantillo, A.Y. & Segar, D.A. (1975). Metal species identification in the environment. A major challenge for the analyst. In: *International Conference. Heavy Metals in the Environment*. Vol. 1, pp. 183-204. Toronto, 1975. Edinburgh, CEP Consultants.

Carefoot, T.H. (1984a). Studies on the nutrition of the supralittoral isopod *Ligia pallasii* using chemically defined artificial diets: assessment of vitamin, carbohydrate, fatty acid, cholesterol and mineral requirements. *Comp. Biochem. Physiol.* **79A**: 655-665.

Carefoot, T.H. (1984b). Nutrition and growth of *Ligia pallasii*. *Symp. Zool. Soc. Lond.* **53**: 455-467.

Carter, A. (1983). Cadmium, copper and zinc in soil animals and their food in a red clover system. *Can. J. Zool.* **61**: 2751-2757.

Carter, A., Kenney, E.A. & Guthrie, T.F. (1980). Earthworms as biological monitors of changes in heavy metal levels in an agricultural soil in British Columbia. In: *Proceedings of the 7th International Conference of Soil Biology*. pp. 344-357. Syracuse, N.Y., July 1979. U.S. Environmental Protection Agency Report No. EPA/560/13-80-038. Washington, Office of Pesticide and

Toxic Substances.

Carter, A., Kenney, E.A., Guthrie, T.F. & Timmenga, H. (1983). Heavy metals in earthworms in non-contaminated and contaminated agricultural soil from near Vancouver, Canada. In: *Earthworm Ecology* (J.E. Satchell, Ed.) pp. 267-274. London & New York, Chapman & Hall.

Cassier, P., Fain-Maurel, M.A. & Alibert, J. (1974). Relation entre la diversité des fonctions intestinales et la structure de l'epithélium mésentérique de *Petrobius maritimus* Leach. *Pedobiologia* **14**: 167-172.

Cataldo, D.A., Garland, T.R. & Wildung, R.E. (1981). Foliar retention and leachability of submicron plutonium and americium particles. *J. Environ. Qual.* **10**: 31-37.

Cavallero, R. & Ravera, O. (1966). Biological indicators of manganese-54 contamination in terrestrial environments. *Nature* **209**: 1259.

Chakrabarti, M.K. & Medda, A.K. (1978). Effect of cobalt chloride on silkworms (*Bombyx mori* L.) Nistari race. *Sci. Cult. (India)* **44**: 406-408.

Chamberlain, W.F. (1977). Amounts of magnesium, manganese, strontium and zinc in horn flies from several geographic areas and cultural conditions. *Southwest. Entomol.* **2**: 11-15.

Chamberlain, W.F., Matter, J.J. & Eschle, J.L. (1977). Trace elements as markers for hornflies, a critical analysis. *Southwest. Entomol.* **2**: 73-78.

Chan, W.H. & Lusis, M.A. (1986). Smelting operations and trace metals in air and precipitation in the Sudbury Basin. *Adv. Environ. Sci. Technol.* **17**: 113-143.

Chandler, J.A. (1977). *X-ray Microanalysis in the Electron Microscope.* Amsterdam, Elsevier North Holland.

Chandler, J.A. (1978). The application of X-ray microanalysis in TEM to the study of ultrathin biological specimens — a review. In: *Electron Probe Microanalysis in Biology* (D.A. Erasmus, Ed.) pp. 37-93. London & New York, Chapman & Hall.

Chaney, W.R., Kelly, J.M. & Strickland, R.L. (1978). Influence of cadmium and zinc on carbon dioxide evolution from litter and soil from a black oak forest. *J. Environ. Qual.* **7**: 115-119.

Chang, S.H., Mergner, W.J., Pendergrass, R.E., Bulger, R.E., Berezesky, I.K. & Trump, B.F. (1980). A rapid method of cryofixation of tissues *in situ* for ultracryomicrotomy. *J. Histochem. Cytochem.* **28**: 47-51.

Chapman, R.F. (1985). Structure of the digestive system. In: *Comprehensive Insect Physiology, Biochemistry and Pharmacology*. Vol. 4, *Regulation, Digestion, Nutrition, Excretion* (G.A. Kerkut & L.I. Gilbert, Eds.) pp. 165-211. Oxford, Pergamon.

Cheng, L. (1980). Incorporation of cadmium into *Drosophila*. *Environ. Pollut.* **21A**: 85-88.

Cherian, M.G. & Goyer, R.A. (1978). Metallothioneins and their role in the metabolism and toxicity of metals. *Life Sci.* **23**: 1-10.

Cheung, W.W.K. & Marshall, A.T. (1973). Studies on water and ion transport in homopteran insects: ultrastructure and cytochemistry of the cicapoid and cercopoid midgut. *Tiss. Cell* **5**: 651-669.

Cisternas, R. & Mignolet, R. (1982). Accumulation of lead in decomposing litter. *Oikos* **38**: 361-364.

Clarkson, J.P., Elmes, M.E., Jasani, B. & Webb, M. (1984). Immunolocalisation of zinc metallothionein in rat tissues. In: *Trace Element Analytical Chemistry in Medicine and Biology*. Vol. 3 (P. Brätter & P. Schramel, Eds.) pp. 291-297. Berlin, Walter De Gruyter.

Clausen, I.H.S. (1984a). Lead (Pb) in spiders: a possible measure of atmospheric Pb pollution. *Environ. Pollut.* **8B**: 217-230.

Clausen, I.H.S. (1984b). Notes on the impact of air pollution (SO_2 and Pb) on spider (Araneae) populations in North Zealand, Denmark. *Entomol. Meddr.* **52**: 33-39.

Clausen, I.H.S. (1986). The use of spiders (Aranaea) as ecological indicators. *Bull. Br. Arachnol. Soc.* **7**: 83-86.

Clifford, B. & Witkus, E.R. (1971). The fine structure of the hepatopancreas of the woodlouse *Oniscus asellus*. *J. Morphol.* **135**: 335-350.

Clubb, R.W., Lords, J.L. & Gaufin, A.R. (1975). Isolation and characterization of a glycoprotein from the stonefly *Pteronarcys californica* which binds cadmium. *J. Insect Physiol.* **21**: 53-60.

Cohen, A.C., Debolt, J.W. & Schreiber, H.A. (1985). Profiles of trace and major elements in whole carcasses of *Lygus hesperus* adults. *Southwest. Entomol.* **10**: 239-243.

Colborn, T. (1982). Measurement of low levels of molybdenum in the environment by using aquatic insects. *Bull. Environ. Contam. Toxicol.* **29**: 422-428.

Cole, J.F. & Volpe, R. (1983). The effect of cadmium on the environment. *Ecotoxicol. Environ. Safety* **7**: 151-159.

Coleman, J.E. (1967). Metal ion dependent binding of sulphonamide to carbonic anhydrase. *Nature* **214**: 193-194.

Collatz, K.G. (1986). Structure and function of the digestive tract. In: *Ecophysiology of Spiders* (W. Nentwig, Ed.) pp. 229-238. Berlin, Springer-Verlag.

Colloff, M.J. & Hopkin, S.P. (1986). The ecology, morphology and behaviour of *Bakerdania elliptica* (Krczal 1959) (Acari: Prostigmata: Pygmephoridae), a mite associated with terrestrial isopods. *J. Zool.* **208A**: 109-123.

Combs, G.F. & Combs, S.B. (1986). *Role of Selenium in Nutrition*. Orlando, Academic Press.

Cooke, M., Jackson, A., Nickless, G. & Roberts, D.J. (1979). Distribution and speciation of cadmium in the terrestrial snail *Helix aspersa*. *Bull. Environ. Contam. Toxicol.* **23**: 445-454.

Cornelis, R. & Wallaeys, B. (1984). Chromium revisited. In: *Trace Element Analytical Chemistry in Medicine and Biology*. Vol. 3 (P. Brätter & P. Schramel, Eds.) pp. 219-233. Berlin, Walter De Gruyter.

Costescu, L.M. & Hutchinson, T.C. (1972). The ecological consequences of soil pollution by metallic dust from the Sudbury smelters. *Proc. Inst. Environ. Sci.* **17**: 540-545.

Coughtrey, P.J. & Martin, M.H. (1976a). The distribution of Pb, Zn, Cd and Cu within the pulmonate mollusc *Helix aspersa* Müller. *Oecologia* **23**: 315-322.

Coughtrey, P.J. & Martin, M.H. (1976b). A comment on the analysis of biological materials for lead using atomic absorption spectroscopy. *Chemosphere* **5**: 183-186.

Coughtrey, P.J. & Martin, M.H. (1977a). The uptake of lead, zinc, cadmium and copper by the pulmonate mollusc *Helix aspersa* Müller, and its relevance to the monitoring of heavy metal contamination of the environment. *Oecologia* **27**: 65-74.

Coughtrey, P.J. & Martin, M.H. (1977b). Cadmium tolerance of *Holcus lanatus* from a site contaminated by aerial fallout. *New Phytol.* **79**: 273-280.

Coughtrey, P.J. & Martin, M.H. (1978a). Cadmium uptake and distribution in tolerant and non-tolerant populations of *Holcus lanatus* grown in solution culture. *Oikos* **30**: 555-560.

Coughtrey, P.J. & Martin, M.H. (1978b). Tolerance of *Holcus lanatus* to lead, zinc and cadmium in factorial combination. *New Phytol.* **81**: 147-154.

Coughtrey, P.J., Jones, C.H., Martin, M.H. & Shales, S.W. (1979). Litter accumulation in woodlands contaminated by Pb, Zn, Cd and Cu. *Oecologia* **39**: 51-60.

Coughtrey, P.J., Martin, M.H., Chard, J. & Shales, S.W. (1980). Microorganisms and metal retention in the woodlouse *Oniscus asellus*. *Soil Biol. Biochem.* **12**: 23-27.

Coughtrey, P.J., Martin, M.H. & Young, E.W. (1977). The woodlouse, *Oniscus asellus*, as a monitor of environmental cadmium levels. *Chemosphere* **6**: 827-832.

Coughtrey, P.J., Martin, M.H. & Unsworth, M.H. (1987). *Pollutant Transport and Fate in Ecosystems*. Oxford, Blackwell.

Coy, C.M. (1984). Control of dust and fume at a primary zinc and lead smelter. *Chemistry in Britain*, May 1984: 418-420.

Crane, D.F. & Cowden, R.R. (1968). A cytochemical study of oocyte growth in four species of millipedes. *Z. Zellforsch. Mikrosk. Anat.* **90**: 414-431.

Crawford, C.S., Minion, G.P. & Bayers, M.D. (1983). Intima morphology, bacterial morphotypes and effects of annual molt on microflora in the hindgut of the desert millipede *Orthoporus ornatus* (Girard) (Diplopoda: Spirostreptidae). *Int. J. Insect Morphol. Embryol.* **12**: 301-312.

Crawford, L.A. (1983). Trace metal transfer from plants to herbivorous insects. PhD Thesis, Liverpool Polytechnic.

Creighton, J.T. (1938). Factors influencing insect abundance. *J. Econ. Entomol.* **31**: 735-739.

Cress, D.C. & Chada, H.L. (1971). Development of a synthetic diet for the greenbug *Schizaphis graminium*. 2. Greenbug development as affected by zinc, iron, manganese and copper. *Ann. Entomol. Soc. Am.* **64**: 1240-1244.

Crichton, R.R. (1982). Ferritin — the structure and function of an iron storage protein. In: *The Biological Chemistry of Iron* (H.B. Dunford, D. Dolphin, K.N. Raymond & L. Sieker, Eds.) pp. 45-61. New York, D. Reidel.

Crossley, D.A. & Van Hook, R.I. (1970). Energy assimilation by the house cricket *Acheta domesticus* measured with radioactive chromium-51. *Ann. Entomol. Soc. Am.* **63**: 512-515.

Crossley, D.A., Reichle, D.E. & Edwards, C.A. (1971). Intake and turnover of radioactive cesium by earthworms (Lumbricidae). *Pedobiologia* **11**: 71-76.

Cunningham, I.J. (1931). Some biochemical and physiological aspects of copper in animal nutrition. *Biochem. J.* **25**: 1267-1294.

Curry, J.P. & Cotton, D.C.F. (1980). Effects of heavy pig slurry contamination on earthworms in grassland. In: *Proceedings of the 7th International Conference of Soil Biology*. pp. 336-343. Syracuse, N.Y., July 1979. U.S. Environmental Protection Agency Report No. EPA 560/13-80-038. Washington, Office of Pesticide and Toxic Substances.

Czarnowska, K. & Jopkiewicz, K. (1978). Heavy metals in earthworms as an index of soil contamination. *Polish J. Soil Sci.* **11**: 57-62.

Dadd, R.H. (1967). Improvement of synthetic diet for the aphid *Myzus persicae* using plant juices, nucleic acid or trace metals. *J. Insect Physiol.* **13**: 763-778.

Dadd, R.H. (1973). Insect nutrition: current developments and metabolic implications. *Ann. Rev. Entomol.* **18**: 381-420.

Dadd, R.H. (1985). Nutrition: Organisms. In: *Comparative Insect Physiology, Biochemistry and Pharmacology*. Vol. 4, *Regulation, Digestion, Nutrition, Excretion* (G.A. Kerkut & L.I. Gilbert, Eds.) pp. 313-390. Oxford, Pergamon.

Dadd, R.H. & Krieger, D.L. (1967). Continuous rearing of aphids of the *Aphis fabae* complex on sterile synthetic diet. *J. Econ. Entomol.* **60**: 1512-1514.

Dadd, R.H. & Mittler, T.E. (1965). Studies on the artificial feeding of the aphid *Myzus persicae* (Sulzer). III. Some major nutritional requirements. *J. Insect Physiol.* **11**: 717-743.

Dadd, R.H. & Mittler, T.E. (1966). Permanent culture of an aphid on a totally synthetic diet. *Experentia* **22**: 832-833.

Dallinger, R. (1977). The flow of copper through a terrestrial food chain. III. Selection of an optimum copper diet by isopods. *Oecologia* **30**: 273-276.

Dallinger, R. & Prosi, F. (1986). Fractionation and identification of heavy metals in hepatopancreas of terrestrial isopods: evidence of lysosomal accumulation and absence of cadmium-thionein. *Proceedings of the Third European Congress of Entomology* (H.H.W. Velthuis, Ed.) p. 328. Amsterdam, August 1986. Amsterdam, Nederlandse Entomologische Vereniging.

Dallinger, R. & Prosi, F. (1988). Heavy metals in the terrestrial isopod *Porcellio scaber* Latr. II. Subcellular fractionation of metal accumulating lysosomes from hepatopancreas. *Cell. Biol. Toxicol.* **4**: 97-109.

Dallinger, R. & Wieser, W. (1977). The flow of copper through a terrestrial food chain. I. Copper and nutrition in isopods. *Oecologia* **30**: 253-264.

Dallinger, R. & Wieser, W. (1984a). Patterns of accumulation, distribution and liberation of Zn, Cu, Cd and Pb in different organs of the land snail *Helix pomatia* L. *Comp. Biochem. Physiol.* **79C**: 117-124.

Dallinger, R. & Wieser, W. (1984b). Molecular fractionation of Zn, Cu, Cd and Pb in the midgut gland of *Helix pomatia* L. *Comp. Biochem. Physiol.* **79C**: 125-129.

Danscher, G. (1984). Autometallography. A new technique for light and electron microscopic visualization of metals in biological tissues (gold, silver, metal sulphides and metal selenides). *Histochemistry* **81**: 331-336.

Daum, R.J. (1965). Agricultural and biocidal applications of organo-metallics. *Ann. New York Acad. Sci.* **125**: 229-241.

Davies, B.E. (1980). Trace element pollution. In: *Applied Soil Trace Elements* (B.E. Davies, Ed.) pp. 287-351. Chichester, John Wiley & Sons.

Davies, B.E. & Ginnever, R.C. (1979). Trace metal contamination of soils and vegetables in Shipham, Somerset. *J. Agric. Sci.* **93**: 753-756.

Davies, B.E. & Holmes, P.L. (1972). Lead contamination of roadside soil and grass in Birmingham, England, in relation to naturally occurring levels. *J. Agric. Sci.* **79**: 479-484.

Davies, B.E. & Houghton, N.J. (1984). Distance decline patterns in heavy metal contamination of soils and plants in Birmingham, England. *Urban Ecol.* **8**: 285-294.

Davies, I.M. & Pirie, J.M. (1980). Evaluation of a "Mussel Watch" project for heavy metals in Scottish coastal waters. *Mar. Biol.* **57**: 87-93.

Davis, G.R.F. & Shah, B.G. (1980). Effect of supplementary zinc on larvae of the yellow mealworm fed rapeseed protein concentrate. *Nutr. Rep. Int.* **22**: 491-495.

Davis, R.D., Carlton-Smith, C.H., Stark, J.H. & Campbell, J.A. (1988). Distribution of metals in grassland soils following surface applications of sewage sludge. *Environ. Pollut.* **49**: 99-115.

De Jonge, W.R.A. & Adams, F.C. (1986). Biogeochemical cycling of organic lead compounds. *Adv. Environ. Sci. Technol.* **17**: 561-594.

De Michele, S.J. (1984). Nutrition of lead. *Comp. Biochem. Physiol.* **78A**: 401-408.

Debry, J.M. & Lebrun, P. (1979). Effets d'un enrichissement en $CuSO_4$ sur le bilan alimentaire de *Oniscus asellus* (Isopoda). *Rev. Ecol. Biol. Sol* **16**: 113-124.

Debry, J.M. & Muyango, S. (1979). Effets du cuivre sur le bilan alimentaire de *Oniscus asellus* (Isopoda) avec référence particulière au cuivre contenu dans le lisier de porcs. *Pedobiologia* **19**: 129-137.

Delkeskamp, E. (1964a). Über den Eisenstoffwechsel bei *Lumbricus terrestris* L. *Z. Vergl. Physiol.* **48**: 332-340.

Delkeskamp, E. (1964b). Über den Porphyrinstoffwechsel bei *Lumbricus terrestris* L. *Z. Vergl. Physiol.* **48**: 400-412.

Dindal, D.L., Newell, L.T. & Moreau, J.P. (1979). Municipal wastewater irrigation: effects on community ecology of soil invertebrates. In: *Utilization of Municipal Sewage Effluent and Sludge on Forest and Disturbed Land* (W.E. Sopper & S.N. Kerr, Eds.) pp. 197-205. Pennsylvania State University Press.

Dirzo, R. (1980). Experimental studies on slug-plant interactions. I. The acceptability of 30 plant species to the slug *Agriolimax carvanae*. *J. Ecol.* **68**: 981-998.

Dismukes, J.F. & Mason, W.H. (1975). Effects of carbaryl and methyl parathion poisoning on the bio-elimination of ^{65}Zn and ^{54}Mn in *Popilius disjunctus*. *Environ. Entomol.* **4**: 221-224.

Dmowski, K. & Karolewski, M.A. (1979). Cumulation of zinc, cadmium and lead in invertebrates and in some vertebrates according to the degree of an area contamination. *Ekol. Pol.* **27**: 333-349.

Dodge, E.E. & Theis, T.L. (1979). Effect of chemical speciation on the uptake of copper by *Chironomus teutans*. *Environ. Sci. Technol.* **13**: 1287-1288.

DoE (1974). *Lead in the Environment and its Significance to Man.* Pollution Paper No. 2. London, HMSO.

DoE (1977). *Report of the Working Party on the Disposal of Sewage Sludge to Land.* London, Department of the Environment.

DoE (1979). *The United Kingdom Environment 1979: Progress of Pollution Control.* Pollution Paper No. 16. London, HMSO.

DoE (1980). *Cadmium in the Environment and its Significance to Man.* Pollution Paper No. 17. London, HMSO.

Doelman, P. & Haanstra, L. (1979a). Effects of lead on soil respiration and dehydrogenase activity. *Soil Biol. Biochem.* **11**: 475-479.

Doelman, P. & Haanstra, L. (1979b). Effects of lead on the decomposition of organic matter. *Soil Biol. Biochem.* **11**: 481-485.

Doelman, P. & Haanstra, L. (1979c). Effects of lead on the soil bacterial microflora. *Soil Biol. Biochem.* **11**: 487-491.

Donadey, C. & Besse, G. (1972). Étude histologique, ultrastructurale et expérimentale des caecums digestifs de *Porcellio dilatatus* et *Ligia oceanica* (Crustacea, Isopoda). *Tethys* **4**: 145-162.

D'Orey, F.L.C. (1975). Contribution of termite mounds to locating hidden copper deposits. *Trans. Inst. Min. Metall.* **84B**: 150-151.

Drummond, R.O., Graham, O.H., Meleney, W.P. & Diamant, G. (1964). Field tests in Mexico with new insecticides and arsenic for the control of *Boophilus* ticks on cattle. *J. Econ. Entomol.* **57**: 340-346.

Dugatova, G. & Podstavkova, S. (1978). Effect of arsenic on *Drosophila melanogaster.* I. Toxic effect and influence on ontogenesis. *Acta Fac. Rerum Nat. Univ. Comenianae, Ser. Genet.* **9**: 63-78.

Dugatova, G., Podstavkova, S. & Trebaticka, M. (1978). Effect of arsenic on *Drosophila melanogaster.* II. Test on recessive lethal and other mutations affecting viability and located in the X chromosome and on the occurrence of chromosomal aberrations. *Acta Fac. Rerum Nat. Univ. Comenianae, Ser. Genet.* **9**: 79-87.

Duncumb, P. (1979). X-ray microanalysis. *J. Microsc.* **117**: 165-174.

Duxbury, T. (1985). Ecological aspects of heavy metal responses in microorganisms. *Adv. Microb. Ecol.* **8**: 185-235.

Duxbury, T. & Bicknell, B. (1983). Metal tolerant bacterial populations from natural and metal polluted soils. *Soil Biol. Biochem.* **15**: 243-250.

Eaton, A.N. & Hutton, R.C. (1988). The use of inductively coupled mass spectrometry for trace metal analysis in natural systems. *Lab. Prac.* **37**: 61-63.

Edney, E.B. (1968). Transition from water to land in isopod crustaceans. *Am. Zool.* **8**: 309-326.

Edwards, C.A. & Heath, G.W. (1963). The role of soil animals in breakdown of leaf material. In: *Soil Organisms* (J. Doeksen & J. Van der Drift, Eds.) pp. 76-84. Amsterdam, Elsevier North Holland.

Edwards, C.A. & Lofty, J.R. (1977). *Biology of Earthworms.* 2nd edition. London, Chapman & Hall.

Ehrhardt, P. (1965a). Speicherung anorganischer Substanzen in den Mitteldarmzellen von *Aphis fabae* Scop. und ihre Bedeutung für die Ernährung. *Z. Vergl. Physiol.* **50**: 293-312.

Ehrhardt, P. (1965b). Magnesium und Calcium enthaltende Einschlusskörper in den Mitteldarmzellen von Aphiden. *Experentia* **21**: 337-338.

Ehrhardt, P. (1968). Die Wirkung verschiedener Spurenelemente auf Wachstum, Reproduktion und Symbioten von *Neomyzus circumflexus* Buckt. (Aphidae, Homoptera, Insecta) bei kunstlicher Ernährung. *Z. Vergl. Physiol.* **58**: 47-75.

Eisenbeis, G. & Wichard, W (1987). *Atlas on the Biology of Soil Arthropods.*

Berlin, Springer-Verlag.

Elinder, C.G. (1985). Cadmium: uses, occurrence and intake. In: *Cadmium and Health: A Toxicological and Epidemiological Appraisal*. Vol. 1, *Exposure, Dose and Metabolism* (L. Friberg, C.G. Elinder, T. Kjellström & G.F. Nordberg, Eds.) pp. 23-63. Florida, CRC Press.

Elinder, C.G. & Lind, B. (1985). Principles and problems of cadmium analysis. In: *Cadmium and Health* (L. Friberg, C.G. Elinder, T. Kjellström & G.F. Nordberg, Eds.) pp. 7-22. Florida, CRC Press.

Elinder, C.G. & Nordberg, M. (1985). Metallothionein. In: *Cadmium and Health* (L. Friberg, C.G. Elinder, T. Kjellström & G.F. Nordberg, Eds.) pp. 65-79. Florida, CRC Press.

Elwood, P.C. (1986). The sources of lead in blood: a critical review. *Sci. Total Environ.* **52**: 1-23.

Engebretson, J.A. & Mason, W.H. (1980). Transfer of ^{65}Zn at mating in *Heliothis virescens*. *Environ. Entomol.* **9**: 119-121.

Engebretson, J.A. & Mason, W.H. (1981). Depletion of trace elements in mated male *Heliothis virescens* and *Drosophila melanogaster*. *Comp. Biochem. Physiol.* **68A**: 523-525.

Enger, M.D., Hildebrand, C.E., Seagrave, J. & Tobey, R.A. (1986). Cellular resistance to cadmium. In: *Cadmium* (E.C. Foulkes, Ed.) pp. 373-380. Handbook of Experimental Pharmacology, Vol. 80. Berlin, Springer-Verlag.

Erasmus, D.A. (1967). Ultrastructural observations in the reserve bladder system of *Cyathocotyle bushiensis* Khan 1962 (Trematoda, Strigeoiden) with special reference to lipid excretion. *J. Parasit.* **53**: 525-536.

Erasmus, D.A. (1974). The application of X-ray analysis in the transmission electron microscope to a study of drug distribution in the parasite *Schistosoma mansoni* (Platyhelminthes). *J. Microsc.* **102**: 56-59.

Esser, H.O. & Moser, P. (1982). An appraisal of problems related to the measurement and evaluation of bioaccumulation. *Ecotoxicol. Environ. Safety* **6**: 131-148.

Esser, J. & El Bassam, N. (1981). On the mobility of cadmium under aerobic soil conditions. *Environ. Pollut.* **26A**: 15-31.

Ettinger, M.J., Darwish, H.M. & Schmidt, R.C. (1986). Mechanisms of copper transport from plasma to hepatocytes. *Fed. Proc.* **45**: 2800-2804.

Everard, L.B. & Swain, R. (1983). Isolation, characterization and induction of metallothionein in the stonefly *Eusthenia spectabilis* following exposure to cadmium. *Comp. Biochem. Physiol.* **75C**: 275-280.

Fain-Maurel, M.A., Cassier, P. & Alibert, J. (1973). Étude infrastructurale et cytochimique de l'intestin moyen de *Petrobius maritimus* Leach en rapport avec ses fonctions excrétrices et digestives. *Tiss. Cell* **5**: 603-631.

Farmer, P. (1987). *Lead Pollution from Motor Vehicles 1974 — 1986: a Select Bibliography*. London, Elsevier Applied Science.

Fergusson, J.E. (1986). Lead: petrol lead in the environment and its contribution to human blood lead levels. *Sci. Total Environ.* **50**: 1-54.

Filshie, B.K., Poulson, D.F. & Waterhouse, D.F. (1971). Ultrastructure of the copper-accumulating region of the *Drosophila* larval midgut. *Tiss. Cell* **3**: 77-102.

Fischer, E. (1977). The function of chloragosomes, the specific age pigment granules of annelids — a review. *Exp. Geront.* **12**: 69-74.

Fischer, E. & Trombitás, K. (1980). X-ray microprobe analysis of chloragosomes of untreated and of EDTA-treated *Lumbricus terrestris* by using fresh air-dried smears. *Acta Histochem.* **66**: 237-242.

Foelix, R.F. (1982). *Biology of Spiders*. Harvard University Press.

Forbes, A.R. (1964). The morphology, histology and fine structure of the gut of the Green Peach Aphid *Myzus persicae* (Sulzer)(Homoptera: Aphididae). *Mem. Entomol. Soc. Can.* **36**: 1-74.

Forester, A.J. (1977). The function of the intestine in the pulmonate mollusc *Helix pomatia* L. *Experentia* **33**: 465-467.

Foulkes, E.C. (1986). Ed. *Cadmium*. Handbook of Experimental Pharmacology, Vol. 80. Berlin, Springer-Verlag.

Fournié, J. & Chétail, M. (1982a). Evidence for a mobilization of calcium reserves for reproduction requirements in *Deroceras reticulatum* (Syn: *Agriolimax reticulatus*)(Gastropoda: Pulmonata). *Malacologia* **22**: 285-291.

Fournié, J. & Chétail, M. (1982b). Accumulation calcique au niveau cellulaire chez les mollusques. *Malacologia* **22**: 265-284.

Fowler, B.A. (1983). Ed. *Biological and Environmental Effects of Arsenic*. Amsterdam, Elsevier.

Fraenkel, G.S. (1958). The effect of zinc and potassium in the nutrition of *Tenebrio molitor*, with observations on the expression of a carnitine deficiency. *J. Nutr.* **65**: 361.

Fraenkel, G. (1959). A historical and comparative survey of the dietary requirements of insects. *Ann. New York Acad. Sci.* **77**: 267-274.

Frankenne, F., Noël-Lambot, F. & Disteche, A. (1980). Isolation and characterization of metallothioneins from cadmium-loaded mussel *Mytilus edulis*. *Comp. Biochem. Physiol.* **66C**: 179-182.

Freedman, B. & Hutchinson, T.C. (1980). Effects of smelter pollutants on forest leaf litter decomposition near a nickel-copper smelter at Sudbury, Ontario. *Can. J. Bot.* **58**: 1722-1736.

Fretter, V. (1952). Experiments with P^{32} and I^{131} on species of *Helix*, *Arion* and *Agriolimax*. *Quart. J. Microsc. Sci.* **93**: 133-146.

Fretter, V. (1953). Experiments with radioactive strontium ^{90}Sr on certain molluscs and polychaetes. *J. Mar. Biol. Ass. U.K.* **32**: 367-384.

Friberg, L. & Kjellström, T. (1981). Cadmium. In: *Disorders of Mineral Metabolism*. Vol. 1 (F. Bronner & J.W. Coburn, Eds.) pp. 317-352. New York, Academic Press.

Friberg, L., Elinder, C.G., Kjellström, T. & Nordberg, G.F. (1985). Eds. *Cadmium and Health: a Toxicological and Epidemiological Appraisal*. Vol. 1, *Exposure, Dose and Metabolism*. Florida, CRC Press.

Friberg, L., Nordberg, G.F. & Vouk, V.B. (1986). *Handbook on the Toxicology of Metals*. 2nd edition. Vol. 1, *General Aspects*. Vol. 2, *Specific Metals*. Amsterdam, Elsevier Science, BV Biomedical Division.

Friberg, L., Piscator, M., Nordberg, G.F. & Kjellström, T. (1974). *Cadmium in the Environment*, 2nd edition. Ohio, CRC Press.

Frieden, E. (1984). *Biochemistry of the Essential Ultratrace Elements*. New York, Plenum.

Friend, J.A. & Richardson, A.M.M. (1986). Biology of terrestrial amphipods.

Ann. Rev. Entomol. **31**: 25-48.

Fuller, H. (1966). Electronenmikroskopische Untersuchungen der malpighischen Gefässe von *Lithobius forficatus* L. *Z. Wiss. Zool.* **173A**: 191-200.

Furr, A.K., Lawrence, A.W., Tong, S.S.C., Grandolfo, M.C., Hofstader, R.A., Bache, C.A., Gutenmann, W.H. & Lisk, D.J. (1976). Multielement and chlorinated hydrocarbon analysis of municipal sewage sludges of American cities. *Environ. Sci. Technol.* **10**: 683-687.

Gadd, G.M. (1981). Mechanisms implicated in the ecological success of polymorphic fungi in metal polluted habitats. *Environ. Technol. Lett.* **2**: 531-536.

Gahukar, R.T. (1975). Effects of dietary zinc sulphate on the growth and feeding behaviour of *Ostrinia nubilalis* Hbn. (Lep. Pyraustidae). *Z. Ang. Entomol.* **79**: 352-357.

Gailey, F.A.Y. & Lloyd, O.L. (1986a). Methodological investigations into low technology monitoring of atmospheric metal pollution: Part 1 — the effects of sampler size on metal concentrations. *Environ. Pollut.* **12B**: 41-59.

Gailey, F.A.Y. & Lloyd, O.L. (1986b). Methodological investigations into low technology monitoring of atmospheric metal pollution: Part 2 — the effects of length of exposure on metal concentrations. *Environ. Pollut.* **12B**: 61-74.

Gailey, F.A.Y. & Lloyd, O.L. (1986c). Methodological investigations into low technology monitoring of atmospheric metal pollution: Part 3 — the degree of replicability of the metal concentrations. *Environ. Pollut.* **12B**: 85-109.

Galloway, J.N., Thornton, J.D., Norton, S.A., Volchok, H.L. & McLean, R.A.N. (1982). Trace metals in atmospheric deposition: a review and assessment. *Atmos. Environ.* **16**: 1677-1700.

Garcia-Miragaya, J., Castro, S. & Paolini, J. (1981). Lead and zinc levels and chemical fractionation in roadside soils of Caracas, Venezuela. *Water Air Soil Pollut.* **15**: 285-297.

Gautney, J.A., Mason, W.H. & Brugh, T.H. (1981). Egg production and bioelimination of ^{59}Fe in *Drosophila melanogaster*. *Environ. Entomol.* **10**: 327-331.

Giashuddin, M. & Cornfield, A.H. (1978). Incubation study on effects of adding varying levels of nickel (as sulphates) on nitrogen and carbon mineralization in soil. *Environ. Pollut.* **15**: 231-234.

Gibbs, P.E., Bryan, G.W. & Ryan, K.P. (1981). Copper accumulation by the polychaete *Melina palmata*: an antipredation mechanism? *J. Mar. Biol. Ass. U.K.* **61**: 707-722.

Giles, F.E., Middleton, S.G. & Grau, J.G. (1973). Evidence for the accumulation of atmospheric Pb by insects in areas of high traffic density. *Environ. Entomol.* **2**: 299-300.

Gill, R., Martin, M.H., Nickless, G. & Shaw, T.L. (1975). Regional monitoring of heavy metal pollution. *Chemosphere* **4**: 113-118.

Gingell, S.M., Campbell, R. & Martin, M.H. (1976). The effect of Zn, Pb and Cd pollution on the leaf surface microflora. *Environ. Pollut.* **11**: 25-37.

Gish, C.D. & Christiensen, R.E. (1973). Cd, Ni, Pb and Zn in earthworms from roadside soils. *Environ. Sci. Technol.* **11**: 1060-1062.

Gist, C.S. & Crossley, D.A. (1975). Feeding rates of some cryptozoa as determined by isotopic half-life studies. *Environ. Entomol.* **4**: 625-631.

Giunti, M. (1879). Ricerche sulla diffusione del rame nel regno animale. *Gazz.*

Chim. Ital. **9**: 546-555.

Godan, D. (1983). *Pest Slugs and Snails.* Berlin, Springer-Verlag.

Goldsmith, C.D. & Scanlon, R.F. (1977). Lead levels in small mammals and selected invertebrates associated with highways of different traffic densities. *Bull. Environ. Contam. Toxicol.* **17**: 311-316.

Goldstein, R.A. & Elwood, J.W. (1971). A two-compartment, three-parameter model for the absorption and retention of ingested elements by animals. *Ecology* **52**: 935-939.

Gordon, H.T. (1959). Minimal nutritional requirements of the German roach *Blattella germanica* L. *Ann. New York Acad. Sci.* **77**: 290-351.

Gould, H.J. (1962). Trials on the control of slugs on arable fields in autumn. *Plant Pathol.* **11**: 125-130.

Gould, J.L., Kirschvink, J.L., Deffeyes, K.S. & Brines, M.L. (1980). Orientation of demagnetized bees. *J. Exp. Biol.* **86**: 1-8.

Gouranton, J. (1967). Accumulation de ferritine dans les noyaux et le cytoplasm de certaines cellules due mésentéron chez des Homopteres Cercopides âgés. *C.R. Acad. Sci. Paris* **264**: 2657-2660.

Gouranton, J. (1968). Composition, structure et mode de formation des concrétions minérales dans l'intestin moyen des Homopteres Cercopides. *J. Cell Biol.* **37**: 316-328.

Gouranton, J. & Folliot, R. (1968). Presence de cristaux de ferritine de grande taille dans les cellules de l'intestin moyen de *Campylenchia latipes* Say (Homoptera, Membracidae). *Rev. Can. Biol.* **27**: 77-81.

Goyffon, M. (1977). Organotropisme du cuivre chez le scorpion. *Rev. Arachnol.* **1**: 9-12.

Graf, F. & Meyran, J.C. (1985). Calcium reabsorption in the posterior caeca of the midgut in a terrestrial crustacean *Orchestia cavimana*. Ultrastructural changes in the postexuvial epithelium. *Cell Tiss. Res.* **242**: 83-95.

Gragam, O.H., Drummond, R.O. & Diamant, G. (1964). The reproductive capacity of female *Boophilus annulatus* collected from cattle dipped in arsenic or coumaphos. *J. Econ. Entomol.* **57**: 409-410.

Grandjean, P. (1981). Blood lead concentrations reconsidered. *Nature* **291**: 188.

Grant, M., Wit, L.C. & Mason, W.H. (1980). Relationship of ^{65}Zn elimination and oxygen consumption in *Acheta domesticus*. *J. Georgia Entomol. Soc.* **15**: 10-19.

Greaves, G.N., Simkiss, K., Taylor, M. & Binsted, N. (1984). The local environment of metal sites in intracellular granules investigated by using X-ray absorption spectroscopy. *Biochem. J.* **221**: 855-868.

Green, D.E., Fry, M. & Blondin, G.A. (1980). Phospholipids as the molecular instruments of ion and solute transport in biological membranes. *Proc. Nat. Acad. Sci.* **77**: 257-261.

Green, L.F.B. (1979). Regional specialization in the Malpighian tubules of the New Zealand glow worm *Arachnocampa luminosa* (Diptera: Mycetophilidae). The structure and function of type I and II cells. *Tiss. Cell* **11**: 673-702.

Greenaway, P. (1985). Calcium balance and moulting in the Crustacea. *Biol. Rev.* **60**: 425-454.

Griffiths, B.S. & Wood, S. (1985). Microorganisms associated with the hindgut of *Oniscus asellus* (Crustacea, Isopoda). *Pedobiologia* **28**: 377-381.

Grodowitz, M.J. & Broce, A.B. (1983). Calcium storage in face fly (Diptera: Muscidae) larvae for puparium formation. *Ann. Entomol. Soc. Am.* **76**: 418-424.

Gueldner, R.C., Hedin, P.A. & Woodard, D.N. (1975). Mineral content of boll weevils, cotton buds and synthetic diets. *J. Econ. Entomol.* **68**: 428-430.

Guha, M.M. & Mitchell, R.L. (1966). Trace and major element composition of the leaves of some deciduous trees. *Pl. Soil* **24**: 90-112.

Gullvag, B.M. (1978). Metal measured in earthworm tissue by X-ray microanalysis. *Proceedings of the Ninth International Congress on Electron Microscopy*. Vol. 2, *Biology* (J.M. Sturgess, Ed.) pp. 114-115. Toronto, 1978. Toronto, Microscopical Society of Canada.

Gunnarsson, T. (1987a). Soil arthropods and their food: choice, use and consequences. PhD Thesis, University of Lund, Sweden.

Gunnarsson, T. (1987b). Selective feeding on a maple leaf by *Oniscus asellus*. *Pedobiologia* **30**: 161-165.

Gunnarsson, T. & Rundgren, S. (1986). Nematode infestation and hatching failure of lumbricid cocoons in acidified and polluted soils. *Pedobiologia* **29**: 165-173.

Gunnarsson, T. & Tunlid, A. (1986). Recycling of faecal pellets in isopods: microorganisms and nitrogen compounds as potential food for *Oniscus asellus* L. *Soil Biol. Biochem.* **18**: 595-600.

Gupta, B.L. & Hall, T.A. (1981). The X-ray microanalysis of frozen-hydrated sections in scanning electron microscopy: an evaluation. *Tiss. Cell* **13**: 623-643.

Gutknecht, J. (1981). Inorganic mercury (Hg^{2+}) transport through lipid bilayer membranes. *J. Membrane Biol.* **6**: 61-66.

Haber, V.R. (1926). The blood of insects with special reference to that of the common household German, or croton, cockroach, *Blatella germanica*, Linn. *Bull. Brook. Entomol. Soc.* **21**: 61-100.

Hall, T.A. (1979). Biological X-ray microanalysis. *J. Microsc.* **117**: 145-163.

Hall, T.A. & Gupta, B.L. (1982). Quantification for the X-ray microanalysis of cryosections. *J. Microsc.* **126**: 333-345.

Hall, T.A. & Gupta, B.L. (1984). The application of EDXS to the biological sciences. *J. Microsc.* **136**: 193-204.

Hallet, J.P., Lardinois, P., Ronneau, C. & Cara, J. (1982). Elemental deposition as a function of distance from an industrial zone. *Sci. Total Environ.* **25**: 99-109.

Hames, C.A.C. & Hopkin, S.P. (in press). The structure and function of the digestive system of terrestrial isopods. *J. Zool.*.

Hamilton, R.S. & Harrison, R.M. (1987). Eds. *Highway Pollution — Proceedings of the Second International Symposium*. London, July 1986. *Sci. Total Environ.* **59**: 1-486.

Haney, A. & Lipsey, R.L. (1973). Accumulation and effects of methyl mercury hydroxide in a terrestrial food chain under laboratory conditions. *Environ. Pollut.* **5**: 305-316.

Hanlon, R.D.G. (1981). Influence of grazing by Collembola on the activity of senescent fungal colonies grown on media of different nutrient concentration. *Oikos* **36**: 362-367.

Hanlon, R.D.G. & Anderson, J.M. (1979). The effects of Collembola grazing on microbial activity in decomposing leaf litter. *Oecologia* **38**: 93-99.

Hanlon, R.D.G. & Anderson, J.M. (1980). The influence of macroarthropod feeding activities on microflora in decomposing oak leaves. *Soil Biol. Biochem.* **12**: 255-261.

Hare, J.D. (1984). Suppression of the Colorado Potato beetle *Leptinotarsa decemlineata* (Say)(Coleoptera: Chrysomelidae) on solanaceous crops with a copper-based fungicide. *Environ. Entomol.* **13**: 1010-1014.

Harman, S.W. & Moore, J.B. (1938). Further studies with lead arsenate substitutes for codling moth control. *J. Econ. Entomol.* **31**: 223-226.

Harmsen, K. (1977). Behaviour of heavy metals in soils. *Agric. Res. Rep. (Versl. Landbouwk. Onderz.)* **866**: 1-171.

Harrison, P.M. (1977). Ferritin: an iron storage molecule. *Sem. Haematol.* **14**: 55-70.

Harrison, P.M. (1985). Ed. *Metalloproteins.* Part 1, *Metal Proteins with Redox Roles.* Part 2, *Metal Proteins with Non-Redox Roles.* Basingstoke, Macmillan.

Harrison, R.M. & Johnston, W.R. (1985). Deposition fluxes of lead, cadmium, copper and polynuclear aromatic hydrocarbons (PAH) on the verges of a major highway. *Sci. Total Environ.* **46**: 121-135.

Harrison, R.M. & Laxen, D.P.H. (1981). *Lead Pollution — Causes and Control.* London, Chapman & Hall.

Harrison, R.M. & Williams, C.R. (1983). Physico-chemical characterization of atmospheric trace metal emissions from a primary zinc-lead smelter. *Sci. Total Environ.* **31**: 129-140.

Harrison, R.M., Laxen, D.P.H. & Wilson, S.J. (1981). Chemical associations of lead, cadmium, copper and zinc in street dusts and roadside soils. *Environ. Sci. Technol.* **15**: 1378-1383.

Harrison, R.M., Johnston, W.R., Ralph, J.C. & Wilson, S.J. (1985). The budget of lead, copper and cadmium for a major highway. *Sci. Total Environ.* **46**: 137-145.

Harrison, R.M., Radojevic, M. & Wilson, S.J. (1986). The chemical composition of highway drainage waters. IV. Alkyllead compounds in runoff waters. *Sci. Total Environ.* **50**: 129-137.

Hartenstein, R. (1964). Feeding, digestion, glycogen and the environmental conditions of the digestive system in *Oniscus asellus. J. Insect Physiol.* **10**: 611-621.

Hartenstein, R. & Hartenstein, F. (1981). Physicochemical changes effected in activated sludge by the earthworm *Eisenia foetida. J. Environ. Qual.* **10**: 377-382.

Hartenstein, R., Leaf, A.L., Neuhauser, E.F. & Bickelhaupt, D.H. (1980a). Composition of the earthworm *Eisenia foetida* (Sav.) and assimilation of 15 elements from sludge during growth. *Comp. Biochem. Physiol.* **66C**: 187-192.

Hartenstein, R., Neuhauser, E.F. & Collier, J. (1980b). Accumulation of heavy metals in the earthworm *Eisenia foetida. J. Environ. Qual.* **9**: 23-26.

Hartenstein, R., Neuhauser, E.F. & Narahara, A. (1981). Effects of heavy metal and other elemental additives to activated sludge on growth of *Eisenia foetida. J. Environ. Qual.* **10**: 372-376.

Hartmann, H.J. & Weser, U. (1984). A structure and function correlation of metal-thiolate proteins. In: *Trace Element Analytical Chemistry in Medicine and Biology*. Vol. 3 (P. Brätter & P. Schramel, Eds.) pp. 267-281. Berlin, Walter De Gruyter.

Hassall, M. (1977). The functional morphology of the mouthparts and foregut in the terrestrial isopod *Philoscia muscorum* (Scopoli, 1763). *Crustaceana* **33**: 225-236.

Hassall, M. & Jennings, J.B. (1975). Adaptive features of gut structure and digestive physiology in the terrestrial isopod *Philoscia muscorum* (Scopoli) 1763. *Biol. Bull.* **149**: 348-364.

Hassall, M. & Rushton, S.P. (1982). The role of coprophagy in the feeding strategies of terrestrial isopods. *Oecologia* **53**: 374-381.

Hassall, M. & Rushton, S.P. (1984). Feeding behaviour of terrestrial isopods in relation to plant defences and microbial activity. *Symp. Zool. Soc. Lond.* **53**: 487-505.

Hassall, M. & Rushton, S.P. (1985). The adaptive significance of coprophagous behaviour in the terrestrial isopod *Porcellio scaber*. *Pedobiologia* **28**: 169-175.

Hassall, M., Turner, J.G. & Rands, M.R.W. (1987). Effects of terrestrial isopods on the decomposition of woodland leaf litter. *Oecologia* **72**: 597-604.

Hayat, M.A. (1980). Ed. *X-ray Microanalysis in Biology*. Baltimore, University Park Press.

Hayes, W.B. (1970). Copper concentrations in the high beach isopod *Tylos punctatus*. *Ecology* **51**: 721-723.

Hays, S.B. (1968). Reproductive inhibition in houseflies with triphenyl tin acetate and triphenyl tin chloride alone and in combination with other compounds. *J. Econ. Entomol.* **61**: 1154-1157.

Heath, G.W. & Arnold, M.K. (1966). Studies on leaf litter breakdown. II. Breakdown of 'sun' and 'shade' leaves. *Pedobiologia* **6**: 238-243.

Heath, G.W., Arnold, M.K. & Edwards, C.A. (1966). Breakdown rates of leaves of different species. *Pedobiologia* **6**: 1-12.

Helmke, P.A., Robarge, W.P., Korotev, R.L. & Schomberg, P.J. (1979). Effects of soil-applied sewage sludge on concentrations of elements in earthworms. *J. Environ. Qual.* **8**: 322-327.

Hemkes, O.J. & Hartmans, J. (1974). Copper content in grass and soil under high voltage lines in industrial and rural areas. In: *Trace Substances in Environmental Health*. Vol. 7 (D.D. Hemphill, Ed.) pp. 175-178. Missouri, University of Missouri.

Henderson, J.F. (1965). Molluscicides. *Rep. Rothamst. Exp. Stat.* (for 1965):173-174.

Henderson, J.F. (1968). Laboratory methods for assessing the toxicity of contact poisons to slugs. *Ann. Appl. Biol.* **62**: 363-369.

Henderson, J.F. (1969). A laboratory method for assessing the toxicity of stomach poisons to slugs. *Ann. Appl. Biol.* **63**: 167-171.

Herms, U. & Brummer, G. (1980). Einfluss der Bodenreaktion auf Loslichkeit und tolerierbare Gesamtgehalte an Nickel, Kupfer, Zink, Cadmium und Blei in Boden und kompostierten Siedlungsabfallen. *Landwirtsch. Forsch.* **33**: 408-423.

Hillerton, J.E. & Vincent, J.F.V. (1982). The specific location of zinc in insect mandibles. *J. Exp. Biol.* **101**: 333-336.

Hillerton, J.E., Robertson, B. & Vincent, J.F.V. (1984). The presence of zinc or manganese as the predominant metal in the mandibles of adult, stored-product beetles. *J. Stored Prod. Res.* **20**: 133-137.

Hirst, J.M., Le Riche, H.H. & Bascomb, C.C. (1961). Copper accumulation in the soils of apple orchards near Wisbech. *Plant Pathol.* **10**: 104-108.

Hiscock, S.A. (1983). Trends in the uses of cadmium (1970-1979). *Ecotoxicol. Environ. Safety* **7**: 25-32.

Hoffmann, P. & Lieser, K.H. (1987). Determination of metals in biological and environmental samples. *Sci. Total Environ.* **64**: 1-12.

Hogg, T.W. (1895). Immunity of some low forms of life from lead poisoning. *Chem. News (Lond.)* **71**: 233-234.

Holdich, D.M. & Mayes, K.R. (1975). A fine-structural re-examination of the so-called 'midgut' of the isopod *Porcellio. Crustaceana* **29**: 186-192.

Honda, K., Nasu, T. & Tatsukawa, R. (1984). Metal distribution in the earthworm *Pheretima hilgendorfi*, and their variations with growth. *Arch. Environ. Contam. Toxicol.* **13**: 427-432.

Hopkin, S.P. (1980). The structure and function of the hepatopancreas of the shore crab *Carcinus maenas* (L.): a study by electron microscopy and X-ray microanalysis. PhD Thesis, University College of North Wales.

Hopkin, S.P. (1986). Ecophysiological strategies of terrestrial arthropods for surviving heavy metal pollution. *Proceedings of the Third European Congress of Entomology* (H.H.W. Velthuis, Ed.) pp. 263-266. Amsterdam, August 1986. Amsterdam, Nederlandse Entomologische Vereniging.

Hopkin, S.P. & Martin, M.H. (1982a). The distribution of zinc, cadmium, lead and copper within the woodlouse *Oniscus asellus* (Crustacea, Isopoda). *Oecologia* **54**: 227-232.

Hopkin, S.P. & Martin, M.H. (1982b). The distribution of zinc, cadmium, lead and copper within the hepatopancreas of a woodlouse. *Tiss. Cell* **14**: 703-715.

Hopkin, S.P. & Martin, M.H. (1983). Heavy metals in the centipede *Lithobius variegatus* (Chilopoda). *Environ. Pollut.* **6B**: 309-318.

Hopkin, S.P. & Martin, M.H. (1984a). Heavy metals in woodlice. *Symp. Zool. Soc. Lond.* **53**: 143-166.

Hopkin, S.P. & Martin, M.H. (1984b). The assimilation of zinc, cadmium, lead and copper by the centipede *Lithobius variegatus* (Chilopoda). *J. Appl. Ecol.* **21**: 535-546.

Hopkin, S.P. & Martin, M.H. (1985a). Transfer of heavy metals from leaf litter to terrestrial invertebrates. *J. Sci. Food Agric.* **36**: 538-539.

Hopkin, S.P. & Martin, M.H. (1985b). Assimilation of zinc, cadmium, lead, copper and iron by the spider *Dysdera crocata*, a predator of woodlice. *Bull. Environ. Contam. Toxicol.* **34**: 183-187.

Hopkin, S.P. & Nott, J.A. (1979). Some observations on concentrically structured, intracellular granules in the hepatopancreas of the shore crab *Carcinus maenas* (L.). *J. Mar. Biol. Ass. U.K.* **59**: 867-877.

Hopkin, S.P. & Nott, J.A. (1980). Studies on the digestive cycle of the shore crab *Carcinus maenas* (L.) with special reference to the B cells in the hepatopancreas. *J. Mar. Biol. Ass. U.K.* **60**: 891-907.

Hopkin, S.P., Martin, M.H., & Moss, S.M. (1985b). Heavy metals in isopods from the supra-littoral zone on the southern shore of the Severn Estuary,

310   *Ecophysiology of Metals in Terrestrial Invertebrates*

U.K. *Environ. Pollut.* **9B**: 239-254.

Hopkin, S.P., Hardisty, G. & Martin, M.H. (1986). The woodlouse *Porcellio scaber* as a 'biological indicator' of zinc, cadmium, lead and copper pollution. *Environ. Pollut.* **11B**: 271-290.

Hopkin, S.P., Watson, K., Martin, M.H. & Mould, M.L. (1985a). The assimilation of heavy metals by *Lithobius variegatus* and *Glomeris marginata* (Chilopoda; Diplopoda). *Bijdragen tot de Dierkunde* **55**: 88-94.

Hopkin, S.P., Gaywood, M.J., Vincent, J.F.V. & Mayes-Harris, E.L.V. (in press, a). Defensive secretion of proteinaceous glues by *Henia* (*Chaetechelyne*) *vesuviana* (Chilopoda, Geophilomorpha). *Proceedings of the Seventh International Congress of Myriapodology*, Vittorio-Veneto, Italy, July 1987.

Hopkin, S.P., Hames, C.A.C. & Bragg, S. (in press, b). Terrestrial isopods as biological indicators of zinc pollution in the Reading area, south-east England. *Mon. Zool. Ital. (N.S.), Monografia 4.*

Horie, Y., Watanabe, K. & Ito, T. (1967). Nutrition of the silkworm *Bombyx mori*. XVIII. Quantitative requirements for potassium, phosphorus, magnesium and zinc. *Bull. Seric. Exp. Stat. Tokyo* **22**: 181-193.

Hosker, R.P. & Lindberg, S.E. (1982). Atmospheric deposition and plant assimilation of gases and particles. *Atmos. Environ.* **16**: 889-910.

Houk, E.J. (1977). Endoplasmic reticular inclusions in the malpighian tubules of the mosquito *Culex tarsalis*. *J. Ultrastruc. Res.* **60**: 63-70.

Howard, B., Mitchell, P.C.H., Ritchie, A., Simkiss, K. & Taylor, M. (1981). The composition of intracellular granules from the metal accumulating cells of the snail *Helix aspersa*. *Biochem. J.* **194**: 507-511.

Hryniewiecka-Szyfter, Z. (1972). Ultrastructure of hepatopancreas of *Porcellio scaber* Latr. in relation to the function of iron and copper accumulation. *Bull. Soc. Amis. Sci. Lett. Poznan (Sér. D)* **12/13**: 135-142.

Hryniewiecka-Szyfter, Z. (1973). Ultrastructure of organelles accumulating copper and iron in hepatopancreas of Crustacea. *Ann. Med. Sect. Polish Acad. Sci.* **18**: 113-115.

Hryniewiecka-Szyfter, Z. & Storch, V. (1986). The influence of starvation and different diets on the hindgut of Isopoda (*Mesidotea entomon*, *Oniscus asellus*, *Porcellio scaber*). *Protoplasma* **134**: 53-59.

Hubbell, S.P., Sikora, A. & Paris, O.H. (1965). Radiotracer, gravimetric and calorimetric studies of ingestion and assimilation rates of an isopod. *Health Phys.* **11**: 1485-1501.

Hubert, M. (1974). Le tissu adipeux de *Cylindroiulus teutonicus* Pocock (*londinensis* C.L.K.) Diplopoda, Iuloidea; étude histologique et ultrastructurale. *C.R. Acad. Sci. Paris* **278D**: 3343-3346.

Hubert, M. (1975). Sur la nature des accumulations minérales et uriques chez *Cylindroiulus teutonicus* Pocock (*londinensis* C.L.K., Diplopoda, Iuloidea). *C.R. Acad. Sci. Paris* **281D**: 151-156.

Hubert, M. (1978a). Données histophysiologiques complémentaires sur les bioaccumulations minérales et puriques chez *Cylindroiulus londinensis* (Leach, 1814)(Diplopode, Iuloidea). *Arch. Zool. Exp. Gén.* **119**: 669-683.

Hubert, M. (1978b). Les cellules hépatiques de *Cylindroiulus londinensis* (Leach, 1814)(Diplopode, Iuloidea). *C.R. Acad. Sci. Paris* **286D**: 627-630.

Hubert, M. (1979). Localization and identification of mineral elements and nitrogenous waste in Diplopoda. In: *Myriapod Biology* (M. Camatini, Ed.) pp. 127-133. London, Academic Press.

Huckabee, J.W. & Blaylock, B.G. (1973). Transfer of mercury and cadmium from terrestrial to aquatic ecosystems. *Adv. Exp. Med. Biol.* **40**: 125-160.

Hueck, H.J. & La Brijn, J. (1973). Mothproofing properties of insecticides. III. Organolead compounds. *Mater. Org.* **8**: 133-143.

Hughes, M.K., Lepp, N.W. & Phipps, D.A. (1980). Aerial heavy metal pollution and terrestrial ecosystems. *Adv. Ecol. Res.* **11**: 218-327.

Hughes, M.N. & Poole, R.K. (1988). *Metals and Micro-Organisms*. London, Chapman & Hall.

Humbert, W. (1974a). Localisation, structure et genèse des concrétions minérales dans le mésentéron des Collemboles Tomoceridae (Insecta, Collembola). *Z. Morph. Tiere* **78**: 93-109.

Humbert, W. (1974b). Étude du pH intestinal d'un Collembole (Insecte, Apterygote). *Rev. Ecol. Biol. Sol* **11**: 89-97.

Humbert, W. (1977). Mineral concretions in midgut of *Tomocerus minor* (Collembola) — microprobe analysis and physiological significance. *Rev. Ecol. Biol. Sol* **14**: 71-80.

Humbert, W. (1978). Cytochemistry and X-ray microprobe analysis of the midgut of *Tomocerus minor* Lubbock (Insecta: Collembola) with special reference to the physiological significance of the mineral concretions. *Cell Tiss. Res.* **187**: 397-416.

Humbert, W. (1979). The midgut of *Tomocerus minor* Lubbock (Insecta, Collembola): ultrastructure, cytochemistry, ageing and renewal during a moulting cycle. *Cell Tiss. Res.* **196**: 39-57.

Hunter, B.A. (1984). The ecology and toxicology of trace metals in contaminated grasslands. PhD Thesis, University of Liverpool.

Hunter, B.A. & Johnson, M.S. (1982). Food chain relationships of copper and cadmium in contaminated grassland ecosystems. *Oikos* **38**: 108-117.

Hunter, B.A., Hunter, L.M., Johnson, M.S. & Thompson, D.J. (1987d). Dynamics of metal accumulation in the grasshopper *Chorthippus brunneus* in contaminated grasslands. *Arch. Environ. Contam. Toxicol.* **16**: 711-716.

Hunter, B.A., Johnson, M.S. & Thompson, D.J. (1984a). Food chain relationships of copper and cadmium in herbivorous and insectivorous small mammals. In: *Metals in Animals* (D. Osbourn, Ed.) pp. 5-10. Institute of Terrestrial Ecology Symposium No. 12. Abbots Ripton, Institute of Terrestrial Ecology.

Hunter, B.A., Johnson, M.S. & Thompson, D.J. (1984b). Cadmium-induced lesions in tissues of *Sorex araneus* from metal refinery grasslands. In: *Metals in Animals* (D. Osbourn, Ed.) pp. 39-44. Institute of Terrestrial Ecology Symposium No. 12. Abbots Ripton, Institute of Terrestrial Ecology.

Hunter, B.A., Johnson, M.S. & Thompson, D.J. (1987a). Ecotoxicology of copper and cadmium in a contaminated grassland ecosystem. I. Soil and vegetation contamination. *J. Appl. Ecol.* **24**: 573-586.

Hunter, B.A., Johnson, M.S. & Thompson, D.J. (1987b). Ecotoxicology of copper and cadmium in a contaminated grassland ecosystem. II. Invertebrates. *J. Appl. Ecol.* **24**: 587-599.

Hunter, B.A., Johnson, M.S. & Thompson, D.J. (1987c). Ecotoxicology of copper and cadmium in a contaminated grassland ecosystem. III. Small mammals. *J. Appl. Ecol.* **24**: 601-614.

Hunziker, P.E. & Kagi, J.H.R. (1985). Metallothionein. In: *Metalloproteins*. Part 2, *Metal Proteins with Non-Redox Roles* (P.M. Harrison, Ed.) pp. 149-181. Basingstoke, Macmillan Press.

Hurley, D.E. (1968). Transition from water to land in amphipod crustaceans. *Am. Zool.* **8**: 327-353.

Hutchinson, T.C. & Whitby, L.M. (1974). Heavy metal pollution in the Sudbury mining and smelting region of Canada. I. Soil and vegetation contamination by nickel, copper and other metals. *Environ. Conserv.* **1**: 123-132.

Hutton, M. (1983). Sources of cadmium in the environment. *Ecotoxicol. Environ. Safety* **7**: 9-24.

Hutton, M. (1984). Impact of airborne metal contamination on a deciduous woodland system. In: *Effects of Pollutants at the Ecosystem Level* (P.J. Sheehan, D.R. Miller, G.C. Butler & P. Bourdeau, Eds.) pp. 365-375. Chichester, John Wiley & Sons.

Icely, J.D. & Nott, J.A. (1980). Accumulation of copper within the 'hepatopancreatic' caeca of *Corophium volutator* (Crustacea: Amphipoda). *Mar. Biol.* **57**: 193-199.

Ihnat, M. (1988). Pick a number — analytical data reliability and biological reference materials. *Sci. Total Environ.* **71**: 85-104.

Ineson, P. & Anderson, J.M. (1985). Aerobically isolated bacteria associated with the gut and faeces of the litter feeding macroarthropods *Oniscus asellus* and *Glomeris marginata*. *Soil Biol. Biochem.* **17**: 843-849.

Ineson, P., Leonard, M.A. & Anderson, J.M. (1982). Effect of collembolan grazing on nitrogen and cation leaching from decomposing leaf litter. *Soil Biol. Biochem.* **14**: 601-605.

Inman, J.C. & Parker, G.R. (1978). Decomposition and heavy metal dynamics of forest litter in northwestern Indiana. *Environ. Pollut.* **17**: 39-51.

Inoue, Y. & Watanabe, T.K. (1978). Toxicity and mutagenicity of cadmium and furylfuramide in *Drosophila melanogaster*. *Jap. J. Genet.* **53**: 183-189.

Ireland, M.P. (1975a). Metal content of *Dendrobaena rubida* (Oligochaeta) in a base metal mining area. *Oikos* **26**: 74-79.

Ireland, M.P. (1975b). The effect of the earthworm *Dendrobaena rubida* on the solubility of lead, zinc and cadmium in heavy metal contaminated soil in Wales. *J. Soil Sci.* **26**: 313-318.

Ireland, M.P. (1975c). Distribution of lead, zinc and cadmium in *Dendrobaena rubida* (Oligochaeta) living in soil contaminated by base metal mining in Wales. *Comp. Biochem. Physiol.* **52B**: 551-555.

Ireland, M.P. (1976). Excretion of lead, zinc and calcium by the earthworm *Dendrobaena rubida* living in soil contaminated with zinc and lead. *Soil Biol. Biochem.* **8**: 347-350.

Ireland, M.P. (1978). Heavy metal binding properties of earthworm chloragosomes. *Acta Biol. Acad. Sci. Hung.* **29**: 385-394.

Ireland, M.P. (1979a). Metal accumulation by the earthworms *Lumbricus rubellus*, *Dendrobaena veneta* and *Eiseniella tetraedra* living in heavy metal polluted sites. *Environ. Pollut.* **19**: 201-206.

Ireland, M.P. (1979b). Distribution of essential and toxic metals in the terrestrial gastropod *Arion ater*. *Environ. Pollut.* **20**: 271-278.

Ireland, M.P. (1981). Uptake and distribution of cadmium in the terrestrial slug *Arion ater* (L.). *Comp. Biochem. Physiol.* **68A**: 37-41.

Ireland, M.P. (1982). Sites of water, zinc and calcium uptake and distribution of these metals after cadmium administration in *Arion ater* (Gastropoda: Pulmonata). *Comp. Biochem. Physiol.* **73A**: 217-221.

Ireland, M.P. (1983). Heavy metal uptake and tissue distribution in earthworms. In: *Earthworm Ecology* (J.E. Satchell, Ed.) pp. 247-265. London & New York, Chapman and Hall.

Ireland, M.P. (1984a). Seasonal changes in zinc, manganese, magnesium, copper and calcium content in the digestive gland of the slug *Arion ater*. *Comp. Biochem. Physiol.* **78A**: 855-858.

Ireland, M.P. (1984b). Effect of chronic and acute lead treatment in the slug *Arion ater* on calcium and delta-aminolaevulinic acid dehydratase activity. *Comp. Biochem. Physiol.* **79C**: 287-290.

Ireland, M.P. (1988). A comparative study of the uptake and distribution of silver in a slug *Arion ater* and a snail *Achatina fulica*. *Comp. Biochem. Physiol.* **90C**: 189-194.

Ireland, M.P. & Fischer, E. (1979). Effect of Pb^{++} on Fe^{++} tissue concentrations and delta-aminolaevulinic acid dehydratase activity in *Lumbricus terrestris*. *Acta Biol. Acad. Sci. Hung.* **29**: 395-400.

Ireland, M.P. & Richards, K.S. (1977). The occurrence and localisation of heavy metals and glycogen in the earthworms *Lumbricus rubellus* and *Dendrobaena rubida* from a heavy metal site. *Histochemistry* **51**: 153-166.

Ireland, M.P. & Richards, K.S. (1981). Metal content, after exposure to cadmium, of two species of earthworms of known differing calcium metabolic activity. *Environ. Pollut.* **26A**: 69-78.

Ireland, M.P. & Wooton, R.J. (1976). Variations in the lead, zinc and calcium content of *Dendrobaena rubida* (Oligochaeta) in a base metal mining area. *Environ. Pollut.* **10**: 201-208.

Irgolic, K.J. (1986). Arsenic in the environment. In: *Frontiers in Bioinorganic Chemistry* (A.V. Xavier, Ed.) pp. 399-408. Hamburg, VCH.

Ito, T. & Niimura, M. (1966). Nutrition of the silkworm *Bombyx mori*. XII. Nutritive effects of minerals. *Bull. Seric. Exp. Stat. Tokyo* **20**: 361-374.

Ivanova, L.V., Brovx, L.Y., Shakhovtseva, T.N., Dolmanova, I.F. & Ugarova, N.N. (1982). Application of *Luciola mingrelica* immobilized luciferase for enzymatic detection of physiologically active metals. *Prikl. Biochim. Microbiol.* **18**: 718-724 (In Russian).

Iyngar, G.V., Kasperek, K. & Feinendegen, L.E. (1978). Retention of the metabolized trace elements in biological tissues following different drying procedures. 1. Antimony, cobalt, iodine, mercury, selenium and zinc. *Sci. Total Environ.* **10**: 1-16.

Izhevskiy, S.S. (1976). The physiological effect of mineral salts on *Leptinotarsa decemlineata* Say. *Ekologiya* **4**: 90-92 (In Russian).

Jackson, D.R. & Watson, A.P. (1977). Disruption of nutrient pools and transport of heavy metals in a forested watershed near a lead smelter. *J. Environ. Qual.* **6**: 331-338.

Jackson, D.R., Selvidge, W.J. & Ausmus, B.S. (1978a). Behaviour of heavy metals in forest microcosms. I. Transport and distribution among components.

Water Air Soil Pollut. **10**: 3-11.

Jackson, D.R., Selvidge, W.J. & Ausmus, B.S. (1978b). Behaviour of heavy metals in forest microcosms. II. Effects on nutrient cycling processes. *Water Air Soil Pollut.* **10**: 13-18.

Jacobson, K.B., Williams, M.W., Turner, J.E. & Christie, N.T. (1985). Cadmium toxicity in *Drosophila*: genetic and physiological parameters. In: *International Conference. Heavy Metals in the Environment.* Vol. 2, pp. 239-241. Athens, 1985. Edinburgh, CEP Consultants.

Jaffre, T., Brooks, R.R., Lee, J. & Reeves, R.D. (1976). *Sebertia acuminata*: a hyperaccumulator of nickel from New Caledonia. *Science* **193**: 579-580.

Jaggy, A. & Streit, B. (1982). Toxic effects of soluble copper on *Octolasium cyaneaum* Sav. (Lumbricidae). *Rev. Suisse Zool.* **89**: 881-890.

Janssen, H.H. (1985). Some histophysiological findings on the mid-gut gland of the common garden snail *Arion rufus* (L.)(Syn. *A. ater rufus* (L.)), *A. empiricorum* (Férrussac), Gastropoda: Stylommatophora. *Zool. Anz.* **215**: 33-51.

Jeantet, A.Y., Martoja, R. & Truchet, M. (1974). Rôle des sphérocristaux de l'epithelium intestinal dans la résistance d'un insecte aux pollutions minérales: données expérimentales obtenues par utilisation de la microsonde électronique et du micro-analyseur par émission ionique secondaire. *C.R. Acad. Sci. Paris* **278D**: 1441-1444.

Jeantet, A.Y., Ballan-Dufrançais, C. & Martoja, R. (1977). Insects resistance to mineral pollution. Importance of spherocrystal in ionic regulation. *Rev. Écol. Biol. Sol* **14**: 563-582.

Jeantet, A.Y., Ballan-Dufrançais, C. & Ruste, J. (1980). Quantitative electron probe analysis on insect exposed to mercury. II. Involvement of the lysosomal system in detoxification processes. *Biol. Cellulaire* **39**: 325-334.

Jeffries, D.S. (1983). Atmospheric deposition of pollutants in the Sudbury area. *Adv. Environ. Sci. Technol.* **15**: 117-154.

Jenkins, T. (1973). Histochemical and fine structure observations of the intestinal epithelium of *Trichuris suis* (Nematoda: Trichinoidea). *Z. Parasitenk.* **42**: 165-183.

Jenkins, T., Erasmus, D.A. & Davies, T.W. (1977). *Trichuris suis* and *T. muris*: elemental analysis of intestinal inclusions. *Exp. Parasit.* **41**: 464-471.

Jennings, T.J. & Barkham, J.P. (1975). Food of slugs in mixed deciduous woodland. *Oikos* **26**: 211-221.

Jennings, T.J. & Barkham, J.P. (1979). Litter decomposition by slugs in mixed deciduous woodland. *Holarct. Ecol.* **2**: 21-29.

Jensen, A. & Jorgensen, S.E. (1984). Analytical chemistry applied to metal ions in the environment. In: *Metal Ions in Biological Systems*. Vol. 18 (H. Sigel, Ed.) pp. 5-59. New York, Marcel Dekker.

John, M.K., Van Laerhoven, C. & Bjerring, J. (1976a). Effect of a smelter complex on the regional distribution of cadmium, lead and zinc in litter and soil horizons. *Arch. Environ. Contam. Toxicol.* **4**: 456-468.

John, M.K., Van Laerhoven, C. & Cross, C. (1976b). Cadmium, lead and zinc accumulation in soils near a smelter complex. *Environ. Lett.* **10**: 23-25.

Jones, D.A., Babbage, P.C. & King, P.E. (1969). Studies on digestion and the fine structure of digestive caeca in *Eurydice pulchra* (Crustacea: Isopoda).

Mar. Biol. **2**: 311-320.

Jones, D.S. & Macfadden, B.J. (1982). Induced magnetization in the monarch butterfly *Danaus plexippus* (Insecta, Lepidoptera). *J. Exp. Biol.* **96**: 1-9.

Jones, K.C. (1987). Honey as an indicator of heavy metal contamination. *Water Air Soil Pollut.* **33**: 179-190.

Jones, R. (1983). Zinc and cadmium in lettuce and radish grown in soils collected near electrical transmission (hydro) towers. *Water Air Soil Pollut.* **19**: 389-395.

Jones, R. & Burgess, M.S.E. (1984). Zinc and cadmium in soils and plants near electrical transmission (hydro) towers. *Environ. Sci. Technol.* **18**: 731-733.

Joosse, E.N.G. & Buker, J.B. (1979). Uptake and excretion of lead by litter dwelling Collembola. *Environ. Pollut.* **18**: 235-240.

Joosse, E.N.G. & Van Vliet, L.H.H. (1982). Impact of blast-furnace plant emissions in a dune ecosystem. *Bull. Environ. Contam. Toxicol.* **29**: 279-284.

Joosse, E.N.G. & Van Vliet, L.H.H. (1984). Iron, manganese and zinc inputs in soil and litter near a blast-furnace plant and the effects on the respiration of woodlice. *Pedobiologia* **26**: 249-256.

Joosse, E.N.G. & Verhoef, H.A. (1987). Developments in ecophysiological research on soil invertebrates. *Adv. Ecol. Res.* **16**: 175-248.

Joosse, E.N.G. & Verhoef, S.C. (1983). Lead tolerance in Collembola. *Pedobiologia* **25**: 11-18.

Joosse, E.N.G., Wulffraat, K.J. & Glas, H.P. (1981). Tolerance and acclimation to zinc of the isopod *Porcellio scaber* Latr. In: *International Conference. Heavy Metals in the Environment*, pp. 425-428. Amsterdam, 1981. Edinburgh, CEP Consultants.

Joosse, E.N.G., Van Capelleveen, H.E., Van Dalen, L.H. & Van Diggelen, J. (1983). Effects of zinc, iron and manganese on soil arthropods associated with decomposition processes. In: *International Conference. Heavy Metals in the Environment*. Vol. 1, pp. 467-470. Heidelberg, 1983. Edinburgh, CEP Consultants.

Jordan, M.J. (1975). Effects of zinc smelter emissions and fire on a chestnut-oak woodland. *Ecology* **56**: 78-91.

Jordan, M.J. & Lechevalier, M.P. (1975). Effects of zinc-smelter emissions on forest floor microflora. *Can. J. Microbiol.* **21**: 1855-1865.

Jorgensen, S.E. & Jensen, A. (1984). Processes of metal ions in the environment. In: *Metal Ions in Biological Systems*. Vol. 18 (H. Sigel, Ed.) pp. 61-103. New York, Marcel Dekker.

Kägi, J.H.R. (1987). *Metallothionein II*. Basel, Birkhauser.

Kägi, J.H.R. & Vallee, B.L.R. (1961). Metallothionein — an acid and zinc containing protein from equine renal cortex. *J. Biol. Chem.* **236**: 2435-2442.

Kaplan, D.L. & Hartenstein, R. (1978). Studies on monoxygenases and dioxygenases in soil macroinvertebrates and bacterial isolates from the gut of the terrestrial isopod *Oniscus asellus* L. *Comp. Biochem. Physiol.* **60B**: 47-50.

Karnak, R.E. & Hamelink, J.L. (1982). A standardized method for determining the acute toxicity of chemicals to earthworms. *Ecotoxicol. Environ. Safety* **6**: 216-222.

Keating, K.I. & Dagbusan, B.C. (1984). Effect of selenium deficiency on cuticle integrity in the Cladocera (Crustacea). *Proc. Nat. Acad. Sci.* **81**: 3433-3437.

Kegley, L.M., Brown, B.W. & Berntzen, A.K. (1969). *Mesocestoides corti*: inorganic components in calcareous corpuscles. *Exp. Parasit.* **25**: 85-92.

Kenaga, E.E. (1965). Triphenyl tin compounds as insect reproduction inhibitor. *J. Econ. Entomol.* **58**: 4-8.

Khan, D.H. & Frankland, B. (1983). Chemical forms of cadmium and lead in some contaminated soils. *Environ. Pollut.* **6B**: 15-32.

Kheirallah, A.M. (1979). Behavioural preferences of *Julus scandinavius* (Myriapoda) to different species of leaf litter. *Oikos* **33**: 466-471.

Kiekens, L. (1984). Behaviour of heavy metals in soils. In: *Utilisation of Sewage Sludge on Land: Rates of Application and Long-Term Effects of Metals* (S. Berglund, R.D. Davis & P. L'Hermite, Eds.) pp. 126-134. New York, D. Reidel.

Kiffer, E. & Benest, G. (1981). Microflora of the digestive tract of *Abax ater* (Coleoptera, Carabidae). *Rev. Écol. Biol. Sol* **18**: 567-578.

Killham, K. & Wainwright, M. (1981). Deciduous leaf litter and cellulose decomposition in soil exposed to heavy atmospheric pollution. *Environ. Pollut.* **26A**: 79-85.

Kimura, M., Seveus, L. & Maramorosch, K. (1975). Ferritin in insect vectors of the maize streak disease agent: electron microscopy and electron microprobe analysis. *J. Ultrastruc. Res.* **53**: 366-373.

King, R.C. (1957). The major inorganic constituents of adult *Drosophila melanogaster*. *Am. Nat.* **91**: 319.

Kirschvink, J.L., Jones, D.S. & Macfadden, B.J. (1985). Eds. *Magnetite Biomineralization and Magnetoreception in Organisms*. Topics in Geobiology, Vol. 5. New York, Plenum.

Kogan, I.G., Grozdova, T.Y. & Kholikova, T.A. (1978). Investigation of the mutagenic effect of $CdCl_2$ on *Drosophila melanogaster* germ cells. *Genetika* **14**: 2136-2140 (In Russian).

Koh, T.S. (1984). An interlaboratory survey of the determination of selenium in blood: a preliminary report. In: *Trace Element Analytical Chemistry in Medicine and Biology*. Vol. 3 (P. Brätter & P. Schramel, Eds.) pp. 201-206. Berlin, Walter De Gruyter.

Koirtyohann, S.R. & Hopkins, C.A. (1976). Loss of trace metals during the ashing of biological materials. *Analyst* **101**: 870-875.

Koirtyohann, S.R., Wallace, G. & Hinderberger, E. (1976). Multi-element analysis of *Drosophila* for environmental monitoring purposes using carbon furnace atomic absorption. *Can. J. Spectrosc.* **21**: 61-64.

Kopp, J.F. (1975). Current status of analytical methodology for trace metals. In: *International Conference. Heavy Metals in the Environment*. Vol. 1, pp. 183-204. Toronto, 1975. Edinburgh, CEP Consultants.

Korte, F. (1983). Ecotoxicology of cadmium: general overview. *Ecotoxicol. Environ. Safety* **7**: 3-8.

Kowal, N.E. (1971). Models of elemental assimilation by invertebrates. *J. Theoret. Biol.* **31**: 469-474.

Kraal, H. & Ernst, W. (1976). Influence of copper high tension lines on plants and soil. *Environ. Pollut.* **11**: 131-135.

Krueger, R.A., Broce, A.B. & Hopkins, T.L. (1987). Dissolution of granules in the Malpighian tubules of *Musca autumnalis* De Geer, during mineralization of the puparium. *J. Insect Physiol.* **33**: 255-263.

Kruse, E.A. & Barrett, G.W. (1985). Effects of municipal sludge and fertilizer on heavy metal accumulation in earthworms. *Environ. Pollut.* **38A**: 235-244.

Kudo, A., Miyahara, S. & Miller, D.R. (1980). Movement of mercury from Minimata Bay into Yatsushiro Sea. *Prog. Water Technol.* **12**: 509-524.

Kukor, J.J. & Martin, M.M. (1986a). The effect of acquired microbial enzymes on assimilation efficiency in the common woodlouse *Tracheoniscus rathkei. Oecologia* **69**: 360-366.

Kukor, J.J. & Martin, M.M. (1986b). Cellulose digestion in *Monochamus marmorator* Kby. (Coleoptera: Cerambycidae): role of acquired fungal enzymes. *J. Chem. Ecol.* **12**: 1057-1070.

Kuterbach, D.A. & Walcott, B. (1986a). Iron-containing cells in the honeybee (*Apis mellifera*). I. Adult morphology and physiology. *J. Exp. Biol.* **126**: 375-387.

Kuterbach, D.A. & Walcott, B. (1986b). Iron-containing cells in the honeybee (*Apis mellifera*). II. Accumulation during development. *J. Exp. Biol.* **126**: 389-401.

Kuterbach, D.A., Walcott, B., Reeder, R.J. & Frankel, R.B. (1982). Iron-containing cells in the honey bee (*Apis mellifera*). *Science* **218**: 695-696.

Lagerwerff, J.V. & Specht, A.W. (1970). Contamination of roadside soil and vegetation with cadmium, nickel, lead and zinc. *Environ. Sci. Technol.* **4**: 583-586.

Lake, D.L., Kirk, P.W.W. & Lester, J.N. (1984). Fractionation, characterization and speciation of heavy metals in sewage sludge and sludge-amended soils: a review. *J. Environ. Qual.* **13**: 175-183.

Lange, W.H. & Macleod, G.F. (1941). Metaldehyde and calcium arsenate in slug and snail baits. *J. Econ. Entomol.* **34**: 321-322.

Lansdown, R. (1985). Ed. *The Lead Debate: The Environment, Toxicology and Health*. Beckenham, Croom Helm.

Lastowski-Perry, D., Otto, E. & Maroni, G. (1985). Nucleotide sequence and expression of a *Drosophila* metallothionein. *J. Biol. Chem.* **260**: 1527-1530.

Lauhachinda, N. & Mason, W.H. (1979). ^{65}Zn excretion as an indirect measurement of Q_{10} in the isopod *Armadillidium vulgare* (Latreille). *J. Alabama Acad. Sci.* **50**: 27-34.

Lawrey, J.D. (1978). Trace metal dynamics in decomposing leaf litter in habitats variously influenced by coal strip mining. *Can. J. Bot.* **56**: 953-962.

Lawton, J.H. (1986). Food shortage in the midst of apparent plenty?: the case for birch-feeding insects. *Proceedings of the Third European Congress of Entomology* (H.H.W Velthuis, Ed.) pp. 219-228. Amsterdam, August 1986. Amsterdam, Nederlandse Entomologische Vereniging.

Le Garff, B. & Barbier, R. (1982). Développment des tubes de Malpighi adultes obtenus par transplantation de disques imaginaux chez *Galleria mellonella* L. (Lepidoptera: Pyralidae). *Int. J. Insect Morphol. Embryol.* **11**: 161-171.

Lee, M.Y., Huebers, H., Martin, A.W. & Finch, C.A. (1978). Iron metabolism in a spider *Dugesiella hentzi. J. Comp. Physiol.* **127B**: 349-354.

Legge, A.H. & Krupa, S.V. (1986). Eds. *Air Pollutants and Their Effects on the Terrestrial Ecosystem*. Advances in Environmental Science and Technology, Vol. 18. Chichester, John Wiley & Sons.

Lennox, F.G. (1940). Studies of the physiology and toxicology of blowflies. 7. A quantitative examination of the iron content of *Lucilia cuprina*. *Bull. Counc. Sci. Ind. Res. Melb.* **102**: 51-65.

Leonhard, S.L., Lawrence, S.G., Friesen, M.K. & Flannagan, J.F. (1980). Evaluation of the acute toxicity of the heavy metal cadmium to nymphs of the burrowing mayfly *Hexagenia rigida*. In: *Advances in Ephemeroptera Biology* (J.F. Flannagan & K.E. Marshall, Eds.) pp. 457-465. New York & London, Plenum Press.

Levy, R. & Cromroy, H.L. (1973). Concentration of some major and trace elements in forty one species of adult and immature insects determined by atomic absorption spectroscopy. *Ann. Entomol. Soc. Am.* **66**: 523-526.

Levy, R., Cromroy, H.L. & Cornell, J.A. (1973). Major and trace elements as bioindicators of acute insect radiosensitivity. *Radiation Res.* **56**: 130-139.

Levy, R., Van Rinsvelt, H.A. & Cromroy, H.L. (1974a). Relative concentration of major and trace elements in adult and immature stages of the red imported fire ant determined by ion-induced X-ray fluorescence. *Florida Entomol.* **57**: 269-273.

Levy, R., Cromroy, H.L. & Cornell, J.A. (1974b). Major and trace elements as biopredictors of radiation-induced insect sterility. *Florida Entomol.* **57**: 303-307.

Levy, R., Cromroy, H.L. & Van Rinsvelt, H.A. (1979). Relative comparisons in major and trace elements between adult and immature stages of two species of fire ants. *Florida Entomol.* **62**: 260-266.

Lewis, H.C. & La Follette, J.R. (1942). Control of brown snail in citrus orchards. *J. Econ. Entomol.* **35**: 359-362.

Lewis, J.G.E. (1981). *Biology of Centipedes*. Cambridge University Press.

Lhonoré, J. (1971). Données cytophysiologiques sur les tubes de Malpighi de *Gryllotalpa gryllotalpa* Latr. (Orthoptère, Gryllotalpidé). *C.R. Acad. Sci. Paris* **272**: 2788-2790.

Licent, E. (1912). Recherches d'anatomie et de physiologie comparées sur le tube digestif des Homoptères supérieurs. *La Cellule (Louvain)* **28**: 1-20.

Little, P. (1973). A study of heavy metal contamination on leaf surfaces. *Environ. Pollut.* **5**: 159-172.

Little, P. (1977). Deposition of 2.75, 5.0 and 8.5 μm particles to plant and soil surfaces. *Environ. Pollut.* **12**: 293-305.

Little, P. & Martin, M.H. (1972). A survey of zinc, lead and cadmium in soil and natural vegetation around a smelting complex. *Environ. Pollut.* **3**: 241-254.

Little, P. & Martin, M.H. (1974). Biological monitoring of heavy metal pollution. *Environ. Pollut.* **6**: 1-19.

Little, P. & Wiffen, R.D. (1977). Emission and deposition of petrol engine exhaust lead. I. Deposition of exhaust Pb to plant and soil surfaces. *Atmos. Environ.* **11**: 437-447.

Little, P. & Wiffen, R.D. (1978). Emission and deposition of lead from motor exhausts. II. Airborne concentration, particle size and deposition of lead near motorways. *Atmos. Environ.* **12**: 1331-1341.

Locke, M. & Leung, H. (1984). The induction and distribution of an insect ferritin — a new function for the endoplasmic reticulum. *Tiss. Cell* **16**: 739-766.

Lodenius, M. (1981). Mercury content of dipterous larvae feeding on macrofungi. *Ann. Entomol. Fenn.* **47**: 63-64.

Lodenius, M. & Tulisalo, E. (1984). Environmental mercury contamination around a chlor-alkali plant. *Bull. Environ. Contam. Toxicol.* **32**: 439-444.

Lotmar, R. (1938). Untersuchungen über den Eisenstoffwechsel der Insekten, besonders der Honigbiene. *Rev. Suisse Zool.* **45**: 237-271.

Ludwig, M. & Alberti, G. (1988). Mineral congregations "spherites" in the midgut gland of *Coelotes terrestris* (Araneae): structure, composition and function. *Protoplasma* **143**: 43-50.

Lund, L.J., Betty, E.E., Page, A.L. & Elliot, R.A. (1981). Occurrence of naturally high cadmium levels in soils and its accumulation by vegetation. *J. Environ. Qual.* **10**: 551-556.

Lyon, G.L., Brooks, R.R., Peterson, P.J. & Butler, G.W. (1970). Some trace elements in plants from serpentine soils. *New Zeal. J. Sci.* **13**: 133-139.

Lytton, D.G., Eastgate, A.R. & Ashbolt, N.J. (1985). Digital mapping of metal-bearing amoebocytes in tissue sections of the Pacific oyster by scanning microphotometry. *J. Microsc.* **138**: 1-14.

Ma, W.C. (1982). The influence of soil properties and worm-related factors on the concentration of heavy metals in earthworms. *Pedobiologia* **24**: 109-120.

Ma, W. (1984). Sublethal toxic effects of copper on growth, reproduction and litter breakdown activity in the earthworm *Lumbricus rubellus* with observations on the influence of temperature and soil pH. *Environ. Pollut.* **33A**: 207-219.

Ma, W.C. (1987). Heavy metal accumulation in the mole *Talpa europea* and earthworms as an indicator of metal bioavailability in terrestrial environments. *Bull. Environ. Contam. Toxicol.* **39**: 933-938.

Ma, W., Edelman, T., Van Beersum, I. & Jans, T. (1983). Uptake of cadmium, zinc, lead and copper by earthworms near a zinc-smelting complex — influence of soil pH and organic matter. *Bull. Environ. Contam. Toxicol.* **30**: 424-427.

Mackay, D. (1988). On low, very low, and negligible concentrations. *Environ. Toxicol. Chem.* **7**: 1-3.

Macrae, E.K. & Bogorad, L. (1958). Conversion of delta-aminolevulinic acid to porphobilinogen in earthworms. *Anat. Rec.* **131**: 577-578.

Madrell, S.H.P. & Gardiner, B.O.C. (1980). The permeability of the cuticular lining of the insect alimentary canal. *J. Exp. Biol.* **85**: 227-237.

Malecki, M.R., Neuhauser, E.F. & Loehr, R.C. (1982). The effect of metals on the growth and reproduction of *Eisenia foetida* (Oligochaeta, Lumbricidae). *Pedobiologia* **24**: 129-137.

Mangum, C.P. (1980). Respiratory function of the hemocyanins. *Am. Zool.* **20**: 19-38.

Marcaillou, C., Truchet, M. & Martoja, R. (1986). Rôle des cellules S de l'épithélium caecal des Crustacés Isopodes dans la capture et la dégradation de protéines hémolymphatiques, et dans le stockage de catabolites (acide urique, sulfure de cuivre, phosphates). *Can. J. Zool.* **64**: 2757-2769.

Maren, T.H. (1967). Carbonic anhydrase: chemistry, physiology and inhibition. *Physiol. Rev.* **47**: 595-781.

Marešova, E., Škrobák, J. & Weismann, L. (1973). Toxische Einwirkung des Kupfers auf die Raupen *Scotia segetum* Den. et Schiff. 1. Inhibitionseffekt der Kupferverbindungen auf die Aktivität der Xanthindehydrogenase. *Biológia (Bratislavia)* **28**: 651-656.

Margoshes, M. & Vallee, B.L. (1957). A cadmium protein from equine kidney-cortex. *J. Am. Chem. Soc.* **79**: 4813.

Marigomez, J.A., Angulo, E. & Moya, J. (1986a). Copper treatment of the digestive gland of the slug *Arion ater* L. 1. Bioassay conduction and histo-chemical analysis. *Bull. Environ. Contam. Toxicol.* **36**: 600-607.

Marigomez, J.A., Angulo, E. & Moya, J. (1986b). Copper treatment of the digestive gland of the slug *Arion ater* L. 2. Morphometrics and histophysiology. *Bull. Environ. Contam. Toxicol.* **36**: 608-615.

Marigomez, J.A., Angulo, E. & Saez, V. (1986c). Feeding and growth responses to copper, zinc, mercury and lead in the terrestrial gastropod *Arion ater* (Linne). *J. Mollusc. Stud.* **52**: 68-78.

Maroni, G. & Watson, D. (1985). Uptake and binding of cadmium, copper and zinc by *Drosophila melanogaster* larvae. *Insect Biochem.* **15**: 55-63.

Maroni, G., Lastowski-Perry, D., Otto, E. & Watson, D. (1986a). Effects of heavy metals on *Drosophila* larvae and a metallothionein cDNA. *Environ. Health Perspec.* **65**: 107-116.

Maroni, G., Otto, E. & Lastowski-Perry, D. (1986b). Molecular and cytogenetic characterization of a metallothionein gene of *Drosophila*. *Genetics* **112**: 493-504.

Maroni, G., Wise, J., Young, J.E. & Otto, E. (1987). Metallothionein gene duplications and metal tolerance in natural populations of *Drosophila melanogaster*. *Genetics* **117**: 739-744.

Marples, A.E. (1979). The occurrence and behaviour of cadmium in soils and its uptake by pasture grasses in industrially contaminated and naturally metal-rich environments. PhD Thesis, Royal School of Mines, Imperial College, London.

Marshall, A.T. (1983). X-ray microanalysis of copper and sulphur-containing granules in the fat body cells of homopteran insects. *Tiss. Cell* **15**: 311-315.

Marshall, A.T. & Cheung, W.W.K. (1973a). Studies on water and iron transport in homopteran insects: ultrastructure and cytochemistry of the cicadoid and cercopoid hindgut. *Tiss. Cell* **5**: 671-678.

Marshall, A.T. & Cheung, W.W.K. (1973b). Calcification in insects. The dwell-ing tube and midgut of machaerotid larvae (Homoptera). *J. Insect Physiol.* **19**: 963-972.

Martin, M.H. & Coughtrey, P.J. (1975). Preliminary observations on the levels of cadmium in a contaminated environment. *Chemosphere* **4**: 155-160.

Martin, M.H. & Coughtrey, P.J. (1976). Comparisons between the levels of lead, zinc, and cadmium within a contaminated environment. *Chemosphere* **5**: 15-20.

Martin, M.H. & Coughtrey, P.J. (1981). Impact of metals on ecosystem function and productivity. In: *Effects of Heavy Metals on Plants*. Vol. 2 (N.W. Lepp, Ed.) pp. 119-158. London & New York, Applied Science.

Martin, M.H. & Coughtrey, P.J. (1982). *Biological Monitoring of Heavy Metal Pollution: Land and Air*. London & New York, Applied Science.

Martin, M.H. & Coughtrey, P.J. (1987). Cycling and fate of heavy metals in a contaminated woodland ecosystem. In: *Pollutant Transport and Fate in Ecosystems* (P.J. Coughtrey, M.H. Martin & M.H. Unsworth, Eds.) pp. 319-336. Oxford, Blackwell.

Martin, M.H., Coughtrey, P.J. & Ward, P. (1979). Historical aspects of heavy metal pollution in the Gordano Valley. *1977 Proceedings Bristol Naturalists Society* **37**: 91-97.

Martin, M.H., Coughtrey, P.J. & Young, E.W. (1976). Observations on the availability of lead, zinc, cadmium and copper in woodland litter and the uptake of lead, zinc and cadmium by the woodlouse *Oniscus asellus*. *Chemosphere* **5**: 313-318.

Martin, M.H., Duncan, E.M. & Coughtrey, P.J. (1982). The distribution of heavy metals in a contaminated woodland ecosystem. *Environ. Pollut.* **3B**: 147-157.

Martin, M.H., Coughtrey, P.J., Shales, S.W. & Little, P. (1980). Aspects of airborne cadmium contamination of soils and natural vegetation. In: *Inorganic Pollution and Agriculture*, pp. 56-69. MAFF Reference Book 326. London, HMSO.

Martin, M.M. (1983). Cellulose digestion in insects. *Comp. Biochem. Physiol.* **75A**: 313-324.

Martin, W.E. & Bils, R.F. (1964). Trematode excretory concretions: formation and fine structure. *J. Parasit.* **50**: 337-344.

Martoja, R. (1972). Données histophysiologiques sur les accumulations minérales et puriques des Thysanoures. *Arch. Zool. Exp. Gén.* **113**: 565-578.

Martoja, R. (1974). Problèms posés par l'adaptation aux Aptérygotes des méthodes histologiques classiques. Application à deux aspects de la physiologie des Thysanoures. *Pedobiologia* **14**: 163-164.

Martoja, R. & Ballan-Dufrançais, C. (1984). The ultrastructure of the digestive and excretory organs. In: *Insect Ultrastructure*. Vol. 2, *The Ultrastructure of Developing Cells* (R.C. King & H. Akai, Eds.) pp. 199-268. New York, Plenum.

Martoja, R. & Martoja, M. (1973). Sur des accumulations naturelles d'aluminium et de silicium chez quelques invertebrés. *C.R. Acad. Sci. Paris* **276D**: 2951-2954.

Martoja, R. & Seureau, C. (1972). Répartition des accumulations de metaux et de déchets puriques chez *Calliphora erythrocephala* (Diptère, Brachycère). *C.R. Acad. Sci. Paris* **274D**: 1534-1537.

Martoja, R. & Truchet, M. (1983). Rôle d'un ommochrome dans l'excretion de metaux essentiels (Cu, Zn) et dans la détoxication à l'égard de contaminants métalliques (Ag, Cd) chez un insecte (*Locusta migratoria*, Orthoptère). *C.R. Acad. Sci. Paris* (Ser. III) **297**: 219-224.

Martoja, R., Alibert, J., Ballan-Dufrançais, C., Jeantet, A.Y., Lhonore, D. & Truchet, M. (1975). Microanalyse et écologie. *J. Micr. Biol. Cell.* **22**: 441-448.

Martoja, R., Lhonore, D. & Ballan-Dufrançais, C. (1977). Bioaccumulation minerale et purique chez les insectes planipennes, *Euroleon nostras*, *Myrmeleon hyalinus*, *Acanthaclisis beaticus*, *Sisyra fuscata*, *Chrysopa* sp. *Arch. Zool. Exp. Gén.* **118**: 441-455.

Martoja, R., Bouquegneau, J.M. & Verthe, C. (1983). Toxicological effects and storage of cadmium and mercury in an insect Locusta migratoria (Orthoptera). J. Invert. Pathol. **42**: 17-32.

Mason, A.Z. (1983). Applications of microincineration in localising biomineralized inorganic deposits in sectioned tissues. In: Biomineralization and Biological Metal Accumulation (P. Westbroek & E.W. De Jong, Eds.) pp. 379-387. New York, D. Reidel.

Mason, A.Z. & Nott, J.A. (1980). The association of the blood vessels and the excretory epithelium in the kidney of Littorina littorea (L.)(Mollusca: Gastropoda). Mar. Biol. Lett. **1**: 355-365.

Mason, A.Z. & Nott, J.A. (1981). The role of intracellular biomineralized granules in the regulation and detoxification of metals in gastropods with special reference to the marine prosobranch Littorina littorea. Aquatic Toxicol. **1**: 239-256.

Mason, A.Z. & Simkiss, K. (1982). Sites of mineral deposition in metal-accumulating cells. Exp. Cell Res. **139**: 383-391.

Mason, A.Z., Simkiss, K. & Ryan, K.P. (1984). The ultrastructural localisation of metals in specimens of Littorina littorea collected from clean and polluted sites. J. Mar. Biol. Ass. U.K. **64**: 699-720.

Mason, W.H. & McGraw, K.A. (1973). Relationship of [65]Zn excretion and egg production in Trichoplusia ni (Hübner). Ecology **54**: 214-216.

Mason, W.H. & Odum, E.P. (1969). The effect of coprophagy on retention and bioelimination of radionuclides by detritus-feeding animals. In: Proceedings of the Second National Symposium on Radioecology (D.J. Nelson & F.O. Evans, Eds.) pp. 721-724. USAEC, Technical Information Service.

Mason, W.H., Wit, L.C. & Blackmore, M.S. (1983a). Bioelimination of [65]Zn in Popilius disjunctus after a dietary zinc supplement. J. Georgia Entomol. Soc. **18**: 246-251.

Mason, W.H., Wit, L.C. & Sherrill, D.E. (1983b). Effect of label level on the bioelimination of [65]Zn in Popilius disjunctus. J. Alabama Acad. Sci. **54**: 94-100.

Massie, H.R., Aiello, V.R. & Williams, T.R. (1981). Cadmium: temperature dependent increase with age in Drosophila. Exp. Geront. **16**: 337-341.

Massie, H.R., Aiello, V.R. & Williams, T.R. (1983a). Chromium levels and aging in mice and Drosophila. Age **6**: 62-65.

Massie, H.R. Williams, T.R. & Aiello, V.R. (1983b). Influence of dietary cadmium and chelators on the survival of Drosophila. Gerontology **29**: 226-232.

Massie, H.R., Williams, T.R. & Aiello, V.R. (1984). Influence of dietary copper on the survival of Drosophila. Gerontology **30**: 73-78.

Masui, H., Suzuki, K. & Matsubara, F. (1986). Distribution and concentration of elements in the tissues and organs of germ-free silkworm larvae fed on diets containing EDTA. J. Seric. Sci. Japan **55**: 23-27.

Mathew, C. & Al-Doori, Z. (1976). The mutagenic effect of the mercury fungicide Ceresan M in Drosophila melanogaster. Mutat. Res. **40**: 31-36.

Matsubara, F., Masui, H. & Suzuki, K. (1986). Concentration of elements in organs and tissues of germ-free silkworm larvae fed on an artificial diet. J. Seric. Sci. Japan **55**: 5-9.

Matsumoto, K. & Fuwa, K. (1986). The estimation of cadmium in biological samples. In: *Cadmium* (E.C. Foulkes, Ed.) pp. 1-31. Handbook of Experimental Pharmacology, Vol 80. Berlin, Springer-Verlag.

Matsumoto, M. (1960). Cytological study of iron of the chloragogen cells in the earthworm. *Sci. Rep. Tohoku Univ.* **24**: 95-105.

Matthewson, M.D. & Baker, J.A.F. (1975). Arsenic resistance in species of multi-host ticks in the Republic of South Africa and Swaziland. *J. S. Afr. Vet. Ass.* **46**: 341-344.

Maurer, R. (1974). Die Vielfalt der Käfer und Spinnenfauna des Wiesenbodens im Einflussbereich von Verkehrsemissionen. *Oecologia* **14**: 327-351.

Maxwell, C.W. (1961). Laboratory tests of some insecticides against adults of the apple maggot *Rhagoletis pomonella* (Walsh). *J. Econ. Entomol.* **54**: 526-528.

Maxwell, C.W. & Parsons, E.C. (1963). Comparing inert ingredients in lead arsenate sprays in laboratory toxicity tests against the apple maggot *Rhagoletis pomonella*. *J. Econ. Entomol.* **56**: 626-628.

Maxwell, C.W., Neilson, W.T.A. & Wood, F.A. (1963). Evaluation of lead arsenate for control of the apple maggot *Rhagoletis pomonella*, in New Brunswick. *J. Econ. Entomol.* **56**: 160-161.

McBrayer, J.F. (1973). Exploitation of deciduous leaf litter by *Apheloria montana* (Diplopoda: Eurydesmidae). *Pedobiologia* **13**: 90-98.

McFarlane, J.E. (1972). Vitamin E, tocopherol quinone and selenium in the diet of the house cricket *Acheta domesticus* L. *Israel J. Entomol.* **7**: 7-14.

McFarlane, J.E. (1974). The function of copper in the house cricket and the relation of copper to vitamin E. *Can. Entomol.* **106**: 441-446.

McFarlane, J.E. (1976). Influence of dietary copper and zinc on growth and reproduction of the house cricket *(Orthoptera: Gryllidae)*. *Can. Entomol.* **108**: 387-390.

McLean, J.A., Stump, I.G., D'Auria, J.M. & Holman, J. (1979). Monitoring trace elements in diets and life stages of the onion maggot, *Hylemya antiqua* (Diptera: Anthomyiidae) with X-ray energy spectrometry. *Can. Entomol.* **111**: 1293-1298.

McNeilly, T., Williams, S.T. & Christian, P.J. (1984). Lead and zinc in a contaminated pasture at Minera, North Wales, and their impact on productivity and organic matter breakdown. *Sci. Total Environ.* **38**: 183-198.

Medici, J.C. & Taylor, M.W. (1966). Mineral requirements of the confused flour beetle. *J. Nutr.* **88**: 181-186.

Medici, J.C. & Taylor, M.W. (1967). Interrelationships among copper, zinc and cadmium in the diet of the confused flour beetle. *J. Nutr.* **93**: 307-309.

Meincke, K.F. & Schaller, K.H. (1974). Uber die Brauchbarkeit der Weinbergschnecke *(Helix pomatia* L.) im Freiland als Indikator für die Belastung der Umwelt durch die Elemente Eisen, Zink und Blei. *Oecologia* **15**: 393-398.

Mello, M.L.S. & Bozzo, L. (1969). Histochemistry, refractometry and fine structure of excretory globules in larval Malpighian tubules of *Melipona quadrifasciata* (Hym. Apoidea). *Protoplasma* **68**: 241-251.

Melvin, R. (1931). A quantitative study of copper in insects. *Ann. Entomol. Soc. Am.* **24**: 485-488.

Meyer, E., Schwazenberger, I., Stark, G. & Wechselberger, G. (1984). Bestand und jahreszeitliche Dynamik der Bodenmakrofauna in einem inneralpinen Eichenmischwald (Tirol, Österreich). *Pedobiologia* **27**: 115-132.

Meyran, J.C., Graf, F. & Nicaise, G. (1986). Pulse discharge of calcium through a demineralizing epithelium in the crustacean *Orchestia*: ultrastructural cytochemistry and X-ray microanalysis. *Tiss. Cell* **18**: 267-283.

Miller, D.R. (1984a). Chemicals in the environment. In: *Effects of Pollutants at the Ecosystem Level* (P.J. Sheehan, D.R. Miller, G.C. Butler & P. Bourdeau, Eds.) pp. 7-14. Chichester, John Wiley & Sons.

Miller, D.R. (1984b). Distinguishing ecotoxic effects. In: *Effects of Pollutants at the Ecosystem Level* (P.J. Sheehan, D.R. Miller, G.C. Butler & P. Bourdeau, Eds.) pp. 15-22. Chichester, John Wiley & Sons.

Miller, W.J. & Neathery, M.W. (1977). Newly recognised trace mineral elements and their role in animal nutrition. *Bioscience* **27**: 674-679.

Mills, C.F. (1979). Trace elements in animals. *Phil. Trans. Roy. Soc. Lond.* **288B**: 51-63.

Mislin, M. & Ravera, O. (1986). Eds. *Cadmium in the Environment*. Basel & Boston, Birkhauser.

Mitchell, R.L. & Burridge, J.C. (1979). Trace elements in soils and crops. *Phil. Trans. Roy. Soc. Lond.* **288B**: 15-24.

Mitra, S. (1986). *Mercury in the Ecosystem*. Switzerland, Trans Tech.

Miyoshi, T., Miyazawa, F. & Shimizu, O. (1971). Effects of heavy metals on the mulberry plant and silkworm. I. Effects of cadmium and zinc on silkworm larvae *Bombyx mori* L. *J. Seric. Sci. Japan* **40**: 323-329.

Miyoshi, T., Miyazawa, F., Shimizu, O. & Machida, J. (1978a). Effect of heavy metals on the mulberry plant and silkworm. II. Effect of dietary lead, copper and arsenite on silkworm larvae *Bombyx mori* L. *J. Seric. Sci. Japan* **47**: 70-76.

Miyoshi, T., Shimizu, O., Miyazawa, F. Machida, J. & Ito, M. (1978b). Effect of heavy metals on the mulberry plant and silkworm. III. Cooperative effects of heavy metals on silkworm larvae *Bombyx mori* L. *J. Seric. Sci. Japan* **47**: 77-84.

Miyoshi, T., Miyazawa, F. & Shimizu, O. (1978c). Effect of heavy metals on the mulberry plant and silkworm. IV. The relation between different compounds of heavy metals and their toxicity and an absorption of heavy metals by the silkworm *Bombyx mori* L. *J. Seric. Sci. Japan* **47**: 101-107.

Mohan, M.S., Zingaro, R.A., Micks, P. & Clark, P.J. (1982). Analysis and speciation of arsenic in herbicide treated soils by DC helium emission spectrometry. *Int. J. Environ. Anal. Chem.* **11**: 175-187.

Mokdad, R., Debec, A. & Wegnez, M. (1987). Metallothionein genes in *Drosophila* constitute a dual system. *Proc. Nat. Acad. Sci.* **84**: 2658-2662.

Moore, P.G. & Rainbow, P.S. (1984). Ferritin crystals in the gut caeca of *Stegocephaloides christianiensis* Boeck and other Stegocephalidae (Amphipoda: Gammaridea): a functional interpretation. *Phil. Trans. Roy. Soc. Lond.* **306B**: 219-245.

Moore, P.G. & Rainbow, P.S. (1987). Copper and zinc in an ecological series of talitroidean Amphipoda (Crustacea). *Oecologia* **73**: 120-126.

Moreton, R.B. (1981). Electron-probe X-ray microanalysis: techniques and recent applications in biology. *Biol. Rev.* **56**: 409-461.

Morgan, A.J. (1979). Non-freezing techniques of preparing biological specimens for electron microprobe X-ray microanalysis. *Scan. Electron Microsc.* **2**: 635-648.

Morgan, A.J. (1980). Preparation of specimens: changes in chemical integrity. In: *X-ray Microanalysis in Biology* (M.A. Hayat, Ed.) pp. 65-165. Baltimore, University Park Press.

Morgan, A.J. (1981). A morphological and electron-microprobe study of the inorganic composition of the mineralized secretory products of the calciferous gland and chloragogenous tissue of the earthworm *Lumbricus terrestris* L. The distribution of injected strontium. *Cell Tiss. Res.* **220**: 829-844.

Morgan, A.J. (1982). The elemental composition of the chloragosomes of nine species of British earthworms in relation to calciferous gland activity. *Comp. Biochem. Physiol.* **73A**: 207-216.

Morgan, A.J. (1985). *X-ray Microanalysis in Electron Microscopy for Biologists.* Royal Microscopical Society Handbook No. 5. London, Royal Microscopical Society.

Morgan, A.J. & Morris, B. (1982). The accumulation and intracellular compartmentation of cadmium, lead, zinc, and calcium in two earthworm species *(Dendrobaena rubida* and *Lumbricus rubellus)* living in highly contaminated soil. *Histochemistry* **75**: 269-287.

Morgan, A.J. & Winters, C. (1982). The elemental composition of the chloragosomes of two earthworm species *Lumbricus terrestris* and *Allolobophora longa)* determined by electron probe X-ray microanalysis of freeze-dried cryosections. *Histochemistry* **73**: 589-598.

Morgan, A.J., Davies, T.W. & Erasmus, D.A. (1975). Analysis of droplets from iso-atomic solutions as a means of calibrating a transmission electron analytical microscope (TEAM). *J. Microsc.* **104**: 271-280.

Morgan, E.H. & Baker, E. (1986). Iron uptake and metabolism by hepatocytes. *Fed. Proc.* **45**: 2810-2816.

Morgan, J.E. (1987). Exogenous and endogenous factors influencing the accumulation of heavy metals by selected earthworm species. PhD Thesis, University College of Wales, Cardiff.

Mori, T. & Kurihara, Y. (1979). Accumulation of heavy metals in earthworms *(Eisenia foetida)* grown in composted sewage sludge. *Sci. Rep. Tohoku Univ.* **37**: 289-297.

Morris, B. & Morgan, A.J. (1986). Calcium-lead interactions in earthworms — observations on *Lumbricus terrestris* L. sampled from a calcareous abandoned lead mine site. *Bull. Environ. Contam. Toxicol.* **37**: 226-233.

Moser, H. & Wieser, W. (1979). Copper and nutrition in *Helix pomatia* (L.). *Oecologia* **42**: 241-251.

Munshower, F. (1972). Cadmium compartmentation and cycling in a grassland ecosystem in the Deer Lodge Valley, Montana. PhD Thesis, University of Montana.

Muntau, H. (1984). Newer essential trace elements and analytical reliability. In: *Trace Element Analytical Chemistry in Medicine and Biology.* Vol. 3 (P. Brätter & P. Schramel, Eds) pp. 563-580. Berlin, Walter De Gruyter.

Muschinek, G., Szentesi, A. & Jermy, T. (1976). Inhibition of oviposition in the bean weevil *(Acanthoscelides obtectus* Say, Col., Bruchidae). *Acta*

Phytopathol. Acad. Sci. Hung. **11**: 91-98.

Muskett, C.J. & Jones, M.P. (1980). The dispersal of lead, cadmium and nickel from motor vehicles and effects on roadside invertebrate macrofauna. *Environ. Pollut.* **23A**: 231-242.

Muttkowski, R.A. (1921). Copper: its occurrence and role in insects and other animals. *Trans. Am. Microsc. Soc.* **40**: 144-157.

Nabholz, J.V. & Crossley, D.A. (1978). Ingestion and elimination of cesium-134 by the spider *Pardosa lapidicina. Ann. Entomol. Soc. Am.* **71**: 325-328.

Nakayama, Y. & Matsubara, F. (1981). Influence of heavy metals on the aseptically reared silkworm larvae. 5. Inhibitory effects of EDTA on cadmium toxicity. *Environ. Control Biol.* **19**: 121-128.

Nation, J.L. & Robinson, F.A. (1971). Concentration of some major and trace elements in honeybees, royal jelly and pollens determined by atomic absorption spectrophotometry. *J. Apic. Res.* **10**: 35-43.

Needham, A.E. (1960). Properties of the connective tissue pigment of *Lithobius forficatus* (L.). *Comp. Biochem. Physiol.* **1**: 72-100.

Neuhauser, E.F. (1978). Phenolic content and palatability of leaves and wood to soil isopods and diplopods. *Pedobiologia* **18**: 99-109.

Neuhauser, E.F., Malecki, M.R. & Loehr, R.C. (1984a). Growth and reproduction of the earthworm *Eisenia foetida* after exposure to sub-lethal concentrations of metals. *Pedobiologia* **27**: 89-98.

Neuhauser, E.F., Meyer, J.A., Malecki, M.R. & Thomas, J.M. (1984b). Dietary cobalt supplements and the growth and reproduction of the earthworm *Eisenia foetida. Soil Biol. Biochem.* **16**: 521-523.

Neuhauser, E.F., Malecki, M.R. & Cukic, Z.V. (1985). Metal content of earthworms in sludge amended soils: uptake and loss. In: *International Conference. Heavy Metals in the Environment.* Vol. 2, pp. 211-213. Athens, 1985. Edinburgh, CEP Consultants.

Nicholson, P.B., Bocock, K.L. & Heal, O.W. (1966). Studies on the decomposition of the faecal pellets of a millipede (*Glomeris marginata* (Villers)). *J. Ecol.* **54**: 755-766.

Nieboer, E. & Richardson, D.H.S. (1980). The replacement of the nondescript term 'heavy metals' by a biologically and chemically significant classification of metal ions. *Environ. Pollut.* **1B**: 3-26.

Nieland, M.L. & Von Brand, T. (1969). Electron microscopy of cestode calcareous corpuscle formation. *Exp. Parasit.* **24**: 279-289.

Nielsen, M.G. & Gissel-Nielsen, G. (1975). Selenium in soil animal relationships. *Pedobiologia* **15**: 65-67.

Nielson, R.L. (1951). Effect of soil minerals on earthworms. *New Zeal. J. Agric.* **83**: 433-435.

Niklas, J. & Kennel, W. (1978). Lumbricid populations in orchards of the Federal Republic of Germany and the influence of fungicides based on copper compounds and benzimidazole derivatives upon them. *Z. Pflanzenkr. Pflanzensch.* **85**: 705-713 (In German).

Nopanitaya, W. & Misch, D.W. (1974). Developmental cytology of the midgut of the fleshfly *Sarcophaga bullata* (Parker). *Tiss. Cell* **6**: 487-502.

Nott, J.A. & Mavin, L.J. (1986). Adaptation of a quantitative programme for the X-ray analysis of solubilized tissue as microdroplets in the transmission electron microscope: application to the moult cycle of the shrimp *Crangon crangon* (L). *Histochem. J.* **18**: 507-518.

Nottrot, F., Joosse, E.N.G. & Van Straalen, N.M. (1987). Sublethal effects of iron and manganese soil pollution on *Orchesella cincta* (Collembola). *Pedobiologia* **30**: 45-53.

Nriagu, J.O. (1978). Ed. *The Biogeochemistry of Lead in the Environment*. Part A, *Ecological cycles*. Part B, *Biological Effects*. Amsterdam, Elsevier/North Holland Biomedical Press.

Nriagu, J.O. (1979). Ed. *Copper in the Environment*. Parts 1 & 2. Chichester, Wiley Interscience.

Nriagu, J.O. (1980a). Ed. *Zinc in the Environment*. Parts 1 & 2. Chichester, Wiley Interscience.

Nriagu, J.O. (1980b). Ed. *Cadmium in the Environment*. Part 1. Chichester, Wiley Interscience.

Nriagu, J.O. (1980c). Ed. *Nickel in the Environment*. Chichester, Wiley Interscience.

Nriagu, J.O. (1981). Ed. *Cadmium in the Environment*. Part 2. *Health Effects*. New York, Wiley Interscience.

Nriagu, J.O. (1988). A silent epidemic of environmental metal poisoning. *Environ. Pollut.* **50**: 139-162.

Nriagu, J.O. & Davidson, C.I. (1986). Eds. *Toxic Metals in the Atmosphere*. *Adv. Environ. Sci. Technol.* **17**. New York, Wiley Interscience.

Nuorteva, P. (1977). Saprophagous insects as forensic indicators. In: *Forensic Medicine, a Study in Trauma and Environmental Hazards*. Vol. 2, *Physical Trauma* (C.G. Tedeschi, W.G. Eckert & L.G. Tedeschi, Eds.) pp. 1072-1095. Philadelphia, Saunders.

Nuorteva, P. & Häsänen, E. (1972). Transfer of mercury from fishes to sarcosaprophagous flies. *Ann. Zool. Fenn.* **9**: 23-27.

Nuorteva, P. & Nuorteva, S.L. (1982). The fate of mercury in sarcosaprophagous flies and in insects eating them. *Ambio* **11**: 34-37.

Nuorteva, P., Häsänen, E. & Nuorteva, S.L. (1978a). Bioaccumulation of mercury in sarcosaprophagous insects. *Norw. J. Entomol.* **25**: 79-80.

Nuorteva, P., Nuorteva, S.L. & Suckcharoen, S. (1980). Bioaccumulation of mercury in blowflies collected near the mercury mine of Idrija, Yugoslavia. *Bull. Environ. Contam. Toxicol.* **24**: 515-521.

Nuorteva, P., Witkowski, Z. & Nuorteva, S.L. (1987). Chronic damage by *Tortrix viridana* L. (Lepidoptera, Tortricidae) related to the content of iron, aluminium, zinc, cadmium and mercury in oak leaves in Niepolomice forest. *Ann. Entomol. Fenn.* **53**: 36-38.

Nuorteva, P., Wuorenrinne, H & Kaistila, M. (1978b). Transfer of mercury from fish carcass to *Formica aquilonia* (Hymenoptera, Formicidae). *Ann. Entomol. Fenn.* **44**: 85-86.

Ode, P.E. & Matthysse, J.G. (1964). Feed additive larviciding to control face fly. *J. Econ. Entomol.* **57**: 637-640.

Olafson, R.W., Sim, R.G. & Boto, K.G. (1979). Isolation and chemical characterization of the heavy metal-binding protein metallothionein from marine invertebrates. *Comp. Biochem. Physiol.* **62B**: 407-416.

Orians, G.H. & Pfeiffer, E.W. (1970). Ecological effects of the war in Vietnam.

Science **168**: 544-554.

Otto, E., Young, J.E. & Maroni, G. (1986). Structure and expression of a tandem duplication of the *Drosophila* metallothionein gene. *Proc. Nat. Acad. Sci.* **83**: 6025-6029.

Otto, E., Allen, J.M., Young, J.E., Palmiter, R.D. & Maroni, G. (1987). A DNA segment controlling metal-regulated expression of the *Drosophila* metallothionein gene *Mtn*. *Mol. Cell Biol.* **7**: 1710-1715.

Pacyna, J.M. (1986a). Emission factors of atmospheric elements. *Adv. Environ. Sci. Technol.* **17**: 1-32.

Pacyna, J.M. (1986b). Atmospheric trace elements from natural and anthropogenic sources. *Adv. Environ. Sci. Technol.* **17**: 33-52.

Page, R.A., Cawse, P.A. & Baker, S.J. (1988). The effect of reducing petrol lead on airborne lead in Wales, U.K. *Sci. Total Environ.* **68**: 71-77.

Parkinson, D., Visser, S. & Whittaker, J.B. (1979). Effects of collembolan grazing on fungal colonization of leaf litter. *Soil Biol. Biochem.* **11**: 529-535.

Patanè, L. (1934). Sulla struttura e la funzione dell'epatopancreas di *Porcellio laevis*. *Arch. Zool. Ital.* **20**: 303-323.

Pattison, S.E. & Cousins, R.J. (1986). Zinc uptake and metabolism by hepatocytes. *Fed. Proc.* **45**: 2805-2809.

Peacock, A.J. (1986). Effects of anions, acetazolamide and copper on diuresis in the tsetse fly *Glossina morsitans morsitans* Westwood. *J. Insect Physiol.* **32**: 157-160.

Pearse, A.G.E. (1972). *Histochemistry* (3rd edition). Baltimore, Williams & Wilkins.

Pearson, R.G. (1963). Hard and soft acids and bases. *J. Am. Chem. Soc.* **85**: 3533-3539.

Peck, W.B. & Whitcomb, W.H. (1968). Feeding spiders on artificial diet. *Entomol. News* **79**: 233-236.

Peirson, D.H. & Cawse, P.A. (1979). Trace elements in the atmosphere. *Phil. Trans. Roy. Soc. Lond.* **288B**: 41-49.

Perel, T.S. (1977). Differences in Lumbricid organization connected with ecological properties. *Ecol. Bull. (Stockholm)* **25**: 56-63.

Perron, J.M., Huot, L. & Smirnoff, W.A. (1966). Toxicité comparée des éléments Al, As, B, Co, Cu, F, Fe, I, Li, Mg, Mo, Zn pour *Plodia interpunctella* (HBN)(Lépidoptère). *Comp. Biochem. Physiol.* **18**: 869-879.

Petit, J. (1970). Sur la nature et l'accumulation de substances minérales dans les ovocytes de *Polydesmus complanatus* (Myriapode: Diplopode). *C.R. Acad. Sci. Paris* **270**: 2107-2110.

Phillips, D.J.H. (1977). The use of biological indicator organisms to monitor trace metal pollution in marine and estuarine environments — a review. *Environ. Pollut.* **13**: 281-317.

Phillips, D.J.H. (1980). *Quantitative Aquatic Biological Indicators*. London & New York, Applied Science.

Pickett, A.D. & Patterson, N.A. (1963). Effect of arsenates on the reproduction of some diptera. *Science* **140**: 493-494.

Piearce, T.G. (1972). The calcium relations of selected lumbricidae. *J. Anim. Ecol.* **41**: 167-188.

Piearce, T.G. (1978). Gut contents of some lumbricid earthworms. *Pedobiologia*

18: 153-157.

Pietz, R.I., Peterson, J.R., Prater, J.E. & Zenz, D.R. (1984). Metal concentrations in earthworms from sewage sludge-amended soils at a strip mine reclamation site. *J. Environ. Qual.* **13**: 651-654.

Pollard, S.D. (1986). Prey capture in *Dysdera crocata* (Araneae : Dysderidae), a long fanged spider. *New Zeal. J. Zool.* **13**: 149-150.

Ponomarenko, A.N., Trufanov, G.V. & Golubev, S.N. (1974). Microelements in soil invertebrates. *Soviet J. Ecol.* **5**: 279-281.

Popham, J.D. & D'Auria, J.M. (1980). *Arion ater* (Mollusca: Pulmonata) as an indicator of terrestrial environmental pollution. *Water Air Soil Pollut.* **14**: 115-124.

Poulson, D.F. (1950a). Chemical differentiation of larval midgut of *Drosophila*. *Genetics* **35**: 130-131.

Poulson, D.F. (1950b). Physiological genetic studies on copper metabolism in the genus *Drosophila*. *Genetics* **35**: 684-685.

Poulson, D.F. & Bowen, V.T. (1951). The copper metabolism of *Drosophila*. *Science* **114**: 486.

Poulson, D.F. & Bowen, V.T. (1952). Organisation and function of the inorganic constituents of nuclei. *Exp. Cell Res.* **Suppl. 2**: 161-180.

Poulson, D.F., Bowen, V.T., Hilse, R.M. & Rubinson, A.C. (1952). The copper metabolism of *Drosophila*. *Proc. Nat. Acad. Sci.* **38**: 912-921.

Prasad, E.A.V. & Dunn, C.E. (1987). Significance of termite mounds in gold exploration. *Curr. Sci.* **56**: 1219-1222.

Prentø, P. (1979). Metals and phosphate in the chloragosomes of *Lumbricus terrestris* and their possible physiological significance. *Cell Tiss. Res.* **196**: 123-134.

Price, P.W., Rathcke, B.J. & Gentry, D.A. (1974). Lead in terrestrial arthropods: evidence for biological concentration. *Environ. Entomol.* **3**: 370-372.

Proctor, J. (1971). The plant ecology of serpentine. III. The influence of a high magnesium/calcium ratio and high nickel and chromium levels in some British and Swedish serpentine soils. *J. Ecol.* **59**: 827-842.

Proctor, J. & Woodell, S.R.J. (1975). The ecology of serpentine soils. *Adv. Ecol. Res.* **9**: 255-365.

Prosi, F. (1983). Storage of heavy metals in organs of limnic and terrestric invertebrates and their effects on the cellular level. *International Conference. Heavy Metals in the Environment.* Vol. 1, pp. 459-462. Heidelberg, September 1983. Edinburgh, CEP Consultants.

Prosi, F. & Back, H. (1985). Indicator cells for heavy metal uptake and distribution in organs from selected invertebrate animals. In: *International Conference. Heavy Metals in the Environment.* Vol. 2, pp. 242-244. Athens, 1985. Edinburgh, CEC Consultants.

Prosi, F. & Dallinger, R. (1988). Heavy metals in the terrestrial isopod *Porcellio scaber* Latr. I. Histochemical and ultrastructural characterization of metal-containing lysosomes. *Cell Biol. Toxicol.* **4**: 81-96.

Prosi, F., Storch, V. & Janssen, H.H. (1983). Small cells in the midgut glands of terrestrial Isopoda: sites of heavy metal accumulation. *Zoomorphol.* **102**: 53-64.

Pszonicki, L. (1984). Reliability of trace analysis of biological materials in the light of IAEA's interlaboratory comparisons. In: *Trace Element Analytical Chemistry in Medicine and Biology*. Vol. 3 (P. Brätter & P. Schramel, Eds.) pp. 583-590. Berlin, Walter De Gruyter.

Pulliainen, E., Lajunen, L.H.J. & Itamies, J. (1986). Lead and cadmium in earthworms (Oligochaeta, Lumbricidae) in northern Finland. *Ann. Zool. Fenn.* **23**: 303-306.

Quarles, H.D., Hanawalt, R.B. & Odum, W.E. (1974). Lead in small mammals, plants and soil at varying distances from a highway. *J. Appl. Ecol.* **11**: 937-949.

Quimby, P.C., Frick, K.E., Wauchope, R.D. & Kay, S.H. (1979). Effects of cadmium on two biocontrol insects and their host weeds. *Bull. Environ. Contam. Toxicol.* **22**: 371-378.

Raeburn, D. (1987). Calcium homeostasis in smooth muscles cells. *Biologist* **34**: 16-19.

Rainbow, P.S. (1985a). Accumulation of Zn, Cu and Cd by crabs and barnacles. *Estuar. Coast. Shelf Sci.* **21**: 669-686.

Rainbow, P.S. (1985b). The biology of heavy metals in the sea. *Int. J. Environ. Stud.* **25**: 195-211.

Rajulu, G.S. (1969). Presence of haemocyanin in the blood of a centipede *Scutigera longicornis* (Chilopoda: Myriapoda). *Curr. Sci.* **38**: 168-169.

Ramel, C. & Magnusson, J. (1968). Genetic effects of organic mercury compounds. II. Chromosome segregation in *Drosophila melanogaster*. *Hereditas* **61**: 231-254.

Raw, F. (1959). Estimating earthworm populations by using formalin. *Nature* **184**: 1661.

Razin, L.V. & Rozhkov, I.S. (1966). Geochemistry of gold in the crust of weathering and in the biosphere in the gold-ore deposits of the Kuranakh type. *Izdatellstvo Nauka Moskva* (In Russian).

Read, H.J. (1988). The effects of heavy metal pollution on woodland leaf litter faunal communities. PhD Thesis, University of Bristol.

Read, H.J. & Martin, M.H. (in press). A study of millipede communities in woodlands contaminated with heavy metals. *Proceedings of the Seventh International Congress of Myriapodology*, Vittorio-Veneto, Italy, July 1987.

Read, H.J., Wheater, C.P. & Martin, M.H. (1986). The effects of heavy metal pollution on woodland communities of surface active Carabidae (Coleoptera). In: *Proceedings of the Third European Congress of Entomology* (H.H.W. Velthuis, Ed.) pp. 295-298. Amsterdam, August 1986. Amsterdam, Nederlandse Entomologische Vereniging.

Read, H.J., Wheater, C.P. & Martin, M.H. (1987). Aspects of the ecology of Carabidae (Coleoptera) from woodlands polluted by heavy metals. *Environ. Pollut.* **48**: 61-76.

Readshaw, J.L. (1972). Failure of lead arsenate in an ecological approach to the control of mites in orchards. *J. Aust. Inst. Agric. Sci.* **38**: 308-309.

Redborg, K.E., Hinesly, T.D. & Ziegler, E.L. (1983). Rearing *Psychoda alternata* (Diptera: Psychodidae) in the laboratory on digested sewage sludge, with some observations on its biology. *Environ. Entomol.* **12**: 412-415.

Reichle, D.E. (1969). Measurement of elemental assimilation by animals from radioisotope retention patterns. *Ecology* **50**: 1102-1104.

Reichle, D.E. (1977). The role of soil invertebrates in nutrient cycling. *Ecol. Bull. (Stockholm)* **25**: 145-156.

Reichle, D.E. & Crossley, D.A. (1965). Radiocesium dispersion in a cryptozoan food web. *Health Phys.* **11**: 1375-1384.

Reinecke, J.P. (1985). Nutrition: Artificial diets. In: *Comparative Insect Physiology, Biochemistry and Pharmacology*. Vol. 4, *Regulation, Digestion, Nutrition, Excretion* (G.A. Kerkut & L.I. Gilbert, Eds.) pp. 391-419. Oxford, Pergamon.

Revel, J.P., Barnard, T. & Haggis, G.H. (1984). Eds. *Science of Biological Specimen Preparation for Microscopy and Microanalysis*. Chicago, Scanning Electron Microscopy Corporation.

Reyes, V.G. & Tiedje, J.M. (1976a). Ecology of the gut microbiota of *Tracheoniscus rathkei* (Crustacea: Isopoda). *Pedobiologia* **16**: 67-74.

Reyes, V.G. & Tiedje, J.M. (1976b). Metabolism of C^{14}-labelled plant materials by woodlice (*Tracheoniscus rathkei* Brandt) and soil microorganisms. *Soil Biol. Biochem.* **8**: 103-108.

Rhodes, S.T. & Mason, W.H. (1971). ^{65}Zn excretion as an index to metabolic rate in the wood-feeding cockroach *Cryptocercus punctulatus*. *Ann. Entomol. Soc. Am.* **64**: 450-452.

Richards, B.N. (1987). *The Microbiology of Terrestrial Ecosystems*. Harlow, Longmans.

Richards, K.S. & Ireland, M.P. (1978). Glycogen-lead relationship in the earthworm *Dendrobaena rubida* from a heavy metal site. *Histochemistry* **56**: 55-64.

Richards, K.S., Rush, A.D., Clarke, D.T. & Myring, W.J. (1986). Soft X-ray contact microscopy, using synchroton radiation, of thin-sectioned, lead-contaminated chloragogenous tissue of the earthworm *Dendrobaena rubida*. *J. Microsc.* **142**: 1-7.

Richardson, B. & Whittaker, J.B. (1982). The effect of varying the reference material on ranking of acceptability indices of plant species to a polyphagous herbivore *Agriolimax reticulatus*. *Oikos* **39**: 237-240.

Ridlington, J.W., Chapman, D.C., Goeger, D.E. & Whanger, P.D. (1981). Metallothionein and Cu-chelatin: characterization of metal-binding proteins from tissues of four marine animals. *Comp. Biochem. Physiol.* **70B**: 93-104.

Robel, R.J., Howard, C.A., Udevitz, M.S. & Curnutte, B. (1981). Lead contamination in vegetation, cattle dung and dung beetles near an interstate highway, Kansas. *Environ. Entomol.* **10**: 262-263.

Roberts, R.D. & Johnson, M.S. (1978). Dispersal of heavy metals from abandoned mine workings and their transference through terrestrial food chains. *Environ. Pollut.* **16A**: 293-310.

Roberts, T.M. (1975). A review of some biological effects of lead emissions from primary and secondary smelters. In: *International Conference. Heavy Metals in the Environment*. Vol. 1, pp. 503-532. Toronto, 1975. Edinburgh, CEP Consultants.

Roberts, T.M. & Goodman, G.T. (1973). The persistence of heavy metals in soils and natural vegetation following closure of a smelter. In: *Trace Substances in Environmental Health*. Vol. 7 (D.D. Hemphill, Ed.) pp. 105-116. Missouri, University of Missouri.

Robson, R.L., Eady, R.R., Richardson, T.H., Miller, R.W., Hawkins, M. & Postgate, J.R. (1986). The alternative nitrogenase of *Azotobacter chroococcum* is a vanadium enzyme. *Nature* **322**: 388-390.

Rolfe, G.L. & Haney, A. (1975). An Ecosystem Analysis of Environmental Contamination by Lead. Institute for Environmental Studies, University of Illinois at Urbana-Champaign.

Ross, D.S., Sjogren, R.E. & Bartlett, R.J. (1981). Behaviour of chromium in soils. IV. Toxicity to microorganisms. *J. Environ. Qual.* **10**: 145-148.

Ross, I.S. (1975). Some effects of heavy metals on fungal cells. *Trans. Br. Mycol. Soc.* **64**: 175-193.

Rossaro, B., Gaggino, G.F. & Marchetti, R. (1986). Accumulation of mercury in larvae and adults *Chironomus riparius* (Meigen). *Bull. Environ. Contam. Toxicol.* **37**: 402-406.

Rühling, Å. & Tyler, G. (1968). An ecological approach to the lead problem. *Bot. Not.* **121**: 321-342.

Rühling, Å. & Tyler, G. (1969). Ecology of heavy metals — a regional and historical study. *Bot. Not.* **122**: 248-259.

Rühling, Å. & Tyler, G. (1973). Heavy metal pollution and decomposition of spruce needle litter. *Oikos* **24**: 402-417.

Russ, J.C. (1978). Electron probe X-ray microanalysis: principles. In: *Electron Probe Microanalysis in Biology* (D.A. Erasmus, Ed.) pp. 5-36. London, Chapman & Hall.

Russell, L.K., Dehaven, J.I. & Botts, R.P. (1981). Toxic effects of cadmium on the garden snail *Helix aspersa*. *Bull. Environ. Contam. Toxicol.* **26**: 634-640.

Ryder, T.A. & Bowen, I.D. (1977a). The slug foot as a site of uptake of copper molluscicide. *J. Invert. Pathol.* **30**: 381-386.

Ryder, T.A. & Bowen, I.D. (1977b). The use of X-ray microanalysis to demonstrate the uptake of the molluscicide copper sulphate by slug eggs. *Histochemistry* **52**: 55-60.

Salama, H.S. (1972). Zinc sulphate induces sterility in the cotton leafworm *Spodoptera littoralis* Boisduval. *Experentia* **28**: 1318.

Salama, H.S. & El-Sharaby, A.F. (1973). Effect of zinc sulphate on the feeding and growth of *Spodoptera littoralis* Boisd. *Z. Ang. Entomol.* **72**: 383-389.

Salomons, W. (1986). Impact of atmospheric inputs on the hydrospheric trace metal cycle. *Adv. Environ. Sci. Technol.* **17**: 409-466.

Salomons, W. & Forstner, U. (1984). *Metals in the Hydrocycle*. Berlin, Springer-Verlag.

Sandford, K.H. (1964). Life history and control of *Atractotomus mali*, a new pest of apple in Nova Scotia (Miridae: Hemiptera). *J. Econ. Entomol.* **57**: 921-925.

Sastry, K.S., Murthy, R.R. & Sarma, P.S. (1958). Studies on the zinc toxicity in the larvae of the rice moth *Corcyra cephalonica*. *Biochem. J.* **69**: 425-428.

Satchell, J.E. & Lowe, D.G. (1967). Selection of leaf litter by *Lumbricus terrestris*. In: *Progress in Soil Biology* (O. Graff & J.E. Satchell, Eds.) pp. 102-119. Amsterdam, North Holland.

Scanlon, P.F. (1979). Ecological implications of heavy metal contamination of roadside habitats. *Proc. Ann. Conf. Southeast Ass. Fish. Wildlife Agencies* **33**: 136-145.

Schmidt, D.J. & Reese, J.C. (1986). Sources of error in nutritional index studies of insects on artificial diet. *J. Insect Physiol.* **32**: 193-198.

Schoetti, G. & Seiler, H.G. (1970). Uptake and localisation of radioactive zinc in the visceral complex of the land pulmonate *Arion rufus*. *Experentia* **26**: 1212-1213.

Schowalter, T.D. & Crossley, D.A. (1982). Bioelimination of ^{51}Cr and [85]Sr by cockroaches *Gromphadorhina portentosa* (Orthoptera: Blaberidae) as affected by mites *Gromphadorholaelaps schaeferi* (Parasitiformes, Laelapidae). *Ann. Entomol. Soc. Am.* **75**: 158-160.

Schreier, H. & Timmenga, H.J. (1986). Earthworm response to asbestos-rich serpentinitic sediments. *Soil Biol. Biochem.* **18**: 85-89.

Schütt, S. & Nuorteva, P. (1983). Metlykvicksilvrets inverkan på aktivitet hos *Tenebrio molitor* (L.)(Col. Tenebrionidae). *Acta Entomol. Fenn.* **42**: 78-81.

Schwarz, K. (1974). New essential trace elements (Sn, V, F, Si): progress report and outlook. In: *Trace Element Metabolism in Animals*. Vol. 2 (W.G. Hoekstra, J.W. Suttie, H.E. Ganther & W. Mertz, Eds.) pp. 355-380. Baltimore, University Park Press.

Schwarz, K. & Spallholz, J.E. (1977). Growth effects of small cadmium supplements in rats maintained under trace element-controlled conditions. In: *Proceedings of the First International Cadmium Conference*. San Francisco, 1977. pp. 105-109. Dorset, Drogher Press.

Scokart, P.O., Meeus-Verdinne, K. & De Borger, R. (1983). Mobility of heavy metals in polluted soils near zinc smelters. *Water Air Soil Pollut.* **20**: 451-463.

Scott, D.B., Nylen, M.U., Von Brand, T. & Pugh, M.H. (1962). The mineralogical composition of the calcareous corpuscles of *Taenia taeniaeformis*. *Exp. Parasit.* **12**: 445-458.

Seastedt, T.R. & Crossley, D.A. (1980). Effects of microarthropods on the seasonal dynamics of nutrients in forest litter. *Soil Biol. Biochem.* **12**: 337-342.

Seastedt, T.R. & Tate, C.M. (1981). Decomposition rates and nutrient contents of arthropod remains in forest litter. *Ecology* **62**: 13-19.

Seitz, K.A. (1986). Excretory organs. In: *Ecophysiology of Spiders* (W. Nentwig, Ed.) pp. 239-248. Berlin, Springer-Verlag.

Sell, D.K. & Bodznick, D.A. (1971). Effects of dietary $ZnSO_4$ on the growth and feeding of the tobacco bud worm *Heliothis virescens*. *Ann. Entomol. Soc. Am.* **64**: 850-855.

Sell, D.K. & Schmidt, C.H. (1968). Chelating agents suppress pupation of the cabbage looper. *J. Econ. Entomol.* **61**: 946-949.

Sevilla, C. & Lagarrigue, J.G. (1979). Oxygen binding characteristics of Oniscoidea hemocyanins (Crustacea; terrestrial isopods). *Comp. Biochem. Physiol.* **64A**: 531-536.

Shamberger, R.J. (1981). Selenium in the environment. *Sci. Total Environ.* **17**: 59-74.

Shamberger, R.J. (1983). *Biochemistry of Selenium*. New York, Plenum.

Sheehan, P.J. (1984a). Effects on individuals and populations. In: *Effects of Pollutants at the Ecosystem Level* (P.J. Sheehan, D.R. Miller, G.C. Butler & P. Bourdeau, Eds.) pp. 23-50. Chichester, John Wiley & Sons.

Sheehan, P.J. (1984b). Effects on community and ecosystem structure and dynamics. In: *Effects of Pollutants at the Ecosystem Level* (P.J. Sheehan, D.R. Miller, G.C. Butler & P. Bourdeau, Eds.) pp. 51-99. Chichester, John Wiley & Sons.

Sheehan, P.J. (1984c). Functional changes in the ecosystem. In: *Effects of Pollutants at the Ecosystem Level* (P.J. Sheehan, D.R. Miller, G.C. Butler & P. Bourdeau, Eds.) pp. 101-145. Chichester, John Wiley & Sons.

Shendrikar, A.D. & Ensor, D.S. (1986). Sampling and measurement of trace element emissions from particulate control devices. *Adv. Environ. Sci. Technol.* **17**: 53-111.

Sherlock, P.L., Bowden, J. & Digby, P.G.N. (1985). Studies of elemental composition as a biological marker in insects. IV. The influence of soil type and host-plant on elemental composition of *Agrotis segetum* (Denis & Schiffermüller)(Lepidoptera, Noctuidae). *Bull. Entomol. Res.* **75**: 675-687.

Sibly, R.M. & Calow, P. (1986). *Physiological Ecology of Animals: An Evolutionary Approach*. Oxford, Blackwell.

Siccama, T.G. & Smith, W.H. (1978). Lead accumulation in a northern hardwood forest. *Environ. Sci. Technol.* **12**: 593-594.

Siegel, S.M., Siegel, B.Z., Puerner, N., Speitel, T. & Thorarinsson, F. (1975). Water and soil biotic relations in mercury distribution. *Water Air Soil Pollut.* **4**: 9-18.

Simkiss, K. (1976a). Cellular aspects of calcification. In: *The Mechanisms of Mineralization in the Invertebrates and Plants* (N. Watanabe & K. Wilbur, Eds.) pp. 1-31. University of South Carolina Press.

Simkiss, K. (1976b). Intracellular and extracellular routes in bio-mineralization. *Symp. Soc. Exp. Biol.* **30**: 423-444.

Simkiss, K. (1977). Biomineralisation and detoxification. *Calcif. Tiss. Res.* **24**: 199-200.

Simkiss, K. (1981). Cellular discrimination processes in metal accumulating cells. *J. Exp. Biol.* **94**: 317-327.

Simkiss, K. (1983a). Lipid solubility of heavy metals in saline solutions. *J. Mar. Biol. Ass. U.K.* **63**: 1-7.

Simkiss, K. (1983b). Trace elements as probes of biomineralization. In: *Biomineralization and Biological Metal Accumulation* (P. Westbroek & E.W. De Jong, Eds.) pp. 363-371. New York, D. Reidel.

Simkiss, K. (1985). Prokaryote-eukaryote interactions in trace element metabolism. *Desulfovibrio sp.* in *Helix aspersa*. *Experentia* **41**: 1195-1197.

Simkiss, K. & Mason, A.Z. (1983). Metal ions: metabolic and toxic effects. In: *The Mollusca*. Vol. 2, *Environmental Biochemistry & Physiology*, pp. 101-164. London & New York, Academic Press.

Simkiss, K. & Mason, A.Z. (1984). Cellular responses of molluscan tissues to environmental metals. *Mar. Environ. Res.* **14**: 103-118.

Simkiss, K. & Taylor, M. (1981). Cellular mechanisms of metal ion detoxification and some new indices of pollution. *Aquatic Toxicol.* **1**: 279-290.

Simkiss, K. & Taylor, M.G. (in press). Metal fluxes across the membranes of aquatic organisms. CRC Critical Reviews in Aquatic Toxicology. Florida, CRC Press.

Simkiss, K., Jenkins, K.G.A., McLellan, J. & Wheeler, E. (1982). Methods of

metal incorporation into intracellular granules. *Experentia* **38**: 333-335.
Sims, R.W. & Gerard, B.M. (1985). *Earthworms*. Linnean Society Synopses of the British Fauna (New Series) No. 31. London & Leiden, E.J. Brill/Dr W. Backhuys.
Singh, B.R. & Narwal, R.P. (1984). Plant availability of heavy metals in a sludge-treated soil. II. Metal extractability compared with plant metal uptake. *J. Environ. Qual.* **13**: 344-349.
Singh, P. (1977). *Artificial Diets for Insects, Mites and Spiders*. New York, Plenum Data Co.
Sivapalan, P. & Gnanapragasam, N.C. (1980). Influence of copper on the development and adult emergence of *Homona coffearia* (Lepidoptera: Tortricidae) reared *in vitro*. *Entomol. Exp. Appl.* **28**: 59-63.
Škrobák, J. & Weismann, L. (1975). Toxic effects of copper on *Scotia segetum* (Den. and Schiff., Lepidoptera). 2. Effects of inorganic copper compounds on mortality of larvae and on weight of pupae. *Biológia (Bratislavia)* **30**: 109-116 (In Czechoslovakian).
Škrobák, J. & Weismann, L. (1979a). Cu, Zn, Mn and Mg contents in the body of *Scotia segetum* kept on a semisynthetic diet in its single postembryonic developmental stages. *Biológia (Bratislavia)* **34**: 107-114 (In Czechoslovakian).
Škrobák, J. & Weismann, L. (1979b). Bioaccumulation of copper and retention of manganese and magnesium in *Scotia segetum* kept on a semisynthetic diet (for larvae) containing an elevated amount of copper. *Biológia (Bratislavia)* **34**: 353-364.
Škrobák, J., Weismann, L. & Škrobáková, E. (1975). Toxic effects of copper on *Scotia segetum* (Den. and Schiff., Lepidoptera). 3. The influence of inorganic copper compounds on adult fertility. *Biológia (Bratislavia)* **30**: 621-631 (In Czechoslovakian).
Škrobák, J., Škrobáková, E. & Weismann, L. (1976). Toxic effects of copper on *Scotia segetum* (Den. and Schiff., Lepidoptera). 4. Oxygen consumption in pupae intoxicated with copper chloride. *Biológia (Bratislavia)* **31**: 615-624 (In Czechoslovakian).
Sminia, T. & Vlugt van Daalen, J.E. (1977). Haemocyanin synthesis in pore cells of the terrestrial snail *Helix aspersa*. *Cell Tiss. Res.* **183**: 299-301.
Smith, D.S., Compher, K., Janners, M., Lipton, C. & Wittle, L.W. (1969). Cellular organization and ferritin uptake in the midgut epithelium of a moth *Ephestia kühniella*. *J. Morphol.* **127**: 41-72.
Smith, K.G.V. (1986). *A Manual of Forensic Entomology*. London, British Museum (Natural History).
Smith, W.H. (1976). Lead contamination of the roadside ecosystem. *J. Air Pollut. Control Ass.* **26**: 758-766.
Sohal, R.S. (1974). Fine structure of the malpighian tubules in the housefly *Musca domestica*. *Tiss. Cell* **6**: 719-728.
Sohal, R.S. & Lamb, R.E. (1977). Intracellular deposition of metals in the midgut of the adult housefly *Musca domestica*. *J. Insect Physiol.* **23**: 1349-1354.
Sohal, R.S. & Lamb, R.E. (1979). Storage-excretion of metallic cations in the adult housefly *Musca domestica*. *J. Insect Physiol.* **25**: 119-124.
Sohal, R.S., Peters, P.D. & Hall, T.A. (1976). Fine structure and X-ray

microanalysis of mineralized concretions in the Malpighian tubules of the housefly *Musca domestica*. *Tiss. Cell* **8**: 447-458.

Sohal, R.S., Peters, P.D. & Hall, T.A. (1977). Origin, structure, composition and age-dependence of mineralized dense bodies (concretions) in the midgut epithelium of the adult housefly *Musca domestica*. *Tiss. Cell* **9**: 87-102.

Somlyo, A.P., Bond, M. & Somlyo, A.V. (1985). Calcium content of mitochondria and endoplasmic reticulum in liver frozen rapidly *in vivo*. *Nature* **314**: 622-625.

Soni, R. & Abbasi, S.A. (1981). Mortality and reproduction in earthworm *Pheretima posthuma* exposed to chromium VI. *Int. J. Environ. Stud.* **17**: 147-149.

Soon, Y.K. (1981). Solubility and sorption of cadmium in soils amended with sewage sludge. *J. Soil Sci.* **32**: 85-95.

Sorsa, M. & Pfeifer, S. (1973a). Response of puffing pattern to *in vivo* treatments with organomercurials in *Drosophila melanogaster*. *Hereditas* **74**: 89-102.

Sorsa, M. & Pfeifer, S. (1973b). Effects of cadmium on developmental time and prepupal puffing pattern of *Drosophila melanogaster*. *Hereditas* **75**: 273-277.

Sposito, G. & Page, A.L. (1984). Cycling of metal ions in the environment. In: *Metal Ions in Biological Systems*. Vol. 18, *Circulation of Metals in the Environment* (H. Sigel, Ed.) pp. 287-332. New York, Marcel Dekker.

Sridhara, S. & Bhat, J.V. (1966a). Trace element nutrition of the silkworm *Bombyx mori* L. I. Effect of trace elements. *Proc. Indian Acad. Sci.* **63B**: 9-16.

Sridhara, S. & Bhat, J.V. (1966b). Trace element nutrition of the silkworm *Bombyx mori* L. II. Relationship between cobalt, vitamin B_{12}, benzinidazole and purines. *Proc. Indian Acad. Sci.* **63B**: 17-25.

Srivastava, P.N. & Auclair, J.L. (1971a). An improved chemically defined diet for the pea aphid, *Acyrthosiphon pisum*. *Ann. Entomol. Soc. Am.* **64**: 474-478.

Srivastava, P.N. & Auclair, J.L. (1971b). Influence of sucrose concentration on the diet, uptake and performance by the pea aphid, *Acyrthosiphon pisum*. *Ann. Entomol. Soc. Am.* **64**: 739-743.

Stafford, E.A. & McGrath, S.P. (1986). The use of acid insoluble residue to correct for the presence of soil-derived metals in the gut of earthworms used as bio-indicator organisms. *Environ. Pollut.* **42A**: 233-246.

Standen, V. (1978). The influence of soil fauna on decomposition by microorganisms in blanket bog litter. *J. Anim. Ecol.* **47**: 25-38.

Stebbing, A.R.D. (1982). Hormesis — the stimulation of growth by low levels of inhibitors. *Sci. Total Environ.* **22**: 213-234.

Steel, C.G.H. (1982). Stages of the intermoult cycle in the terrestrial isopod *Oniscus asellus* and their relation to biphasic cuticle secretion. *Can. J. Zool.* **60**: 429-437.

Steeves, H.R. (1968). Presence of compounds containing iron in the digestive system. *Nature* **218**: 393-394.

Steudel, A. (1913). Absorption und secretion im darm von insecten. *Zool. Jahrb. Zool. Physiol. Tiere* **33**: 165-224.

Stevenson, C.D. (1985). Analytical advances and changing perceptions of environmental heavy metals. *J. Roy. Soc. New Zeal.* **15**: 355-362.

Storch, V. (1982). Der einfluss der Ernährung auf die Ultrastruktur der grossen Zellen in den Mitteldarmdrüsen terrestrischer Isopoda *(Armadillidium vulgare, Porcellio scaber)*. *Zoomorphol.* **100**: 131-142.

Storch, V. (1984). The influence of nutritional stress on the ultrastructure of the hepatopancreas of terrestrial isopods. *Symp. Zool. Soc. Lond.* **53**: 167-184.

Stott, D.E., Dick, W.A. & Tabatabai, M.A. (1985). Inhibition of pyrophosphatase activity in soils by trace elements. *Soil Sci.* **139**: 112-117.

Streit, B. (1985). Effects of high copper concentrations on soil invertebrates (earthworms and orabatid mites). Experimental results and a model. *Oecologia* **64**: 381-388.

Strojan, C.L. (1978a). Forest leaf litter decomposition in the vicinity of a zinc smelter. *Oecologia* **32**: 203-212.

Strojan, C.L. (1978b). The impact of zinc smelter emissions on forest litter arthropods. *Oikos* **31**: 41-46.

Strufe, R. (1968). Problems and results of residue studies after application of molluscicides. *Residue Rev.* **24**: 79-168.

Stubbs, R.L. (1977). Cadmium — the metal of benign neglect. In: *Proceedings of the First International Cadmium Conference*. San Francisco, 1977. pp. 7-12. Dorset, Drogher Press.

Sumi, Y. Suzuki, T., Yamamura, M., Hatakeyama, S., Sugaya, Y & Suzuki, K.T. (1984). Histochemical staining of cadmium taken up by the midge larva *Chironomus yoshimatsui* (Diptera, Chironomidae). *Comp. Biochem. Physiol.* **79A**: 353-357.

Sumner, A.T. (1983). X-ray microanalysis — a histochemical tool for elemental analysis. *Histochem. J.* **15**: 501-542.

Sunderland, K.D. & Sutton, S.L. (1980). A serological study of arthropod predation on woodlice in a dune grassland ecosystem. *J. Anim. Ecol.* **49**: 987-1004.

Suzuki, K.T., Yamamura, M. & Mori, T. (1980). Cadmium-binding proteins induced in the earthworm. *Arch. Environ. Contam. Toxicol.* **9**: 415-424.

Suzuki, K.T., Aoki, Y., Nishikawa, M., Masui, H. & Matsubara, F. (1984). Effect of cadmium-feeding on tissue concentrations of elements in a germ-free silkworm *(Bombyx mori)* larvae and distribution of cadmium in the alimentary canal. *Comp. Biochem. Physiol.* **79C**: 249-253.

Swaine, D.J. & Mitchell, R.L. (1960). Trace element distribution in soil profiles. *J. Soil Sci.* **11**: 347-368.

Swift, M.J., Heal, O.W. & Anderson, J.M. (1979). *Decomposition in Terrestrial Ecosystems*. Oxford, Blackwell.

Szyfter, Z. (1966). The correlation of moulting and changes occurring in the hepatopancreas of *Porcellio scaber* Latr. (Crustacea; Isopoda). *Bull. Soc. Amis. Sci. Lett. Poznan* **7D**: 95-114.

Tanaka, K., Kobayashi, M. & Ichikawa, Y. (1982). Intracellular distribution of zinc in *Bombyx mori* larvae (Lepidoptera: Bombycidae) treated with $ZnCl_2$. *Jap. J. Appl. Entomol. Zool.* **26**: 103-111.

Tapp, R.L. (1975). X-ray microanalysis of the midgut epithelium of the fruit fly *Drosophila melanogaster*. *J. Cell Sci.* **17**: 449-459.

Tapp, R.L. & Hockaday, A. (1977). Combined histochemical and X-ray microanalytical studies on the copper accumulating granules in the midgut of larval *Drosophila. J. Cell Sci.* **26**: 201-215.

Tatsuyama, K., Egawa, H., Yamamoto, H. & Senmaru, H. (1975). Cadmium resistant micro-organisms in the soil polluted by metals. *Trans. Mycol. Soc. Japan* **16**: 69-78.

Taylor, C.W. (1984). Calcium distribution during egg development in *Calliphora vicina. J. Insect Physiol.* **30**: 905-910.

Taylor, M.G. & Simkiss, K. (1984). Inorganic deposits in invertebrate tissues. *Environ. Chem.* **3**: 102-138.

Taylor, M., Greaves, G.N. & Simkiss, K. (1983). Structure of granules in *Helix aspersa* by EXAFS and other physical techniques. In: *Biomineralization and Biological Metal Accumulation* (P. Westbroek & E.W. De Jong, Eds.) pp. 373-377. New York, D. Reidel.

Taylor, M., Simkiss, K. & Greaves, G.N. (1986). Amorphous structure of intracellular mineral granules. *Biochem. Soc. Trans.* **14**: 549-552.

Taylor, M.G., Simkiss, K., Greaves, G.N. & Harries, J. (1988). Corrosion of intracellular granules and cell death. *Proc. Roy. Soc.* **234B**: 463-476.

Terra, W.R. & Ferreira, C. (1981). The physiological role of the peritrophic membrane and trehalase: digestive enzymes in the midgut and excreta of starved larvae of *Rhynchosciara. J. Insect Physiol.* **27**: 325-331.

Terriere, L.C. & Rajadhyakasha, N. (1964). Reduced fecundity of the two-spotted spider mite on metal chelate-treated leaves. *J. Econ. Entomol.* **57**: 95-99.

Terwilliger, N.B. (1982). Effect of subunit composition on quaternary structure of isopod *(Ligia pallasii)* hemocyanin. *Biochemistry* **21**: 2579-2586.

Terwilliger, R.C. & Terwilliger, N.B. (1985). Molluscan hemoglobins. *Comp. Biochem. Physiol.* **81B**: 255-261.

Thomas, P.G. & Ritz, D.A. (1986). Growth of zinc granules in the barnacle *Elminius modestus. Mar. Biol.* **90**: 255-260.

Thompson, K.C. & Reynolds, R.J. (1978). *Atomic Absorption, Fluorescence and Flame Emission Spectroscopy — a Practical Approach.* London, Charles Griffin.

Timm, F. (1958). Zur Histochemie der Schwermetalle, das Sulfid-Silber-Verfahren. *Dtsch. Z. Ges. Gerichtl. Med.* **46**: 706-711.

Toshkov, A.S., Shabanov, M.M. & Ibrishimov, N.I. (1974). Attempts to use bees to prove impurities in the environment. *Dokl. Bolg. Akad. Nauk.* **27**: 699-702.

Tsalev, D.L. & Zaprianov, Z.K. (1983). *Atomic Absorption Spectrometry in Occupational and Environmental Health Practice.* Vol. 1, *Analytical Aspects of Health Significance.* Vol. 2, *Determination of Individual Elements.* Florida, CRC Press.

Turbeck, B.O. (1974). A study of the concentrically laminated concretions, spherites, in the regenerative cells of the midgut of lepidopterous larvae. *Tiss. Cell* **6**: 627-640.

Turner, D.R. (1984). Relationships between biological availability and chemical measurements. In: *Metal Ions in Biological Systems.* Vol. 18, *Circulation of Metals in the Environment* (H. Sigel, Ed.) pp. 137-164. New York, Marcel Dekker.

Turnock, W.J., Gerber, G.H. & Sabourin, D.U. (1980). An evaluation of the use of elytra and bodies in X-ray energy-dispersive spectroscopic studies of the red turnip beetle, *Entomoscelis americana* (Coleoptera: Chrysomelidae). *Can. Entomol.* **112**: 609-614.

Tyler, G. (1972). Heavy metals pollute nature; may reduce productivity. *Ambio* **1**: 52-59.

Tyler, G. (1974). Heavy metal pollution and soil enzymatic activity. *Pl. Soil* **41**: 303-311.

Tyler, G. (1975). Heavy metal pollution and mineralisation of nitrogen in forest soils. *Nature* **255**: 701-702.

Tyler, G. (1976). Influence of vanadium on soil phosphatase activity. *J. Environ. Qual.* **5**: 216-217.

Tyler, G. (1978). Leaching rates of heavy metal ions in forest soil. *Water Air Soil Pollut.* **9**: 137-148.

Tyler, G. (1984). The impact of heavy metal pollution on forests: a case study of Gusum, Sweden. *Ambio* **13**: 18-24.

Tyler, G., Mörnsjö, B. & Nilsson, B. (1974). Effects of Cd, Pb and sodium salts on nitrification in a mull soil. *Pl. Soil* **40**: 237-242.

Tyler, G., Bengtsson, G., Folkeson, L., Gunnarsson, T., Rundgren, S., Rühling, Å. & Söderström, B. (1984). *Metallfororening I Skogsmark — Biologiska Effekter*. Lund, Statens Naturvardsverk PM 1910.

Tyler, L.D. & McBride, M.B. (1982). Mobility and extractability of cadmium, copper, nickel and zinc in organic and mineral soil columns. *Soil Sci.* **134**: 198-205.

Udevitz, M.S., Howard, C.A., Robel, R.J. & Curnutte, B. (1980). Lead contamination in insects and birds near an interstate highway, Kansas. *Environ. Entomol.* **9**: 35-36.

Underwood, E.J. (1977). *Trace Elements in Human and Animal Nutrition* (4th edition). New York, Academic Press.

Underwood, E.J. (1979). Trace elements and health: an overview. *Phil. Trans. Roy. Soc. Lond.* **288**: 5-14.

Usher, M.B. & Ocloo, J.K. (1975). Testing the termite resistance of small, treated with water-borne preservatives wood blocks. *Holzforschung* **29**: 147-151.

Van Capelleveen, H.E. (1983). Effects of iron and manganese on isopods. In: *International Conference. Heavy Metals in the Environment*. Vol. 1, pp. 666-669. Heidelberg, September 1983. Edinburgh, CEP Consultants.

Van Capelleveen, H.E. (1985). The ecotoxicity of zinc and cadmium for terrestrial isopods. In: *International Conference. Heavy Metals in the Environment*. Vol. 2, pp. 245-247. Athens, 1985. Edinburgh, CEP Consultants.

Van Capelleveen, H.E. (1987). Ecotoxicity of heavy metals for terrestrial isopods. PhD Thesis, Amsterdam, Free University Press.

Van Capelleveen, H.E., Van Straalen, N.M., Van den Berg, M. & Van Wachem, E. (1986). Avoidance as a mechanism of tolerance for lead in terrestrial arthropods. In: *Proceedings of the Third European Congress of Entomology* (H.H.W. Velthuis, Ed.) pp. 251-254. Amsterdam, August 1986. Amsterdam, Nederlandse Entomologische Vereniging.

Van Hook, R.I. (1974). Cadmium, lead and zinc distributions between earthworms and soils: potential for biological accumulation. *Bull. Environ. Contam. Toxicol.* **12**: 509-512.

Van Hook, R.I. & Yates, A.J. (1975). Transient behaviour of cadmium in a grassland arthropod food chain. *Environ. Res.* **9**: 76-83.

Van Hook, R.I., Blaylock, B.G., Bondietti, E.A., Francis, C.W., Huckabee, J.W., Reichle, D.E., Sweeton, F.H. & Witherspoon, J.P. (1976). Radioisotope techniques in the delineation of the environmental behaviour of cadmium. *Environ. Qual. Safety* **5**: 166-182.

Van Hook, R.I., Harris, W.F. & Henderson, G.S. (1977). Cadmium, lead and zinc distributions and cycling in a mixed deciduous woodland. *Ambio* **6**: 281-286.

Van Raaphorst, J.G., Van Weers, A.W. & Haremaker, H.M. (1974). Loss of zinc and cobalt during dry ashing of biological material. *Analyst* **99**: 523-527.

Van Rhee, J.A. (1967). Development of earthworm populations in orchard soils. In: *Progress in Soil Biology* (J.E. Satchell & O. Graff, Eds.) pp. 360-369. Amsterdam, North Holland.

Van Rhee, J.A. (1969). Effects of biocides and their residues on earthworms. *Mededelingen Rijksfaculteit Landbouwwetenschappen (Gent)* **34**: 682-689.

Van Rhee, J.A. (1975). Copper contamination effects on earthworms by disposal of pig waste in pastures. In: *Proceedings of the 5th International Colloquium on Soil Zoology* (J. Vanek, Ed.) pp. 451-456. Prague, September 1973. The Hague, Dr W. Junk/B.V. Publishers.

Van Rhee, J.A. (1977). Effect of soil pollution on earthworms. *Pedobiologia* **17**: 201-208.

Van Rinsvelt, H.A., Duerkes, R., Levy, R. & Cromroy, H.L. (1973). Major and trace element detection in insects by ion induced X-ray fluorescence. *Florida Entomol.* **56**: 286-290.

Van Straalen, N.M. & De Goede, R.G.M. (1987). Productivity as a population performance index in life-cycle toxicity tests. *Water Sci. Technol.* **19**: 13-20.

Van Straalen, N.M. & Van Meerendonk, J.H. (1987). Biological half-lives of lead in *Orchesella cincta* (L.) Collembola. *Bull. Environ. Contam. Toxicol.* **38**: 213-219.

Van Straalen, N.M. & Van Wensem, J. (1986). Heavy metal content of forest litter arthropods as related to body-size and trophic level. *Environ. Pollut.* **42A**: 209-221.

Van Straalen, N.M., Burghouts, T.B.A. & Doornhof, M.J. (1985). Dynamics of heavy metals in populations of Collembola in a contaminated pine forest soil. In: *International Conference. Heavy Metals in the Environment*. Vol. 1, pp. 613-615. Athens, 1985. Edinburgh, CEP Consultants.

Van Straalen, N.M., Groot, G.M. & Zoomer, H.R. (1986a). Adaptation of Collembola to heavy metal soil contamination. In: *Proceedings of the International Conference Environmental Contamination*, Amsterdam, 1986. pp. 16-20. Edinburgh, CEP Consultants.

Van Straalen, N.M., Van Zalinge, J. & Doucet, P.G. (1986b). Life history theory on accumulation and turnover of pollutants. In: *Proceedings of the Third European Congress of Entomology* (H.H.W. Velthuis, Ed.) pp. 299-302. Amsterdam, August 1986. Amsterdam, Nederlandse Entomologische

Vereniging.

Van Straalen, N.M., Burghouts, T.B.A., Doornhof, M.J., Groot, G.M., Janssen, M.P.M., Joosse, E.N.G., Van Meerendonk, J.H., Theeuwen, J.P.J.J., Verhoef, H.A. & Zoomer, H.R. (1987). Efficiency of lead and cadmium excretion in populations of *Orchesella cincta* (Collembola) from various contaminated forest soils. *J. Appl. Ecol.* **24**: 953-968.

Van de Westeringh, W. (1972). Deterioration of soil structure in worm free orchard soils. *Pedobiologia* **12**: 6-15.

Varma, A. (1984). *Handbook of Atomic Absorption Analysis.* Vols. 1 & 2. Florida, CRC Press.

Vašák, M. (1986). The spatial structure of metallothionein — a feat of spectroscopy. In: *Zinc Enzymes* (I. Bertini, C. Luchinat, W. Maret & M. Zeppezauer, Eds.) pp. 595-606. Boston, Birkhauser.

Vasudev, V. & Krishnamurthy, N.B. (1979). Dominant lethals induced by cadmium chloride in *Drosophila melanogaster. Curr. Sci.* **48**: 1007-1008.

Vergnano-Gambi, O., Gabbrielli, R. & Pancaro, L. (1982). Nickel, chromium and cobalt in plants from Italian serpentine areas. *Acta Oecologia Oecologia Plant.* **3**: 291-306.

Vernon, G.M., Herold, L. & Witkus, E.R. (1974). Fine structure of the digestive tract epithelium in the terrestrial isopod *Armadillidium vulgare. J. Morphol.* **144**: 337-360.

Viggiani, G., Castronuovo, N. & Borrelli, C. (1972). Effeti secondari di 40 fitofarmaci su *Leptomastidea abnormis* Grit (Hym. Encyrtidae) e *Scymus includens* Kirsch (Col. Coccinellidae), importanti nemeci naturali del *Planococcus citri* (Risso). *Bull. Lab. Entomol. Agrar. Portico* **30**: 88-103.

Volbeda, A. & Hol, W.G.J. (1986). The structure of the copper-containing oxygen-transporting hemocyanins from arthropods. In: *Frontiers in Bioinorganic Chemistry* (A.V. Xavier, Ed.) pp. 584-593. Hamburg, VCH.

Von Brand, T., Mercado, T.I., Nylen, M.U. & Scott, D.B. (1960). Observations on function, composition and structure of calcareous corpuscles. *Exp. Parasit.* **9**: 205-214.

Von Brand, T., Weinbach, E.C. & Claggett, C.E. (1965). Incorporation of phosphate into the soft tissues and calcareous corpuscles of larval *Taenia taeniaeformis. Comp. Biochem. Physiol.* **14**: 11-20.

Wade, K.J., Flanagan, J.T., Currie, A. & Curtis, O.J. (1980). Roadside gradients of Pb and Zn concentrations in surface dwelling invertebrates. *Environ. Pollut.* **1B**: 87-93.

Wade, S.E., Bache, C.A. & Lisk, D.J. (1982). Cadmium accumulation by earthworms inhabiting municipal sludge-ammended soil. *Bull. Environ. Contam. Toxicol.* **28**: 557-560.

Wakayama, E.J., Dillwith, J.W., Howard, R.W. & Blomquist, G.J. (1984). Vitamin B_{12} levels in selected insects. *Insect Biochem.* **14**: 175-179.

Waku, Y. & Sumimoto, K.I. (1971). Metamorphosis of midgut epithelial cells in the silkworm (*Bombyx mori*, L.) with special regard to the calcium salt deposits in the cytoplasm. I. Light microscopy. *Tiss. Cell* **3**: 127-136.

Waku, Y. & Sumimoto, K.I. (1974). Metamorphosis of midgut epithelial cells in the silkworm (*Bombyx mori*, L.) with special regard to the calcium salt deposits in the cytoplasm. II. Electron microscopy. *Tiss. Cell* **6**: 127-136.

Walker, G. (1970). The cytology, histochemistry and ultrastructure of the cell types found in the digestive gland of the slug *Agriolimax reticulatus* (Müller). *Protoplasma* **71**: 91-109.

Walker, G. (1972). The digestive system of the slug *Agriolimax reticulatus* (Müller): experiments on phagocytosis and nutrient absorption. *Proc. Malac. Soc. Lond.* **40**: 33-43.

Walker, G. (1977). "Copper" granules in the barnacle *Balanus balanoides*. *Mar. Biol.* **39**: 343-349.

Wallace, B. (1982). *Drosophila melanogaster* populations selected for resistance to NaCl and CuSO$_4$ in both allopatry and sympatry. *J. Hered.* **73**: 35-42.

Wallwork, J.A. (1983). *Earthworm Biology*. London, Edward Arnold.

Walther, V.H., Klausnitzer, B. & Richter, K. (1984). Beeinflussung der Natalität von *Aphis fabae* Scopoli (Insecta, Homoptera) durch anthropogene Noxen, insbesondere Schwermetalle. *Zool. Anz.* **212**: 26-34.

Walton, K.C. (1986). Fluoride in moles, shrews and earthworms near an aluminium reduction plant. *Environ. Pollut.* **42A**: 361-371.

Walton, K.C. (1987). Factors determining amounts of fluoride in woodlice *Oniscus asellus* and *Porcellio scaber*, litter and soil near an aluminium reduction plant. *Environ. Pollut.* **46**: 1-9.

Wang, T.H. & Wu, H.W. (1948). On the structure of the Malpighian tubules of centipedes and their excretion of uric acid. *Sinensia* **18**: 1-11.

Warburg, M.R. (1987). Isopods and their terrestrial environment. *Adv. Ecol. Res.* **17**: 187-242.

Waterhouse, D.F. (1940). Studies of the physiology and toxicology of blowflies. 6. The absorption and distribution of iron. *Bull. Counc. Sci. Ind. Res. Melb.* **102**: 28-50.

Waterhouse, D.F. (1945a). Studies of the physiology and toxicology of blowflies. 10. A histochemical examination of the distribution of copper in *Lucilia cuprina*. *Bull. Counc. Sci. Ind. Res. Melb.* **191**: 5-20.

Waterhouse, D.F. (1945b). Studies of the physiology and toxicology of blowflies. 11. A quantitative investigation of the copper content of *Lucilia cuprina*. *Bull. Counc. Sci. Ind. Res. Melb.* **191**: 21-39.

Waterhouse, D.F. (1950). Studies on the physiology and toxicology of blowflies. XIV. The composition, formation and fate of the granules in the Malpighian tubules of *Lucilia cuprina* larvae. *Aust. J. Sci. Res.* **3B**: 76-112.

Waterhouse, D.F. (1952). Studies on the digestion of wool by insects. IV. Absorption and elimination of metals by lepidopterous larvae with special reference to the clothes moth, *Tineola biselliella* (Humm.). *Aust. J. Sci. Res.* **5B**: 143-168.

Waterhouse, D.F. & Stay, B. (1955). Functional differentiation in the mid-gut epithelium of blowfly larvae as revealed by histochemical tests. *Aust. J. Biol. Sci.* **8**: 253-277.

Waterhouse, D.F. & Wright, M. (1960). The fine structure of the mosaic midgut epithelium of blowfly larvae. *J. Insect Physiol.* **5**: 230-239.

Watkins, B. & Simkiss, K. (1988a). The effect of oscillating temperatures on the metal ion metabolism of *Mytilus edulis*. *J. Mar. Biol. Ass. U.K.* **68**: 93-100.

Watkins, B. & Simkiss, K. (1988b). Effects of temperature oscillations on the distribution of ^{65}Zn on cytoplasmic proteins. *Comp. Biochem. Physiol.* **89C**: 53-55.

Watkinson, J.H. & Dixon, G.M. (1979). Effect of applied selenate on ryegrass and on larvae of soldier fly, *Inopus rubriceps* Macquart. *New Zeal. J. Exp. Agric.* **7**: 321-325.

Watson, A.P. (1975). Trace element impact on forest floor litter in the new lead belt region of South Eastern Missouri. *Trace Elements Environ. Health* **9**: 227-236.

Watson, A.P., Van Hook, R.I. & Reichle, D.E. (1976). Toxicity of organic and inorganic arsenicals to an insect herbivore. *Environ. Sci. Technol.* **10**: 356-359.

Watson, J.P. (1970). Contribution of termites to development of zinc anomaly in Kalahari sand. *Trans. Inst. Min. Metall.* **79B**: 53-59.

Webb, J.S., Thornton, I., Thompson, M., Howarth, R.J. & Lowenstein, P.L. (1978). *The Wolfson Geochemical Atlas of England and Wales*. Oxford, Oxford University Press.

Webb, M. (1979). Ed. *The Chemistry, Biogeochemistry and Biology of Cadmium*. Amsterdam, Elsevier/North Holland Biomedical Press.

Weinberg, E. (1984). Iron withholding; a defense against infection and neoplasia. *Physiol. Rev.* **64**: 65-102.

Weismann, L. & Svatarakova, L. (1981). The influence of lead on some vital manifestations of insects. *Biológia (Bratislavia)* **36**: 147-151.

Welz, B. (1985). *Atomic Absorption Spectrometry* (2nd edition). Hamburg, VCH.

Westermarck, T.W. (1984). Consequences of low selenium intake for man. In: *Trace Element Analytical Chemistry in Medicine and Biology*. Vol. 3 (P. Brätter & P. Schramel, Eds.) pp. 49-70. Berlin, Walter De Gruyter.

White, J.J. (1968). Bioenergetics of the woodlouse *Tracheoniscus rathkei* Brandt in relation to litter decomposition in a deciduous forest. *Ecology* **49**: 694-704.

White, S.L. & Rainbow, P.S. (1985). On the metabolic requirements for copper and zinc in molluscs and crustaceans. *Mar. Environ. Res.* **16**: 215-229.

Whitehead, G.B. (1961). Investigation of the mechanism of resistance to sodium arsenite in the blue tick *Boophilus decoloratus* Koch. *J. Insect Physiol.* **7**: 177-185.

Wielgus-Serafinska, E. & Kawka, E. (1976). Accumulation and localization of lead in *Eisenia foetida* (Oligochaeta) tissues. *Folia Histochem. Cytochem.* **14**: 315-320.

Wiersma, G.B. (1986). Trace metals in the atmosphere of remote areas. *Adv. Environ. Sci. Technol.* **17**: 201-266.

Wieser, W. (1961). Copper in isopods. *Nature* **191**: 1020.

Wieser, W. (1965a). Untersuchungen über die Ernährung und den Gesamtstoffwechsel von *Porcellio scaber* (Crustacea: Isopoda). *Pedobiologia* **5**: 304-331.

Wieser, W. (1965b). Über die Häutung von *Porcellio scaber* Latr. *Verh. Dt. Zool. Ges. Kiel.* 1964: 178-195.

Wieser, W. (1965c). Electrophoretic studies on blood proteins in an ecological series of isopod and amphipod species. *J. Mar. Biol. Ass. U.K.* **45**: 507-523.

Wieser, W. (1966). Copper and the role of isopods in the degradation of organic

344   *Ecophysiology of Metals in Terrestrial Invertebrates*

matter. *Science* **153**: 67-69.

Wieser, W. (1967). Conquering terra firma: the copper problem from the isopod's point of view. *Helgoländer Wiss. Meer.* **15**: 282-293.

Wieser, W. (1968). Aspects of nutrition and the metabolism of copper in isopods. *Am. Zool.* **8**: 495-506.

Wieser, W. (1978). Consumer strategies of terrestrial gastropods and isopods. *Oecologia* **36**: 191-201.

Wieser, W. (1979). The flow of copper through a terrestrial food web. In: *Copper in the Environment*. Part 1, *Ecological Cycling* (J.O. Nriagu, Ed.) pp. 325-355. Chichester, Wiley Interscience.

Wieser, W. (1984). Ecophysiological adaptations of terrestrial isopods: a brief review. *Symp. Zool. Soc. Lond.* **53**: 247-265.

Wieser, W. & Klima, J. (1969). Compartmentalization of copper in the hepatopancreas of isopods. *Mikroscopie* **24**: 1-9.

Wieser, W. & Makart, H. (1961). Der Sauerstoffverbrauch und der Gehalt an Ca, Cu und einigen anderen Spurenelementen bei terrestrischen Asseln. *Z. Naturforsch.* **16B**: 816-819.

Wieser, W. & Wiest, L. (1968). Ökologische aspekte des Kupferstoffwechsels terrestrischer Isopoden. *Oecologia* **1**: 38-48.

Wieser, W., Busch, G. & Büchel, L. (1976). Isopods as indicators of the copper content of soil and litter. *Oecologia* **23**: 107-114.

Wieser, W., Dallinger, R. & Busch, G. (1977). The flow of copper through a terrestrial food chain. II. Factors influencing the copper content of isopods. *Oecologia* **30**: 265-272.

Wigglesworth, V.B. & Salpeter, M.M. (1962). Histology of the Malpighian tubules in *Rhodnius prolixis*. *J. Insect Physiol.* **8**: 299-307.

Wigham, H., Martin, M.H. & Coughtrey, P.J. (1980). Cadmium tolerance of *Holcus lanatus* L. collected from soils with a range of cadmium concentrations. *Chemosphere* **9**: 123-125.

Wignarajah, S. & Phillipson, J. (1977). Numbers and biomass of centipedes (Lithobiomorpha: Chilopoda) in a *Betula-Alnus* woodland in N.E. England. *Oecologia* **31**: 55-66.

Wild, H. (1975a). Termite Ni uptake on the serpentines of the Great Dyke of Rhodesia. *International Conference. Heavy Metals in the Environment*. Vol. 2, pp. 73-74. Toronto, 1975. Edinburgh, CEP Consultants.

Wild, H. (1975b). Termites and the serpentines of the Great Dyke of Rhodesia. *Trans. Rhod. Sci. Ass.* **57**: 1-11.

Williams, M.W., Hoeschele, J.D., Turner, J.E., Jacobson, K.B., Christie, N.T., Paton, C.L., Smith, L.H., Witschi, H.R. & Lee, E.H. (1982). Chemical softness and acute metal toxicity in mice and *Drosophila*. *Toxicol. Appl. Pharmacol.* **63**: 461-469.

Williams, R.J.P. (1981). Natural selection of the chemical elements. *Proc. Roy. Soc. Lond.* **213B**: 361-397.

Williamson, P. (1975). Use of ^{65}Zn to determine the field metabolism of the snail *Cepaea nemoralis* L. *Ecology* **56**: 1185-1192.

Williamson, P. (1979a). Opposite effects of age and weight on cadmium concentrations of a gastropod mollusc. *Ambio* **8**: 30-31.

Williamson, P. (1979b). Comparison of metal levels in invertebrate detrivores and their natural diets: concentration factors reassessed. *Oecologia* **44**: 75-79.

Williamson, P. (1980). Variables affecting body burdens of lead, zinc and cadmium in a roadside population of the snail *Cepaea hortensis* Müller. *Oecologia* **44**: 213-220.

Williamson, P. & Evans, P.R. (1972). Lead: levels in roadside invertebrates and small mammals. *Bull. Environ. Contam. Toxicol.* **8**: 280-288.

Willis, J.B. (1975). Atomic spectroscopy in environmental studies: fact and artifact. *International Conference. Heavy Metals in the Environment.* Vol. 1, pp. 69-91. Toronto, 1975. Edinburgh, CEP Consultants.

Wit, L.C., Mason, W.H. & Blackmore, M.S. (1984). The effects of crowding on the bioelimination of ^{65}Zn in *Popilius disjunctus. J. Georgia Entomol. Soc.* **19**: 8-14.

Wolburg, H., Hevert, F., Wessing, A. & Porstendoerfer, J. (1973). Die Konkremente des larvalen Primärharnes von *Drosophila hydei.* 1. Struktur. *Cytobiologie* **8**: 25-38.

Wolfenbarger, A.D., Guerra, A.A. & Lowry, W.L. (1968). Effect of organometallic compounds on Lepidoptera. *J. Econ. Entomol.* **61**: 78-81.

Wolff, E.W. & Peel, D.A. (1985). The record of global pollution in polar snow and ice. *Nature* **313**: 535-540.

Wood, J.M. (1974). Biological cycles for toxic elements in the environment. *Science* **183**: 1049-1058.

Wood, J.M. (1984a). Evolutionary aspects of metal ion transport through cell membranes. In: *Metal Ions in Biological Systems.* Vol. 18, *Circulation of Metals in the Environment* (H. Sigel, Ed.) pp. 223-237. New York, Marcel Dekker.

Wood, J.M. (1984b). Microbiological strategies in resistance to metal ion toxicity. In: *Metal Ions in Biological Systems.* Vol. 18, *Circulation of Metals in the Environment* (H. Sigel, Ed.) pp. 333-351. New York, Marcel Dekker.

Wright, K.A. & Newell, I.M. (1964). Some observations of the fine structure of the midgut of the mite *Anystis sp. Ann. Entomol. Soc. Am.* **57**: 684-693.

Wright, M.A. & Stringer, A. (1980). Lead, zinc and cadmium content of earthworms from pasture in the vicinity of an industrial complex. *Environ. Pollut.* **23A**: 313-321.

Wróblewski, R., Roomans, G.M., Ruusa, J. & Hedberg, B. (1979). Elemental analysis of histochemically defined cells in the earthworm *Lumbricus terrestris. Histochemistry* **61**: 167-176.

Xavier, A.V. (1986). Ed. *Frontiers in Bioinorganic Chemistry.* Hamburg, VCH.

Yamamura, M., Mori, T. & Suzuki, K.T. (1981). Metallothionein induced in the earthworm. *Experentia* **37**: 1187-1189.

Yates, L.R. & Crossley, D.A. (1981). Cesium 134 and strontium 85 turnover rates in the Chilopoda *Scolopocryptops nigridia* (Myriapoda). *Pedobiologia* **21**: 145-151.

Young, R.S. (1979). *Cobalt in Biology and Biochemistry.* London & New York, Academic Press.

Zatta, P. (1984). Zinc transport in the haemolymph of *Carcinus maenas* (Crustacea, Decapoda). *J. Mar. Biol. Ass. U.K.* **64**: 801-807.

Zelenayova, E. & Weismann, L. (1983). The effect of CdCl$_2$ in the semisynthetic food of caterpillars upon the gonads of imagos of *Scotia segetum* (Den.

and Schiff.)(Lepidoptera, Noctuidae). *Biológia (Bratislavia)* **38**: 941-948 (In Czechoslovakian).

Zhulidov, A.V. & Emets, V.M. (1979). Accumulation of lead in the bodies of beetles in contaminated environments associated with automobile exhausts. *Dokl. Akad. Nauk. SSSR* **6**: 1515-1516 (In Russian).

Zhulidov, A.V., Poltavskii, A.N. & Emets, V.M. (1982). Method for studying the migrations of nocturnal lepidopterous insects in geochemically heterogeneous regions. *Soviet J. Ecol.* **13**: 398-400.

Species Index

Subject Index

Acari. *See* Mites; Ticks
Accuracy
  cf. precision, 30; Fig. 4.1
  of published metal concentration data, 30
Acids, purity of, 27
Aluminium
  as a 'non-heavy metal', 1
  possible essentiality of, 10
  toxicity of to moth larvae, 160, 161
Americium, retention of on leaf surfaces, 48
Amphipods
  metal-containing granules in, 238, 262
  metals in, 140–1
  terrestriality of, 120
Annelida. *See* Earthworms
Antagonism
  of cadmium, copper and zinc in beetles, Table 7.18
  of copper and molybdenum, 18, 200
  of metal ions at plant root uptake sites, 50
  of metals in invertebrates, 84
Antimony
  as a metalloid, 1; Fig. 1.1
  as sulphide in gut of clothes moth larvae, 270
  in drugs for treatment of trematode infection, 280
Ants
  as biological monitors of pollution, 210
  cadmium and copper in, Merseyside, 171; Table 5.9

Ants—*contd.*
  density of around Gusum brass mill, 67
  effects of metal-containing pesticides on, 277
  lead and copper in, Gusum, 171; Table 5.6
  mercury in, 172
  metal-containing granules in, 276
  metals in, 171–2
Aphids
  diet of, 83
  essentiality of metals in, 155; Fig. 7.14
  metal-containing granules in, 238, 270
  metals in, 154–7
  symbiotic gut microorganisms in, 91, 155
Aranae. *See* Spiders
Arsenic
  as a metalloid, 1; Fig. 1.1
  as an essential element, 10
  as sulphide in gut of clothes moth larvae, 270
  effect of on chromosomes of *Drosophila*, 169
  in pesticides, 20, 163, 171, 277
  in termites, 154
  methylation of, 12
  tolerance to, in ticks and flies, 92, 170, 192
  toxicity of, in katydids, 151–2
  toxicity of, in moth larvae, 160
Artificial diets
  for cockroaches, 153
  for crickets, 150
  for insects, 94–5

X-ray fluorescence
  methodology, 40
  sensitivity of, Table 4.4
X-ray microanalysis
  detection limit of, 233
  development of, 6–7, 218
  methodology of, 226–33; Figs. 9.2,
    9.3
  spectra from, Figs. 9.6B, 9.8, 9.11,
    9.14B,C

Zinc
  ^{65}Zn in beetles, 95, 175; Fig. 7.16
  ^{65}Zn in cockroaches, 153
  ^{65}Zn in isopods, 136; Fig. 7.10
  ^{65}Zn in moths, transfer of during
    mating, 158
  ^{65}Zn in slugs, 226
  ^{65}Zn in snails, 243–6; Tables 9.1,
    9.2
  ^{65}Zn in spiders, 187
  as an essential element, 10
  as sulphide in gut of clothes moth
    larvae, 270
  availability to early life, 4
  'budget' of in Haw Wood, Table
    5.4
  essentiality of to aphids, 155; Fig.
    7.14B
  essentiality of to moth larvae, 162–
    3
  in amphipods, 140–1
  in bees, 170
  in beetles, 173–6; Table 7.18; Fig.
    7.17

Zinc—*contd.*
  in carbonic anhydrase, 10, 222,
    224; Tables 2.1, 2.2
  in centipedes, 180–5; Table 7.20;
    Fig. 7.20
  in cockroaches, 153
  in earthworms, 107–20, 251–6;
    Tables 7.4, 7.5, 7.7; Fig. 9.11
  in enzymes, 220
  in flies, 164, 168–70; Table 7.16;
    Fig. 7.15
  in grasshoppers, 149, 150, 151
  in haemocyanin, 223
  in insect mandibles, 142, 149, 158
  in isopods, 121–40, 256–62; Tables
    7.9, 7.10; Figs. 4.6, 4.7, 7.8A,
    7.9, 8.2, 9.14B,C
  in metal-containing granules, 234–
    41
  in metallothioneins, 220–2
  in millipedes, 178–80; Fig. 7.18C
  in molluscicides, 99
  in moth larvae, 160–2; Table 7.14
  in red blood cells, 9
  in sewage sludges, Table 3.1
  in slugs and snails, 100–7; Tables
    7.1, 7.3; Fig. 7.2A
  in spiders, 187–91; Table 7.21
  in super-oxide dismutase, 4
  leaching rate of from sewage
    sludge, 50
  requirement for in Crustacea, 120–
    1, 127
  toxicity of to flies, 168; Fig. 7.15
  toxicity of to isopods, 127, 140;
    Fig. 4.6